NATURE'S CROSSROADS

HISTORY OF THE URBAN ENVIRONMENT

MARTIN V. MELOSI AND JOEL A. TARR, EDITORS

NATURE'S CROSSROADS

The Twin Cities and Greater Minnesota

EDITED BY

George Vrtis and
Christopher W. Wells

UNIVERSITY OF PITTSBURGH PRESS

Published by the University of Pittsburgh Press, Pittsburgh, Pa., 15260
Copyright © 2022, University of Pittsburgh Press
Manufactured in the United States of America
Printed on acid-free paper
10 9 8 7 6 5 4 3 2 1

Cataloging-in-Publication data is available from the Library of Congress

ISBN 13: 978-0-8229-4738-7
ISBN 10: 0-8229-4738-2

Cover photo: iStockPhoto
Cover design: Joel W. Coggins

For our students at Carleton College and Macalester College,
and for our children

CONTENTS

ACKNOWLEDGMENTS

THE IDEA FOR THIS BOOK EMERGED just over a decade ago, not long after both of us had moved to Minnesota and taken up our positions at Carleton College and Macalester College. We began talking to one another about our shared interests and the rather limited amount of work done on the state's environmental history. We were both committed to trying to incorporate local history and scholarship into our courses, and as newcomers to Minnesota with young families we both had a strong desire to connect to the region's stories and landscapes. Those two intersecting forces inspired the genesis of this book, and so we wish to begin by thanking the many students we have taught over the years in our environmental history courses, as well as our respective children for launching us on the illuminating and lengthy odyssey that has led to this book.

We are also grateful to several institutions and their staffs for supporting this book. The Minnesota Historical Society has been a steadfast partner in all of the work we have undertaken for this book and for our earlier work on Minnesota's environmental past. We will always be thankful to the many staff members who provided expert consultations, taught us about their collections; waived duplication and copyright use fees; published our bibliography and research guide on Twin Cities environmental history, *Twin Cities Environmental History: A Bibliography of Published and Unpublished Sources* (2012); hosted our conference on Minnesota environmental history; helped us develop and launch our app, Minnesota Environments; and encouraged us at every turn to complete this book. We are especially grateful to Robert Horton and Lesley Kadish for their inspiring and creative approach to thinking about history and for always offering their enthusiastic and unwavering support. Thanks also to Debbie Miller, who was particularly generous in guiding us into the society's collections.

The Minnesota Historical and Cultural Heritage Grant Program (also known as Legacy Grants) was another mainstay in helping make this book possible. Since 2008 this state-funded program has been providing financial support for projects focused on preserving Minnesota's history and culture, and we were fortunate to receive one of these grants in 2010. We feel very

lucky to live in a state that is committed to preserving and enhancing historical education, and we hope that our book contributes to that worthy goal in a
small way.

Our home institutions—Carleton College and Macalester College—were
also essential to the completion of this work. Both colleges contributed funding
at key moments in the development of this project and provided good, colle-
gial, and supportive working environments. A series of grants supported our
conference, the development of our app, the hiring of research assistants, and
the production of the book's map and index. Among the many colleagues who
supported this project on our own campuses, we wish to single out a few for
special thanks. At Carleton, our warm thanks to Beverly Nagel, Carly Bjorn,
Janet Russell, and Kim Smith. And at Macalester, our equally warm thanks to
Dan Hornbach, Anne Esson, Helen Warren, Lynn Hertz, and Fritz Vandover.
Additionally, several of our campus colleagues read parts of the manuscript and
gave us insightful feedback that greatly improved this book, particularly Con-
stanza Ocampo-Raeder, Katrina Phillips, and Rebecca Wingo. We also wish
to thank the three Carleton and Macalester students who served as research
assistants on this project, Natalie Locke, Callie Millington, and Jenni Rogan, as
well as Jerome Cookson, who provided invaluable cartographic advice on our
map of the Twin Cities and Greater Minnesota.

In addition to the generous support of all of these institutions, we wish to
thank our fellow authors who joined us in this effort to better understand Minnesota's environmental history. Each of them undertook research on important
and innovative topics, and each has made the book far richer for their insights.
We are particularly grateful to Kathleen Brosnan for serving as our conference
chair and for providing the Afterword for this book, and to Steven Hoffman
who was working on a chapter examining the links between the Twin Cities
and Alberta's tar sands but passed away before it was completed.

Thanks too to our editors at the University of Pittsburgh Press, Sandy
Crooms and Josh Shanholtzer, for their encouragement, critical suggestions,
and patience with this book. It has been a privilege to work with them and
to join the History of the Urban Environment series. We also remain indebt-
ed to the two anonymous peer reviewers whom the press selected to review
our manuscript. They made thoughtful, challenging suggestions that have improved this book in important ways.

We also wish to offer our profound thanks to the land itself. At a time of
significant ferment in American society and culture, Minnesota's beautiful
and complicated landscape—the ancestral and contemporary homeland of
the Dakota and Ojibwe peoples—has sustained us, inspired us, and taught us
about the intermingling of ecological and social forces. During the time we
have worked on this book, important conversations about racism, settler colo-

nialism, and injustice have risen to the top of the agenda in American society and in our field of environmental history, helping us understand Minnesota's environmental past and present with greater insight and clarity. And so, to the Twin Cities and Greater Minnesota, with all their varied landscapes, peoples, and rich and complicated histories, our sincere thanks.

Finally, we wish to thank our families for supporting us during the long journey this book required. Few know the travails scholars go through in their research and writing like their loved ones, and so to our families—Anne, Meadow, and Henry; and Marianne, Jack, Annie, and Meg—our heartfelt thanks for your love and understanding.

NATURE'S CROSSROADS

INTRODUCTION

Unearthing
Nature's Crossroads

George Vrtis and Christopher W. Wells

ON THE MORNING OF JUNE 12, 2018, a raccoon started climbing up the side of the twenty-five-story UBS Plaza tower in downtown Saint Paul, Minnesota. Just after noon, it reached the twenty-second floor and settled down for a nap on a window ledge. By then the raccoon had a following. Across the street in the Minnesota Public Radio (MPR) newsroom, a reporter was chronicling the raccoon's every move, and #MPRraccoon had begun to trend on Twitter, Facebook, and media outlets across the country. On the street below a crowd had gathered to cheer on the raccoon, while Saint Paul's animal control and fire department officials plotted possible capture and rescue strategies.[1]

After its midday nap, the raccoon started climbing again. It meandered up, down, and sideways around the building all afternoon before "settling in for the night" some two hundred feet above street level.[2] The story was captured on the evening news, with a reporter at the local CBS affiliate likening the raccoon to a "smaller, hairier version of Spiderman."[3] Celebrity followers joined in over Twitter as well, including James Gunn, the director of the hit Marvel movie *Guardians of the Galaxy*, which features a raccoon hero named Rocket: "I'll donate a thousand bucks," Gunn wrote, "to the political charity of choice to anyone who saves the raccoon. I can't handle this. Poor dude."[4]

But the raccoon wasn't quite ready to call it a night. In the early morning hours of the following day, the raccoon finished its ascent of the building, only to end up moments later trapped in a wire cage, eating cat food out of a can. From there animal control experts carried the raccoon across the building's roof to a freight elevator and on a far less daring trip back down to street level. They then loaded the raccoon into a truck for a quick ride to the suburbs and set it free.[5]

As the raccoon began its new suburban life, its story became something of a national and international sensation. That evening, the story brought a touch of levity to news and late-night broadcasts across the country and around the world. On *NBC Nightly News*, Lester Holt introduced the story before turning it over to another reporter who likened the raccoon to something of a superhero: "Move over, Spiderman. Step aside, Tom Cruise. Meet the world's newest daredevil, MPR Raccoon, no stunt-double needed."[6] The BBC began their segment on the raccoon by simply saying, "Now I can't believe you haven't heard about it."[7] And on *The Daily Show*, Trevor Noah opened with the raccoon, humorously reflecting on our need for an uplifting story: "Oh what wondrous news! A raccoon is climbing something taller than usual! Life is beautiful again!"[8] By the end of the day the MPR raccoon story had circled the globe and was so widespread that Minnesota's leading newspaper, the Minneapolis *Star Tribune*, observed, "The saga was the most talked-about story online, even usurping the news of the historic meeting between political adversaries US president Donald Trump and North Korean leader Kim Jong Un."[9]

The interest in and response to the MPR raccoon story is revealing. Most urban Americans have long assumed that nothing wilder than robins and squirrels, and perhaps the occasional gopher or fox, share their carefully partitioned concrete and wooden worlds. Those beliefs are not new. They are embedded in long-standing conceptions of cities and nature as separate and opposing worlds.[10] On the one side is the city, the place where we try to wall ourselves off from the natural world, where we look to conquer and control nature in order to build our urban homes. To use the environmentalist Bill McKibben's provocative book title, the city symbolizes *The End of Nature*.[11] It is the place where human forces have become so powerful, so pervasive, that they have come to rival, if not utterly overwhelm, the ancient forces of the natural world. On the other side of this divide is the natural world. It is the world of first things, the place where the autonomous, uncontrolled, and sometimes unruly forces of a pristine natural world still run free. In this dichotomous view, cities and nature lie at opposite poles, human and nonhuman worlds, with an unmistakable line separating the two.

Nice and comforting as straight lines may be in keeping things well organized, they have little to do with most environmental and social realities. The

MPR raccoon reveals this. Though it was certainly not the first wild creature to venture into an American city, it serves as a good example of the outmoded vision of cities and nature as separate and opposing worlds. The history of *wildlife and the city* is profoundly entangled, just as is the larger category of *nature and the city*. Cities are built in particular regional environments, each with its own climate, topography, hydrology, and flora and fauna. How we use and change those landscapes—whether through bulldozing, building, controlling the flow of water, or gardening—affects natural systems, which then respond according to their own logics. But as much as cities are tied up with the physical world, they are also deeply tied up with a cultural and social one—a human one. How we value the natural world, envision and understand it, layer it with cultural and social constructs like race, class, and gender—all of these phenomena (and many more) influence our relationships with nature. To see cities and nature as separate worlds is to miss the countless ways they shape and reshape one another, as well as the complicated, layered, messy ways they inscribe privilege, power, and meaning into environmental thought, policies, institutions, personal interactions, and material realities. By collapsing these worn-out and unhelpful dichotomies, all kinds of possibilities open for improving how we understand the nature of cities and the environmental and social relationships that lie at their core. Indeed, these are some of the critical insights that environmental historians have recently uncovered in their studies of New York, Boston, Philadelphia, Chicago, New Orleans, Houston, Denver, Los Angeles, San Francisco, Sacramento, Seattle, and other cities in North America and around the world.[12]

Whether cities are imagined as deeply enmeshed in nature or somehow rigidly separated from it, environmental historians have yet to focus much of their imaginative energy on Minnesota's two largest cities—the Twin Cities of Minneapolis and Saint Paul—or "the Cities," as locals tend to call them. All historians who study the Cities eventually find their way to the Mississippi River, and this is reflected in a range of works that examine such fundamental issues as the way political debates, economic initiatives, and cultural developments have all interfaced with America's largest river in some way. Other environmental aspects that have received some attention from historians include the major industries that have shaped the Cities' evolution and development such as wheat and lumber milling, environmental concerns like sanitation and pollution, and the work of nature writers and park advocates, including the architect of the Minneapolis park system, Theodore Wirth, and the prominent and influential wilderness advocate, Sigurd Olson, who helped establish both Voyageurs National Park and the Boundary Waters Canoe Area Wilderness in far northern Minnesota.[13]

This book therefore aims to broaden and deepen our understanding of the

Twin Cities' environmental history, and to contribute new research that offers new stories and perspectives to ongoing conversations in the field of urban environmental history generally. In the seventeen original chapters that follow, a multidisciplinary collection of scholars and public agency officials explore some of the significant issues and developments that have shaped the environmental history of the Cities and their hinterlands since Minneapolis and Saint Paul emerged fourteen miles apart in the middle of the nineteenth century. While authors were given the freedom to focus on the topics and questions that interested them most, they were also asked to consider several major focal points that shape the book's empirical and historiographical ambitions. First among these was a consideration of the mutual reshaping that the Cities and their hinterlands brought to one another. Drawing specifically on William Cronon's path-breaking book, *Nature's Metropolis: Chicago and the Great West* (1991), authors were asked to think hard about the environmental significance of the ways that people, resources, ideas, values, and social realities flow back and forth between the Cities and their hinterlands. Other focal points included changes in the regions' physical environments, peoples' understandings of and attitudes toward the environmental changes that took shape, and the cultural realities, social policies, and governmental regulations that influenced developments. By drawing attention to these common focal points, our intention was to illuminate important and potentially distinctive aspects of the environmental history of the Twin Cities and greater Minnesota, bring clarity and coherence to the book as a whole, and contribute new findings, new perspectives, and new stories to urban environmental history generally.

The book is organized into three thematic sections, each unfolding in a roughly chronological sequence, followed by a reflective Afterword. Those thematic sections, in order, are titled "The Dynamics of Environmental Change: Cities, Commodities, Hinterlands," "The Twin Cities and the Built Environment," and "Environmental Politics, Thought, and Justice." While many of the chapters speak to more than one of these themes, we have positioned them in the book based on the focus of their argument. Taken all together, the chapters in this book offer an expansive vision of the Twin Cities and greater Minnesota's environmental past. The Cities have long been powerful engines of environmental change, shaping and reshaping their urban landscapes and the vast hinterlands drawn into their orbit. And those changes have always been tied up with every other aspect of our human and nonhuman existence. Coursing through Twin Cities environmental history are large, complex forces like capitalism and settler colonialism, industrialization and urbanization, and the numerous complicated realms of interaction and life that we describe as politics, racism, gender, labor relations, intellectual thought, society, and culture. These are nature's crossroads, the place we call the Twin Cities and the environmental

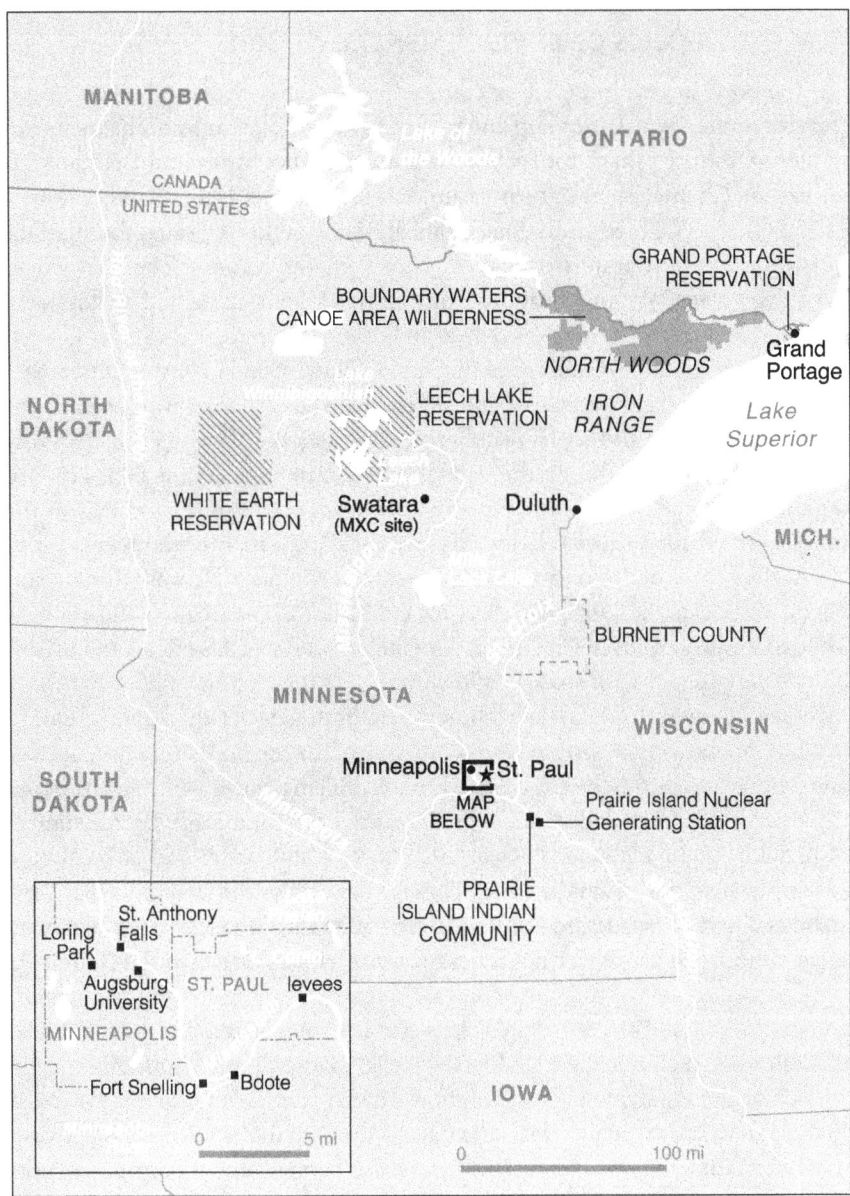

Map 1. Map of the Twin Cities and Greater Minnesota: Major Locations Examined in This Book

and cultural relationships that have shaped them and their hinterlands. After reading through this book, we hope that readers will find it hard to think about the Twin Cities—and about urban environmental history and many other fields of American history—in the same way.

THE DYNAMICS OF ENVIRONMENTAL CHANGE

The book begins in Part I by considering the environmental relationships that wove the Twin Cities and their hinterlands into a common history with nature at its core. That process began, as Christopher Wells and George Vrtis observe in Chapter 1, with the pressured treaty cessions and military conquest of Dakota and Ojibwe lands that established the settler-colonial foundations of two new remote waterfront cities, Saint Paul and Minneapolis, that would eventually transform much of the world around them. Imagined as dots on a commercial map, the two cities shared an important strategic location between the abundant natural resources of the upper Midwest and growing settlements with hungry markets to the east and south. But a closer look reveals geographical, technological, and social peculiarities that together forged each city's distinctive relationship with nature and with its urban twin. As the Cities rose to regional, national, and international prominence over the second half of the nineteenth century, those relationships would change in important ways.

At the center of these early developments in Minneapolis was wheat, a staple in American and European diets for centuries. Between 1850 and 1900, as Thomas Finger argues in Chapter 2, demand for wheat in American and British markets combined with a series of technological innovations, financial investments, and environmental developments on both sides of the Atlantic Ocean to simultaneously make Minneapolis into one of the top flour-milling centers—and Minnesota into one of the top wheat-producing states—in the nation. As these processes unfolded, farmers, millers, investors, and everyday consumers of bread brought a new ecological order to city and hinterland alike. By the 1870s Minneapolis's Saint Anthony Falls, once restless and dynamic, had been stabilized and encased, and the Red River Valley's once vast prairies had been plowed under and planted in endless rows of wheat. The two developments were opposite sides of the same coin, and together they gathered the nutrients and energy from Minnesota landscapes and sent them eastward to feed industrial laborers in places like Buffalo, New York, Liverpool, and London.

While the emergence and evolution of Minneapolis's flour mills and the Red River Valley's wheat farms were tightly knotted, the relationships between the Twin Cities and their hinterlands were also complicated by competing economic and political interests. These dynamics are clearly evident in Chapter 3, where David Lanegran reveals how the land-use history of Burnett County, Wisconsin, was shaped by the ebbs and flows of economic, technological, and political forces emanating from the neighboring state capitals of Saint Paul and Madison, Wisconsin. Burnett County's earliest developments reflect Madison's influence, which promoted the region's agricultural potential and led to a combination of subsistence and commercial farming in the second half of the nineteenth century. The pendulum then swung to Saint Paul for the first three decades of the twentieth century, when Saint Paul's Crex Carpet Company bought up large tracts of Burnett County marshland and began harvesting wiregrass and sedge to make twine and carpets. The pendulum would continue to swing back and forth between Madison and Saint Paul through the twentieth century, shaping Burnett County and its landscapes into the shared cultural and economic hinterland of both cities.

Beneath the flows of economic and political power that tied the Twin Cities to places like the Red River Valley and Burnett County were differing peoples, each with their own reasons and ideas for living as they did. These themes are centered in Chapter 4, where Michael McNally explores the construction of a series of dams in the Upper Mississippi watershed in the 1880s and the ways these dams channeled the commercial dreams of millers and the concerns of US policy makers at the expense of Anishinaabe communities. Damming the outlet streams of lakes such as Lake Winnibigoshish and Leech Lake led to higher water levels, which secured reliable flows of the Mississippi River at Saint Anthony Falls. In a clear case of environmental injustice, those higher lake levels also drowned nearly two hundred thousand acres of reservation land, devastating fishing and gathering areas, inundating burial grounds, forcing relocations, and undercutting cultural relationships that were rooted in the very landscapes that were being destroyed. The ensuing controversy over damages, McNally shows, was never entirely settled. The very concept of damages deployed by the US government was unable to capture the deeper cultural and spiritual importance that the Anishinaabeg vested in the land, a significance that went far beyond the kinds of economic losses and assimilationist policies that organized the approach of federal officials.

As the dam controversy unfolded and the Anishinaabeg struggled to be heard, Minnesota's North Woods were undergoing a series of environmental transformations that were never far removed from the influence of the Twin Cities. Visualized as a series of four simplified time-lapsed photographs, the North Woods would appear as a heavily forested region at the time of statehood in 1858, a stump-riddled cutover in 1890, a thin scattering of small farms amid a sea of stumps, brush, and stones in 1910, and finally, an emerging rustic vacationland and patchwork woodlands in 1930. The stories embedded in these images, and the passages between them, are the subjects of the next two chapters.

Each phase in the logging and farming of northern Minnesota, Kevin Brown explains in Chapter 5, was not due solely to market demand emanating from the Twin Cities and elsewhere. Rather, the changes were the result of what Brown calls "working environments," the day-to-day interactions among laborers, environmental conditions, and most important, differing forms and amounts of capital that all met, mingled, and remade northern Minnesota. Though conditions were dangerous for loggers and farmers alike, timber company investments in the form of saws, horses, steam engines, and management allowed them to overcome the region's harsh environmental conditions and level the northern forest in the late nineteenth century. In contrast, independent farmers were notoriously poor and lacked access to capital in any form, be it tools, horses, or machinery. This dearth of capital limited the success of farmers in the region in the early twentieth century, and ultimately helped lead to the regrowth of the area's vast forests.

By the early twentieth century, as the timber industry fell into decline and agricultural failures mounted, the North Woods were slowly transformed into a vacation landscape for residents of the Twin Cities and beyond. The redevelopment of the North Woods into a place of outdoor leisure rather than labor, Aaron Shapiro points out in Chapter 6, commodified nature in new and important ways. In place of felling the forest or trying to wrestle vegetables from rocky soils, tourism advocates promoted the region's scenic beauty, tranquility, and rugged wilderness characteristics as a respite from the increasingly industrial, urban world. Though the region's natural features certainly helped make this transition possible, the process was fueled by local, state, and national developments that together proved crucial in positioning vacations as part of American consumer culture and as responsive to conservationist efforts in the Progressive Era. Even today, the rugged imagery of Minnesota's North Woods and the common summer refrain of "going up north" owe their origins to these late nineteenth- and early twentieth-century developments.

THE TWIN CITIES AND THE BUILT ENVIRONMENT

Part II of *Nature's Crossroads* narrows attention to the Twin Cities' urban confines and the changing environmental relationships that residents crafted to sustain their lives. With the development of ever-more intricate and technologically sophisticated transportation and communication systems, energy and water delivery services, and waste collection and disposal practices, each of these "city services" created new webs of environmental relationships that underpinned urban life. They also reflected, as John Anfinson observes in his study of the Twin Cities' water supply systems in Chapter 7, contemporary ideas about nature, science, economics, and especially public health. Differing mixes of those ideas led city officials to take divergent paths in building the uneven experience with typhoid and other water-borne diseases until 1900. By then the early stirrings of the bacteriological revolution were undercutting the older miasma theory of disease in medical thought, giving city officials new tools for combating the recurrent epidemics that had periodically ravaged their communities over the past four decades.

A similar process was at work in the Cities' former farmland and open spaces. Changing ideas about nature and society, Karen Wellner argues in Chapter 8, led to the nearly continuous reimagining of Minneapolis's first urban park, Loring Park. Established in 1883 on several hundred acres of soggy land, the park's initial design reflected the influence of Frederick Law Olmsted, the designer of New York City's Central Park and the preeminent landscape architect of the day. Olmsted and other park advocates, including Horace Cleveland, who designed Loring Park, believed that pastoral open spaces in cities helped alleviate the strains of the industrial world, inspired genteel moral values, and promoted a more egalitarian social order. In response to these and other views, Cleveland began work on Loring Park, excavating two ponds, adding gentle slopes, and mixing in an assortment of trees, grassy areas, and flower beds, all connected by winding pathways meant to inspire a kind of refined nature-society harmony. From those beginnings, Loring Park would be reinvented several times over the coming century, each time responding to the changing ideas about nature and society that influenced Minneapolis's Board of Park Commissioners.

Among the most enthusiastic users of Loring Park, and the many other Twin Cities parks that followed, have been bicyclists. At the turn of the twentieth century, as James Longhurst reveals in Chapter 9, the Twin Cities were leaders in what he calls "the golden age of cycling." By 1902 Saint Paul had a network of 115 miles of cycle paths, while Minneapolis had 75 miles in the city and more beyond its city limits. The paths connected the two cities while providing cyclists with connections to parks, other recreational sites, and the cities' hinterlands. Both cities were also part of a much larger nationwide movement that envisioned a network of interstate bicycle paths stretching across the country. The movement faded around 1910, falling victim to weaknesses in the mixed private-public model of funding bike paths, as well as the sudden rise of the automobile and the Good Roads movement. Though many of the old paths were paved over for automobiles in the twentieth century, some were maintained and today are part of Minneapolis's Grand Scenic Byway, an award-winning 51-mile bike path that encircles the city.

Bicycle paths, roadways, and other transportation systems have had an enormous influence on the shape of urban development in the Twin Cities, as well as in many other cities across the country. As Robert Thompson shows in Chapter 10, Minneapolis's very form—defined by its extensive spread of single-family homes radiating outward from the city center—is part of the legacy of streetcar lines that were developed in the late nineteenth century. By the 1920s, and accelerating sharply after World War II, that model was challenged as city officials and developers responded to a complicated mix of developments that included a surge in automobile ownership, population declines in the city, and environmental initiatives like Sustainable Urbanism and Smart Growth. The direction of change led first to incorporating the kind of automobile-centered planning that was becoming commonplace in the suburbs in the 1960s. And then in the 1990s the city reversed course, turning its back on suburban models steeped in car culture and low-density sprawl, and instead focused on fostering high-density, pedestrian-friendly, and mixed-use neighborhoods. What this means for the future of the city remains a vital, if unanswered, question.

Decisions about how to organize land use and respond to changing environmental and cultural influences have also been shaped by the countless landowners who have called the Twin Cities home. On the campus of Augsburg College in Minneapolis, for instance, as Joseph Underhill explores in Chapter 11, the college's water management regime has reflected a mix of national, local, and institutional influences and commitments that have shifted significantly over the college's more than 150-year history. When the college was founded in 1869, students drew water from campus wells and distributed it to the entire campus along footpaths. Those relationships began to change in 1890, when the campus was connected to the municipal water supply system and students began playing sports rather than hauling well water across campus. But the larger changes, Underhill notes, were seeded after World War I and accelerated in the 1960s. During these periods, the college transitioned from a Lutheran seminary to a modern liberal arts college, adding the social and natural sciences to its earlier emphasis on religious training. These curricular changes aligned with the environmental and civil rights movements of the 1960s, as well as the college's historic Lutheran traditions, to reposition a concern for the local Mississippi watershed and sustainability on campus for the first time. The campus is now focused on rebuilding wetlands and rainwater harvesting systems, and helping students see the campus, Minneapolis, and the world as interconnected and interdependent watersheds.

ENVIRONMENTAL POLITICS, THOUGHT, AND JUSTICE

Whether it was rainwater gardens, parks, or bicycle paths, the evolution of the Twin Cities urban landscapes and their hinterlands have been part of long-running debates about how best to use nature, about who benefits and who does not, about whose ideas are privileged and whose are not, and about what all of our many different ways of connecting to nature mean. These diverse and complicated debates are steeped in environmental politics, environmental thought, and environmental justice, which are the focus of the third and final part of the book. The section begins by turning to the northeastern tip of Minnesota, where Twin Cities' conservationists and historians engaged in a lengthy effort to preserve an eighteenth-century British fur trade fort on the Ojibwe's Grand Portage Indian Reservation between 1922 and 1958. Drawing on the aims and policies of the Progressive Era conservation movement, Chantal Norrgard argues in Chapter 12, the campaign to preserve Grand Portage marshaled the logic of both conservation and historical preservation to extend the state's political and cultural influence over both the Grand Portage Ojibwes and the landscape. After a series of negotiations among Native, local, state, and federal officials that stretched over three decades, the land was finally ceded to the federal government, and the Grand Portage National Monument was created in 1958. It stands today as a complicated place: one that commemorates Euro-American history, that collapses the fur trade post into a romanticized wilderness past, and that embodies the tension-riddled politics of historical preservation, conservation, and the very ownership of Grand Portage's history.

Northeastern Minnesota is also home to the Iron Range, where tensions between iron-ore mining communities and the environmental movement have been simmering since World War II. Those tensions, Jeffrey Manuel writes in Chapter 13, ran deeper than the classic jobs-versus-the-environment calculus that pits resource extractive industries like mining against environmental groups. For many Iron Range residents, the region's enormous mine pits and mountains of overburden are symbols of progress, the proud work of generations who turned an isolated northern forest into one of the greatest mining districts the world has ever known. For many Twin Cities environmentalists, the same landscape is seen as scarred and battered, embodying a worrisome overreach of industrial production that threatens the region's lakes, forests, and human communities. Those simplified visions of the region help explain decades of shifting political positions within Minnesota's historically dominant Democratic-Farmer-Labor (DFL) party, growing rifts between Iron Range residents and distant lawmakers in Saint Paul over the state's increasingly proenvironmental policies, and a deepening sense of resentment across the Iron Range based on efforts to turn the region into a playground for nature-seeking tourists from the Twin Cities and elsewhere.

The proenvironmental policies that agitated some Iron Range residents over the last half-century were part of Minnesota's progressive stance toward environmental protection generally. Since the emergence of the American environmental movement in the 1960s, Minnesota has been at the forefront of the nation's struggle to embrace environmental initiatives and build a more sustainable society. This is evident in such signature developments as the creation of the Minnesota Pollution Control Authority in 1967, passage of the Minnesota Clean Indoor Air Act in 1975, and the establishment of the Boundary Waters Canoe Area Wilderness in 1978. It is also evident, as Todd Wildermuth points out in Chapter 14, in the creation of the Minnesota Experimental City Authority in 1971. The goal of the MXC, as it was commonly known, was to create "the greenest modern city ever." The MXC would eliminate polluting industries, recycle its own fresh water, and build housing with views of nature. Cars and trucks would run underground, the population would be capped at 250,000 people, and the city would be built from modular components, allowing for its continuous reinvention like a set of Legos. Although the whole thing might sound rather otherworldly today, the idea was part of a national trend in land-use and urban planning that was being debated alongside other cutting-edge environmental initiatives in the 1960s and 1970s. Although the idea faltered for many reasons and was abandoned by 1973, it represents a rather extraordinary commitment to the ideals and optimism that animated the heart of the environmental movement in its day.

Minnesota's long and deep commitment to environmental protection is also evident in the enactment of the Minnesota Acid Deposition Control Act in 1982. As Gregory Pratt observes in Chapter 15, the Acid Deposition Control Act and subsequent administrative rulemaking outcomes were as remarkable in their own ways as the MXC. Concerned about the acid rain threat to Minnesota's thousands of lakes, the legislature took action to reduce the main source of acid deposition—sulfur emissions from coal-burning power plants—in the face of local opposition and reluctance at the national level. The act overcame resistance from electric utilities, the mining industry, out-of-state coal producers, and scientific uncertainties, and it even overcame an acknowledgment that Minnesota could not solve its acid rain problem on its own. Eight years before national legislation emerged to address the issue for the country, and a decade before any other state took steps to address the problem, Minnesota lawmakers drew on the state's deep heritage of environmental protection to lead the nation in protecting freshwater ecosystems.

Each of the environmental innovations discussed so far fit rather neatly within the traditional focus of American environmentalism. The conservation of resources, the preservation of wildland, and the reduction of harmful pollutants have all been central to the environmental agenda since the turn-of-the-century conservation movement, and these issues and the problems they represent have been regularly cast as universal risks for all Americans. Left out of this frame have been issues of inequality and power. Environmental harm and environmental decision making have always fallen unevenly across American society, further structuring inequities produced by racial, class, and gender divisions in American society. This fusion of social justice and environmental concern began to take root in Minnesota (and elsewhere) in the late 1960s, more than a decade before it took shape as a national environmental justice movement. Those stories are the subject of the next two chapters.

In the late 1960s and early 1970s the American Indian Movement (AIM) made national headlines with a series of militant protests and occupations of federal land. Perhaps the most famous was their participation in the nineteen-month occupation of the former federal penitentiary on Alcatraz Island in the middle of San Francisco Bay. But the group's origins, William Barnett reveals in Chapter 16, lie far from the national spotlight, and grew out of landscapes of racial discrimination and police brutality that are deeply rooted in the Twin Cities. In 1968 Dennis Banks, Clyde Bellecourt, and several other Native Americans living in Minneapolis founded AIM to protest the grim housing and health conditions, poverty and discrimination, and rampant police brutality that Native Americans routinely experienced in the Twin Cities. In response AIM launched wide-ranging efforts to improve living conditions, demand equal treatment, and embrace Indigenous languages and traditions. From their base in Minneapolis, AIM expanded to other cities, built bridges to reservations, and grew during the 1970s into a national movement that pressed for land restoration and tribal sovereignty, among other key issues.

The linked social and environmental inequalities articulated by AIM are also evident in the ongoing battle over the storage of radioactive wastes at the Prairie Island Nuclear Generating Station. Located along the west bank of the Mississippi River just thirty miles from the Twin Cities, Prairie Island is a place of striking contrasts. The northern end of the island is home to the Prairie Island Indian Community (a federally recognized Mdewakanton Sioux Indian reservation), while the southern half is home to the nuclear generating station. Even before construction began on the station in 1969, as James Feldman explains in Chapter 17, debates over its risks and hazards were well underway locally and nationally. By the mid-1970s those debates narrowed to the storage of radioactive waste, which held the power to shut down nuclear power production if it could not be resolved. As the controversy unfolded, it increasingly turned on the principles of environmental justice. The Prairie Island Indian Community, other local residents (including those in the Twin Cities), AIM, and the environmental community put the disproportionate risks of environmental harm, the unequal power to participate in environmental decision making, and the importance of procedural and participatory democracy at the center of their arguments. While the debate over storing hazardous radioactive waste at Prairie Island continues today, it is important to recognize that the nature of the arguments deployed foreshadowed the national environmental justice movement that emerged in the 1980s.

The quest for environmental justice at Prairie Island, and the many other stories that make up this book, reveal the complex ways that culture and nature have combined in the making and constant remaking of the Twin Cities and their hinterlands. Far from being places separated from nature—places where the MPR raccoon and all of nature are banished once and for all—the Cities have always been inextricably connected with the natural world. Those connections blur the defiant environmental binaries that people have long used to structure the meanings of the words *city* and *nature*, while also masking the insights that can be realized by collapsing the all too rigid boundary between them. Once that boundary is wiped away, however, the relationships among the Twin Cities, their hinterlands, and the natural world come into view in far more clear, meaningful, and revealing ways.

We begin to see how a nick point in the sandstone bed of the Mississippi River at Saint Anthony Falls drew settlers to the site of what was to become Minneapolis, eventually tying them to the distant prairie soils of the Red River Valley and to hungry stomachs as far away as London. Fourteen miles downstream where the Mississippi's rapid descent through a narrow rock-filled gorge finally flattens out, Saint Paul organized itself along the riverbank's two natural levees and tied the emergent Twin Cities to steamboats and national markets. As the Twin Cities grew and evolved, so did the nature of their environmental relationships and the meanings assigned to them. Commodity flows like wheat and lumber would continue to shape the Twin Cities and their hinterlands, just as the search for clean water, open space, and other environmental amenities would influence the Cities' built environment and urban form. Running hard through all of these developments were growing social and cultural tensions over how nature was used, whose views dominated environmental decision making, and how prevailing social inequalities magnified the unequal and unjust distribution of environmental benefits and harms in Minnesota and in American society.

Those deep and pervasive connections are the central message of this book. The connections among the Twin Cities and their hinterlands' hydrology and climate, industries and hazardous wastes, poverty and wealth, open spaces and schools, public policies and zoning requirements, racial and ethnic divisions— these are nature's crossroads. Unearthing them helps us to see just how complicated and consequential those interrelationships have become, and just how much our urban lives have always been tied to and dependent on the natural world. Many more stories remain to be told, and we hope this volume helps others to reimagine the many crossroads that run through Minnesota's environmental history.

PART I

The Dynamics of Environmental Change

Cities, Commodities, Hinterlands

A TALE OF TWO WATERFRONTS

Commerce, Industry, and the Environmental Transformation of Minnesota's Twin Cities

Christopher W. Wells and George Vrtis

IN THE SECOND HALF OF THE NINETEENTH CENTURY, the Twin Cities of Saint Paul and Minneapolis grew from tiny, remote outposts on the margins of a rapidly expanding country into large, thriving cities with regional, national, and international significance. The two cities, whose downtown business districts are separated by just fourteen meandering miles of the Mississippi River, shared an important location, standing at the crossroads between the land and natural resources of the areas now known as Minnesota, the Dakotas, and Canadian Manitoba on one hand and the markets of a growing United States on the other. Although both cities rose to prominence in the same half-century of rapid growth by absorbing a flood of new settlers and capitalizing on the surrounding region's natural wealth, the economies of each city connected Minnesota's ecosystems to expanding national markets in different ways.

As a result, the Twin Cities of Minnesota developed as distinct and independent river cities, each with its own local and regional environmental relationships that can be seen in the evolving ways that residents shaped and interacted with their downtown waterfronts. Saint Paul, located at the head of navigation on the Mississippi River just below its confluence with the Minnesota River, built and rebuilt its waterfront around trade during its period of rapid growth, focusing on developing the transportation networks—first steamboats and then railroads—that tied the surrounding region to the city's merchants and delivered the region's natural resources and farm produce to distant markets. Minneapolis, on the other hand, grew to prominence by straddling Saint Anthony Falls, the only major waterfall on the Mississippi River. By constantly tinkering with how they managed the waterfall and harnessed its power, the city's millers turned the waterfront into a center for processing the region's two most valuable natural resources—timber and grain—for regional, national, and international markets. The environmental history of Saint Paul and Minneapolis during their rise to prominence is thus the tale of two waterfronts, one commercial and the other industrial, each built around a different strategy of cashing in on the region's natural abundance during the period when American settlers seized control of the region's land and natural resources.

BEFORE THERE WERE CITIES

Before there were cities called Saint Paul and Minneapolis, before there was a state called Minnesota, there was Dakota land. Both Saint Paul and Minneapolis became cities during the second half of the nineteenth century as the United States pushed its borders westward past the Mississippi. As part of this westward expansion, settlers ultimately wrested land away from the Dakota and Ojibwe (also known as Anishinaabe) peoples who called Minnesota home in order to knit the region's land and abundant natural resources into the fabric of a growing United States. Without controlling the land and its resources and without the ability to remake the region around new economic relationships with the rest of the United States, neither Minnesota nor the cities of Saint Paul and Minneapolis could have emerged as they did. Yet the processes of settler colonialism and colonial conquest did not follow a straight line.[1]

At the site that eventually became the city of Saint Paul, the land began to take its current shape roughly twelve thousand years ago under the powerful forces of retreating glaciers. The Glacial River Warren, which drained the meltwaters of Glacial Lake Agassiz, carved the valleys of the Minnesota and Mississippi Rivers. At the site that is now Saint Paul, a giant waterfall formed, stretching roughly 2,700 feet across with a 175-foot drop, similar in width and height to the modern horseshoe falls on the Canadian side of Niagara Falls. Traversing bedrock composed of three different strata—a thick layer of soft Saint Peter sandstone at the bottom, followed by a thin layer of soft Glenwood shale and a cap of hard Saint Peter limestone—the waterfall slowly receded as falling water eroded away the softer bottom layers at the foot of the falls, undercutting the limestone and causing its unsupported lip to collapse. As the glacial meltwaters receded, the waterfall shrank in size as it eroded its way slowly upstream. The waterfall split at the confluence of the Mississippi and Minnesota

Rivers, devolving on the Minnesota side into rapids two miles upstream, where the limestone layer disappeared. On the Mississippi side, where the Dakota named the waterfall Owamniyomni, it continued to retreat upstream, carving a gorge with 90-foot bluffs through the bedrock.[2]

According to Dakota spiritual beliefs, it was at the place they called Bdote, where the Minnesota River (Mni Sota Wakpa) flowed into the Mississippi River (Haha Wakpa), that the Dakota people were brought from the stars to live over the center of the earth. Beyond Bdote, they called the land Mni Sota Makoce, and eventually spread out in four bands across much of what are now the states of Minnesota, North and South Dakota, and Nebraska, as well as parts of Iowa, Wisconsin, and the Canadian provinces of Manitoba, Saskatchewan, and Alberta.[3] This homeland incorporated a number of different ecosystems, stretching from rich deciduous and coniferous woodlands in the east to prairie grasslands in the west. The Dakota moved around from season to season, living some of the year in portable lodges covered with bison hides and the rest in more substantial structures. By moving their homes with the seasons, they were able to take advantage of the ample sources of food that each ecosystem provided—staying put on fertile agricultural soils to grow corn, squash, and beans for some parts of the year, and during others moving to the places where food was most seasonally abundant, hunting bison, deer, and waterfowl; fishing in lakes and rivers; harvesting wild rice; foraging for nuts, berries, and roots; tapping maples to make sugar; and trapping smaller animals.[4]

Outside forces eventually intruded on the Dakota and their way of life, a process that began even before the arrival of Euro-American settlers. The most disruptive early foreign arrivals in the region, for example, were not settlers but European microbes like smallpox, to which the Dakota lacked immunity. Those European-introduced diseases killed so many that the Dakota population fell by one third between 1680 and 1805.[5] The initial effects of expanding Euro-American settlement also came indirectly, when Ojibwe migrants, themselves pushed from their homelands by white settlers, began to move west onto Dakota land. Initially the Ojibwe acted as important intermediaries between the Dakota and French voyageurs, helping to establish a thriving fur trade. In the late seventeenth century, however, Ojibwe migrants reached the edge of Dakota land, and by the early eighteenth century were beginning to encroach into eastern Minnesota. In the 1730s tensions flared into war.[6] Meanwhile, French and British traders also competed for access to the region, not with the aim of seizing land for colonial settlement but to determine who would control the profits from the region's lucrative fur trade.[7]

In the late eighteenth century, as European empires jockeyed for advantage in North America, war and revolution combined to realign imperial politics. First, at the end of the Seven Years' War in 1763, France surrendered its colonial holdings in North America to Britain. Then, soon after the United States emerged victorious from the American Revolution in 1783, the new nation began to compete vigorously with British traders from Canada for control of the region. Its first important foray came in 1805, shortly after the Louisiana Purchase gave the United States claim to the area west of the Mississippi, when Lieutenant Zebulon Pike visited Bdote as part of an ultimately unsuccessful US expedition to find the headwaters of the Mississippi River.[8] While in the area he brokered an unauthorized treaty with local Dakota leaders, securing some one hundred thousand acres around the confluence up to Saint Anthony Falls for the construction of a military fort. The first American troops did not arrive until 1819, however, and did not finish work on the fort that came to be known as Fort Snelling until 1825.[9]

Built on the bluffs overlooking the confluence of the Minnesota and Mississippi Rivers, Fort Snelling quickly became the center of American activity in the region. The fort's primary purposes were to secure American control over the fur trade by protecting American interests against British Canadian competitors, to try to keep the peace between competing Dakota and Ojibwe groups, and to prevent settlers from encroaching on the area and disrupting the fur trade.[10] In 1825 the American Fur Company opened a headquarters in Mendota, directly across from Fort Snelling on the far bank of the Minnesota and just above Pike Island and the confluence with the Mississippi. It attracted lively activity and helped the company achieve dominance over the fur trade by the end of the 1820s. By the mid-1830s Fort Snelling stood at the center of a fur trade network that extended north into Canada and as far west as the Missouri River.[11]

By the mid-1830s, however, the system of credit on which the fur trade depended had become overextended and structurally unsound, with fur traders' account books brimming, in the words of their primary historian, with "accumulated losses masquerading as bad debts."[12] As a result, leading fur traders began to assess the steady western push of American settlers with a new eye. Until then they had staunchly opposed opening new territory for settlement because settlers threatened the conditions—social, economic, and environmental—on which the fur trade depended. As the credit underpinnings of the fur trade became increasingly unstable, however, its leaders began to develop a new goal: leveraging the federal treaty-making process for personal advantage. In 1836 the fur traders at Mendota saw their chance when the new territorial governor of Wisconsin Territory, Henry Dodge, began to push for a land cession treaty in what is now western Wisconsin and eastern Minnesota. The fur traders inserted themselves into the process, which produced two treaties in 1837, one with the Ojibwe and another with the Dakota, that together ceded millions of acres of valuable timberlands in current-day Wisconsin and Minnesota, including the portion of Minnesota known as the Saint Croix Triangle, which occupies the area between the Mississippi and Saint Croix Rivers below a line running east from near current-day Brainerd, Minnesota, into Wisconsin. In addition to the mix of immediate and deferred annual payments (called "annuities") that went to the Ojibwe and Dakota in exchange for the land cessions—and in a windfall for traders—the federal government also agreed to make direct cash payments to fur traders to cover significant portions of the recorded debts they carried on their books.[13]

The 1837 treaties opened a new era in the region's history, marking a shift in American interests away from a single-minded focus on exploiting furs and toward a much grander objective: seizing direct control of Dakota and Ojibwe land. That land was rich with natural resources, especially timber, the potential worth of which made the profits of the fur trade pale by comparison. It also had rich soil, which in the long run would make the region attractive for settlement and agriculture. Upon ratification of the Dakota treaty in 1838, land east of the Mississippi in the Saint Croix Triangle officially opened to settlers, who immediately began to make land claims. This set in motion the processes that transformed and divided Dakota land into the privately deeded parcels that the US legal system recognized.[14] It was at this point that two ramshackle towns began to develop within the ceded area on the east bank of the Mississippi: Saint Paul, located at a break in the bluffs just below the Fort Snelling military reserve's boundary, and Saint Anthony (later part of Minneapolis), located alongside the waterfall that the Dakota called Owamniyomni but that settlers called Saint Anthony Falls.[15] Adopting a commercial orientation from the start, Saint Paul's population grew in the 1840s in proportion to its steamboat traffic, which began regular packet service to Saint Louis in 1847. Steamboat arrivals rose steadily through 1849, when ninety-five steamboats made their way up the Mississippi to dock at Saint Paul. In Saint Anthony, by contrast, upon receiving word that the 1837 treaties had been ratified, Franklin Steele, an enterprising young storekeeper at Fort Snelling, managed to outrace the fort's commander, Major Joseph Plympton, to stake the crucial claim to the land along the east bank of Saint Anthony Falls. Lacking the capital required to develop the site, however, it took Steele almost a decade to erect a milldam and open his first sawmill.[16]

The 1837 treaties created an opening for settlers, but before Americans could fully exploit the region's natural wealth and before they could create the conditions that would propel the transformation of these two tiny new towns into the large, flourishing cities of Saint Paul and Minneapolis, the new settlers had to make good on their designs to gain firmer control over the Dakota and Ojibwe land beyond the boundaries of the Saint Croix Triangle. The first step in this direction came in 1848 when Wisconsin became a state, prompting Congress in early 1849 to organize Minnesota Territory from the area running west from the Saint Croix River to the Missouri River and north from Iowa to the Canadian border. Congress named Saint Paul as the territorial capital, sparking a building boom and prompting the town's population to double within three weeks.[17] Among other things, organizing Minnesota as a territory meant introducing new forms of sovereignty to the region, including naming a territorial governor, creating a territorial legislature, introducing a new set of legal relationships with the United States, and providing a clear path to statehood.[18]

Almost immediately after the creation of the Minnesota Territory, its political leaders began to push Dakota and Ojibwe leaders to sign new land cession treaties that would legally open the rest of Minnesota for settlement and natural resource extraction. In 1851, after two years of careful planning, American negotiators pressured Dakota leaders into signing two separate treaties—one with the Wahpetun and Sisseton bands at Traverse des Sioux and another with the Mdewakanton and Wahpekute bands at Mendota. The treaties together ceded nearly all remaining Dakota lands in Minnesota for a mix of annuities and onetime payments while confining the Dakota to a long, narrow reservation on the upper Minnesota River. In addition, the Americans were able to trick some Dakota leaders and pressure others into signing separate "trader's papers," which obligated them to repay all of the recorded debts accumulated by individual members of their bands—some of which stretched back as far as 1819. It was, in the words of one of the episode's major historians, "a sorry spectacle of deceit, coercion, and promises broken almost before they were recorded."[19] The state's leaders then turned their attention to the Ojibwe, whose land was less attractive for farming but contained vast pinelands that promised ample riches for American timber operators. As a result, the treaties of La Pointe (1854) and Washington (1855), which ceded all but the northernmost portions of Ojibwe lands in exchange for a mix of immediate payments and long-term annuities, focused primarily on opening the territory for logging rather than settlement. As lumbering operations slowly turned the region's forests to stumplands, however, the environment could no longer support and sustain traditional Ojibwe ways of life.[20]

Once these land cession treaties opened the rich ecosystems of Minnesota Territory to the claims of settlers and timber operators, immigrants began to pour in. In 1850, just after the formation of Minnesota Territory, the territory's population stood at just 6,077. By 1857, following the key treaties in 1851, 1854, and 1855, its population had swelled to more than 150,000, with farmers spreading across its fertile soils and loggers advancing into its vast pinelands.[21] Political leaders used this rapid growth to launch a campaign for statehood that culminated in Minnesota's entry into the union as its thirty-second state on May 11, 1858.[22]

As settlers rushed to claim the land as their own, the United States finalized its conquest of Dakota land in the short, brutal US-Dakota War of 1862. The war began in August, after a summer of crop failures, hunger, and the failure of Indian agents to issue rations provoked tensions that flared into violence. The fighting, which ended before the end of September, provided the state's leaders with all the reason they needed to banish the Dakota from Minnesota. First, a military commission held perfunctory trials of Dakota combatants that ultimately resulted in thirty-eight hangings, the largest mass execution in American history. Next, they confined some 1,600 Dakota noncombatants in a camp through the winter on Pike Island, where overcrowding, disease, and exposure killed several hundred people. That spring Congress used the war as a pretext to extinguish all former treaties with the Dakota, dismantle their reservation, and relocate the survivors of the Pike Island camp to the Crow Creek Reservation in Dakota Territory. As far as the US government was concerned, this was no longer Mni Sota Makoce, land of the Dakota, but Minnesota, part of the United States of America.[23]

MARKETING NATURE IN SAINT PAUL

With this transformation the new state's fertile soils, valuable pinelands, and rich mineral deposits stood ripe for the taking, and over the next several decades a wave of new settlers rushed in to take advantage of the bounty. The state's population more than doubled between 1860 and 1870 from 172,023 to 439,706, nearly doubled again to 780,773 in 1880, and reached 1.3 million in 1890. In its first decades of rapid growth, Saint Paul's leaders maximized the city's advantages as the head of navigation on the Mississippi River by orienting their activities toward the waterfront, building Saint Paul around the steamboats that linked Minnesota to larger national markets. In doing so they established the city's status as the region's primary commercial center, and later cemented that status by becoming the region's railroad hub. As a result they ensured that as the state's population grew and as settlers remade the state's environment, much of the countryside's trade would flow through Saint Paul as Minnesotans interacted with the broader American economy.[24]

The founders of Saint Paul picked a good place to build a city around steamboats. Located at a break in the bluffs, the site had ready access to the waterfront from the east bank of the river (Figure 1.1). In addition, it had two natural levees where steamboats could land and unload, each of which adopted a distinct commercial focus that capitalized on a different aspect of the city's crossroads location within the developing region. Located just below the Mississippi's confluence with the Minnesota River, the Upper Landing cultivated trade up the Minnesota River and then across a short portage to the Red River, which flows north to Manitoba. The larger Lower Landing, on the other hand,

Figure 1.1. In this 1857 view of Saint Paul, a steamboat chugs upstream past Harriet Island, headed toward the Mississippi's confluence with the Minnesota River, while another
steamboat docks at the Upper Landing. Behind the landing, the burgeoning town's buildings crowd around the break in the bluffs. Courtesy of the Minnesota Historical Society.

devoted itself to the steamboats plying the Mississippi River, where Saint Paul's
location at the river's head of navigation gave it a clear strategic advantage:
nearly all of the new settlers streaming into Minnesota during its early period
of growth came up the river and disembarked in Saint Paul. Initially most arrived via a seven-hundred-mile steamboat journey from Saint Louis, but after
1854 Chicago became the major way station for settlers. By then railroad tracks
radiated outward from Chicago in all directions, allowing settlers to travel by
rail to Rock Island, Illinois, before making the final 350-mile trip to Saint Paul
by steamboat. As Minnesota's population swelled, so did trade in Saint Paul,
where both of its levees teemed with activity. In 1857, for example, 216 boats
from the Minnesota River docked at the Upper Landing, while 1,026 traveled
up the Mississippi to the Lower Landing.[25]

Through the 1850s and into the 1860s Saint Paul built its commercial dominance by capitalizing on its position at the center of regional transportation
networks during the period when settlers were fanning across Minnesota to
establish new towns and farms. For every steamboat that traveled up the Mississippi to Saint Paul, a new group of passengers disembarked to find hotels
and boardinghouses to house them, banks and loan offices standing ready to
extend credit, land agencies and real estate firms offering to facilitate land ac-
quisitions, and, for those who wanted to stay, plenty of jobs. For those who
left Saint Paul to establish farms in other parts of Minnesota, or to join the

timber companies operating in Minnesota's vast pineries, the city's merchants built a lively trade equipping them with tools and supplies. Just as important, the city's overland transportation firms (known as express companies) built a robust overland transportation system, with Saint Paul at its center, that specialized in moving goods, supplies, and passengers between the city and new interior settlements. Meanwhile Saint Paul's merchants were well positioned to distribute commodities from the surrounding region, including both farm products and timber, to both local and downriver markets.[26] Because of these relationships, Saint Paul grew along with the region, sprouting from a town of less than a thousand residents in 1849, the year it became the capital of Minnesota Territory, to a population of eleven thousand on the eve of the Civil War, a rate of growth all the more impressive given that the city lost half of its population in the aftermath of the deep nationwide economic recession known as the Panic of 1857.[27]

Through this period both Saint Paul and its surrounding hinterland depended on the free-flowing Mississippi River to connect with the outside world. At the time the Mississippi was both wider and shallower than it is today, with marshy, somewhat indistinct banks that varied considerably as the river rose and fell. Islands and sandbars braided the channel, making it tricky to navigate by steamboat without running aground, especially during low water. Fallen trees created snags that could (and sometimes did) do serious damage to boats. Above Saint Paul, where the river became narrower and faster above its confluence with the Minnesota River, and where huge chunks of limestone left by the retreat of Saint Anthony Falls lined the riverbed, the Mississippi quickly became unsafe to navigate, reinforcing Saint Paul's status as the river's last port of call. In Saint Paul people filled, leveled, and extended the levees to serve the commercial needs of expanding steamboat operations. The river itself, however, remained unchanged until 1868, when the Army Corps of Engineers began operating two boats on the Upper Mississippi that focused on scraping sandbars, clearing snags and overhanging trees, and removing wrecked ships from the channel. Each change of the seasons cast the river's importance to Saint Paul in sharp relief. Every winter the river froze, causing the city's economy to slow to a standstill, and every spring the ice cleared, reawakening the economy as soon as steamboats could once again make their way upstream to the city.[28]

As central a role as steamboats played in Saint Paul's early development, the city's heyday as a steamboat city was short-lived. Even as Minnesota's territorial leaders were preparing for statehood, they already had begun to view the city's future through the prism of railroads, envisioning a network of four trunk railways radiating outward from Saint Paul. The Panic of 1857 wrecked these initial plans, but the Civil War revived both the economy and Minnesota's railroad fever; leaders then concentrated on building an intrastate network of railroads focused on the Twin Cities. In June 1862 the Saint Paul–headquartered Saint Paul and Pacific Railroad (SP&P) began to operate the state's first railroad, a ten-mile link between Saint Paul and Saint Anthony. Within a few years its tracks reached north across the state into the fertile Red River Valley. Meanwhile, the Minnesota Valley Railroad (MVR) began to stretch southward into Iowa and Nebraska, facilitating the transition of Commodore William Davidson (who dominated Saint Paul's steamboat trade) and James C. Burbank (who owned Saint Paul's dominant express company) into railroading.

The pace of construction accelerated after the Civil War, reinforcing Saint Paul's central location as the state's total mileage of railroad tracks jumped from 210 in 1865 to 560 in 1868. Often built to follow the transportation routes that express companies had already established, the railroads quickly absorbed trade that formerly belonged to steamboats and express wagons. But even as railroads captured this trade, causing steamboat arrivals in Saint Paul to fall steadily after 1858, the early railroad network was still focused on Saint Paul's waterfront, where the river remained the only outlet to eastern markets. As a result, the Lower Landing's warehouses, distribution activities, and freight forwarding operations continued to hum with activity. Even after the first railroad connections to Chicago opened—including a roundabout connection in 1869 and a direct one in 1872—Saint Paul retained both its position at the center of the region's transportation system and its regional commercial dominance.[29]

As railroads built outward from the Twin Cities, federal land policies attracted immigrants and sped the transformation of southern and western Minnesota's grasslands—until only recently Dakota land—into privately deeded farmland. The Homestead Act of 1862, for example, offered 160-acre plots to US citizens and intended citizens in exchange for a modest filing fee and the promise to live on the land for five years. Immigrants responded immediately, filing ten thousand Homestead claims in Minnesota by the end of the Civil War; by 1880 the number of claims swelled to sixty-two thousand, accounting for nearly one-seventh of the state's land.[30] The federal government also granted extensive land holdings to railroads to spur their expansion, attracting an additional wave of settlers. In 1856, for example, Congress designated 12 million of Minnesota's 53.5 million acres for land grants to encourage railroad construction, following the general template established earlier in the decade which stipulated that unless lines were completed within a designated period, the lands would revert back to the United States. In 1864 Congress chartered the Northern Pacific Railroad Company (NPR), ensuring that a transcontinental line would extend west through Minnesota on its way to western ports on Puget Sound. After the Civil War—while Saint Paul's leaders maneuvered to ensure that the NPR would share tracks with the SP&P and thereby stop in Saint Paul—railroad companies raced to construct their lines before the land grants elapsed. Upon completing their routes and securing their land grants, the railroads sold the land to immigrants seeking farmland and directed the proceeds toward constructing new routes (and securing new land claims).[31] By the mid-1880s, as railroads completed the construction of their trunk lines and Minnesota's population climbed past 1.1 million, all of the state's best soils had been settled, transformed into farms, and tied by rail to the Twin Cities.[32]

As the state grew so did Saint Paul, which undertook extensive renovations to remake its waterfront around railroads. In doing so, however, it began to marginalize the once central place of the river in the city's everyday life. Around the Lower Levee, which became the focus of railroad activity even before the first Union Depot railroad station opened in 1880, the railroads created the space they needed to operate by extending the shoreline into the river. They began by constructing their lines on trestles that spanned sandbars and seven small islands in the river itself. Later they razed local hills and used them to fill in the area between the old shoreline and their tracks. In places where the river was too narrow to accommodate new railroad tracks, such as at Dayton's Bluff and in parts of downtown Saint Paul, they blasted the bluffs back to make room. Soon the entire stretch of waterfront between the Upper and Lower Levees had been filled and extended into the river, shrinking the river's width by as much as a thousand feet as it flowed through Saint Paul.

With an economy less dependent on an ice-free river and the daily movements of steamboats, and with a formidable barrier of railroad tracks rising between downtown and the waterfront, the city's daily life and commerce reoriented around the railroads. During the same years that the city's population burgeoned from 20,030 in 1870 to 111,397 in 1885, residential growth also turned away from the river. Aided by a horse-drawn streetcar system that began operating in 1872, the city's elites and middle-class residents relocated into new neighborhoods along the bluffs to the west of downtown, leaving the lower-lying industrial corridors along the river to the city's poor and working-class residents.[33]

Ironically, the most ambitious plans to tame the Mississippi to serve navigation interests came to fruition only after railroads replaced steamboats as the region's dominant transportation system. Facing political pressure from farmers who chafed at their growing dependence on railroads and what seemed like their capricious rate-setting practices, in 1878 Congress authorized the Army Corps of Engineers to undertake the wholesale transformation of the Upper Mississippi by creating a continuous four-and-a-half-foot-deep navigation channel from Saint Paul to the mouth of the Illinois River, just above Saint Louis. For the next two decades, the Army Corps transformed the river with two primary tools: "wing dams," which narrowed the river by creating new shoreline, and "closing dams," which forced the main river channel to stay on one side of each of the many islands dividing the river. Together these changes constricted the flow of water, ensuring that even at low water the river's current would carve through sandbars and maintain a navigable channel with a consistent minimum depth. In addition to ensuring that the steamboat trade in Saint Paul, no matter how diminished, would not entirely disappear, the project also caused what one of its leading historians has called "dramatic changes in the physical and ecological character" of the river.[34]

The career of James J. Hill, Saint Paul's most famous and powerful resident, illustrates Saint Paul's evolving relationships with the river and with the surrounding region. Hill migrated to Saint Paul in 1856 from Canada when he was eighteen, finding work on the docks before becoming a clerk for a steamboat line. In 1866 he built a warehouse and launched a company focused on the efficient transfer of freight between steamboats and railroads. Soon he began to specialize in handling coal, which fueled both steamboats and railroads, and became wealthy by dominating the region's fuel trade. In the early 1870s he expanded into the steamboat business, focusing on the Red River Valley trade. In 1877 he made a momentous leap into railroads, joining a group known as "the Associates" in the purchase of the SP&P, which had fallen victim to the Panic of 1873. Renamed as the Saint Paul, Minneapolis, and Manitoba railroad (Manitoba), Hill served as its general manager and quickly turned it into a thriving enterprise by hastening the transformation of the Red River Valley. In the booming real estate market of the early 1880s, the Manitoba sold off its land grants to eager settlers, who soon transformed the valley into prosperous farmland. Meanwhile, the Manitoba connected Saint Paul to Winnipeg and became a profitable line, carrying people, goods, and agricultural products—especially wheat—between the Red River Valley and the Twin Cities.

As his wealth grew, Hill's interests shifted away from the original waterfront focus that had animated his rise to power. Hill helped cement Saint Paul's residential shift away from the river in the early 1880s, for example, by leaving his Canada Street mansion near the Lower Levee to build an opulent 36,500-square-foot mansion on the bluffs overlooking the city. By the mid-1880s Hill's personal and business interests also no longer focused on the city's waterfront, but on the vast regional transportation system that he controlled. Then in the late 1880s his ambitions shifted even further afield as he began to earn the nickname by which he is known today, the "Empire Builder." The Manitoba became the Great Northern Railway in 1887, and Hill began to build westward toward the coast, completing a transcontinental line to Seattle in January 1893. Its completion symbolized a new, more national and interconnected period in Saint Paul's history.[35]

By the time Hill began to enact his transcontinental dreams in the late 1880s, Saint Paul's initial flush period of runaway development—based on rapid population growth and a steadily expanding trade with the settlers who were busy remaking Dakota and Ojibwe land into a patchwork of farms and timber operations—was coming to an end. After 1880 commercial Saint Paul fell steadily behind industrial Minneapolis in size and influence, despite more than tripling its population from 41,473 in 1880 to 133,156 in 1890. In the next decade, however, Saint Paul's growth slowed, adding a more modest 30,000 residents by 1900. Even so, Saint Paul remained a thriving city situated at the center of a wealthy region's transportation system. In the last two decades of the nineteenth century, Saint Paul embarked on a new era in its environmental history, turning its attention toward modernizing its infrastructure: building a water system, installing sewers, paving streets, and electrifying its streetcar network.[36] New residential development continued to spread farther and farther away from the waterfront that had been central to the city's early development, and the river became less and less central to its overall economy and civic identity. Having sprung to life as a town built around its steamboat landings, by the end of the nineteenth century Saint Paul had become a railroad city, all but turning its back on the river as the central, defining feature of its everyday life.

ENGINEERING NATURE IN MINNEAPOLIS

Unlike Saint Paul, which began to grow immediately after the 1837 treaties opened the Saint Croix Triangle for legal settlement, growth at Saint Anthony Falls was constrained by limited access to land and capital in the decades before the Civil War. Franklin Steele claimed the valuable east bank of Saint Anthony Falls in 1838—the same year that Saint Paul began to grow—but it took him a decade to attract investors, secure formal title to the land, and finish the initial construction of his dam, sawmill, and related operations there. Not until 1849 was he able to subdivide his land, file plats, and formally found the town of Saint Anthony, prompting its first growth spurt. In 1852 four sawmills on the east bank milled 8 million board feet per year, and in 1857 Saint Anthony's population reached 4,689.[37] Across the falls, where the land remained part of the Fort Snelling Military Reserve until 1852, the many squatters there were not able to purchase clear title to their claims until 1855–1856. Only after securing title were they able to found the town of Minneapolis in 1856, the same year that the territorial legislature chartered two waterpower companies—the Saint Anthony Falls Water Power Company on the east bank and the Minneapolis Mill Company on the west bank—to develop the waterfall's power.[38]

Despite identical charters, the two companies diverged significantly in their approaches, which explains Minneapolis's surge ahead of Saint Anthony. Short of capital, the Saint Anthony Company suffered from chaotic management, contentious relationships among a churning group of investors, and the loss of a key waterpower site on Nicollet Island. In Minneapolis, by contrast, a small, unified group of wealthy owners modeled their operations on the leading milling center in the United States at the time, the mills at Lowell, Massachusetts. In addition to adopting Lowell's leasing system for waterpower, in 1857 they dug a 40-foot-wide, 215-foot-long, 14-foot-deep power canal into the soft sandstone of the river's west bank to supply their mills, along with an extensive network of head- and tailraces. By 1858 Minneapolis had seven sawmills to Saint Anthony's six. Over the next decade Saint Anthony's population stagnated, numbering just 5,013 in 1870, while Minneapolis—which added mills after extending its power canal to 600 feet in 1866—surged ahead of Saint Anthony to 13,066.[39]

As the leaders of Minneapolis and Saint Anthony developed an industrial waterfront around Saint Anthony Falls, they confronted a variety of environmental challenges. Perhaps most significantly, the sawmills at the waterfall depended on an untamed river whose most troublesome characteristic, from the perspective of the millers, was its variability. In a typical year the river ran high with meltwater in the spring, lost half its volume or more by late summer and into the fall, and then fell by half again after icing over during the winter. But not every year was typical, and the river could behave unpredictably: both flooding and droughts were ever-present possibilities. In addition to shaping the river, seasonal fluctuations also shaped the logging industry on which the sawmills depended. Each winter crews spread into the forests to cut timber, and each spring, after the ice broke, they floated it downstream to the sawmills.[40] Three boom companies formed to handle the log drives, including one that operated in the field at the site of the cuts, another that operated above the waterfall, and a third that operated below it.

The variability of the river, together with certain timber industry practices, created headaches for the sawmill operators at the falls. Floods could be particularly damaging, for example, threatening not only the sawmill platforms that lined the waterfall, which floodwaters occasionally swept away, but also the falls themselves. Driven by the high water, both ice and logs routinely pounded the thin limestone lip of the waterfall, accelerating natural processes of erosion. By 1866 leaders of the Minneapolis Mill Company had become so concerned by this problem that they invested $35,000 to erect a wooden apron across the western channel of the falls, only to see a flood promptly batter it beyond repair in June 1867. The episode underscored both the river's destructive force and the growing focus of the waterpower companies on finding engineering solutions to their problems.[41]

The desire to engineer the falls came into dramatic conflict with natural processes of erosion and with the river's destructive power in October 1869. First, rising floodwaters brought work on a new, half-built replacement apron to a temporary halt. Then the river dramatically escaped its bounds, opening a break into an underground tunnel being excavated as a tailrace. As the river plunged into the tunnel, it soon created an expanding whirlpool that eroded away big chunks of Hennepin Island on the east side, threatening the falls themselves with collapse. Frantic efforts through the winter managed to seal the hole with masonry walls, but another flood in spring 1870 reopened the hole, sucking a mill and warehouse into the river. Again crews leapt into action, and Congress, watching with a wary eye from Washington, appropriated funding for the Army Corps of Engineers to take over the rescue operation.

Things soon seemed to be in hand, but in spring 1871 yet another break opened in the east-side millpond at a different point along the tunnel, prompting a strategic reassessment that ultimately led to a more thorough reengineering of the falls.[42] Further study revealed problems much bigger than the leaking tunnel: the protective top layer of limestone thinned and disappeared just 1,200 feet upstream of the falls. Worse, water was already infiltrating through the limestone and into the soft lower layer of sandstone, creating an existential threat to the waterfall. The best way to proceed, the engineers decided, was to build a vast underground dike above the falls—a thick concrete wall sunk 40 feet deep into the riverbed and spanning its entire width—to prevent further erosion. With funding from Congress, the Army Corps spent the remainder of the decade reengineering the falls. Corps crews began with the dike, followed by a new apron across the falls, low dams above them, and a sluiceway to control the passage of logs.[43] Construction crews also filled the problematic tunnel with tens of thousands of cubic feet of gravel to prevent any future collapses.[44] The result was a heavily engineered waterfall that bore only passing similarity to the one that had initially drawn residents to the area. Just as Saint Paul had remade its waterfront to serve the steamboats and railroads that made it a commercial hub, Minneapolis's newly remade waterfall clearly reflected the industrial purposes to which the city's residents put the river.

At the time of the tunnel break, the primary (though not exclusive) industrial function of Saint Anthony Falls was to provide power to sawmills, whose soaring output through the 1850s and 1860s drove population growth and economic expansion on both sides of the river and increasingly gave the two towns a sense of common identity. Sawmills at the falls were the leading employers in both towns from the start, steadily expanding their production from 12 million board feet in 1856 to more than 90 million in 1869—enough lumber to build a plank road eight feet wide and four inches thick stretching across 532 miles, or the distance between Minneapolis and Saint Louis. Meanwhile the total value of forest products swelled from $357,900 in 1860 to $1.73 million in 1870.[45] Connected to seemingly inexhaustible northern forests by the Mississippi River and its major tributaries, the sawmills at Saint Anthony and Minneapolis sang the song of prosperity and together propelled the two towns into position as one of the nation's leading sawmilling centers. As Saint Anthony and Minneapolis rose in stature, the common ground between them increasingly came to seem more important than their rivalry, especially given the prominence of their common downstream competitor, Saint Paul. In 1872 the two towns became one, with Minneapolis legally absorbing Saint Anthony into its boundaries, forming a united city with a population of twenty-eight thousand residents.[46]

During the same decade that engineers remade the falls, equipping them with a new dike, apron, and dams, a quiet technological revolution catapulted the flour industry ahead of the timber industry at Saint Anthony Falls, changing the city's environmental relationships both with the falls themselves and with the hinterland that kept its mills humming with activity. Before the 1870s traditional flour milling techniques produced a relatively undesirable flour from the region's primary variety of wheat, hard spring wheat. The low-quality flour had a limited market, and in 1870 the city's flour mills produced just 193,000 barrels, making them a distant second to sawmills as the most important industry at the falls. Beginning in the early 1870s, however, millers introduced a series of new technologies that allowed them to grind hard spring wheat, at low cost and in steadily increasing quantities, into a fine, white flour that fetched a premium price. Production grew quickly, and by 1873 Minneapolis's flour production jumped to 585,000 barrels of superior flour. The newfound capacity to engage in large-scale, profitable flour production benefited not just millers, but also wheat farmers, spurring a rapid expansion of wheat cultivation in western Minnesota and the Dakotas that sent a rising torrent of grain flowing to Minneapolis's mills.[47] As a result of these changes, the city's millers increased both the range of natural resources that they processed for the market and the geographical extent of the environmental relationships that sustained industrial operations at the falls. After being dependent primarily on timber operations in the North Woods through the 1860s, in the 1870s the city's waterfront industries began to diversify to make use of a new natural resource: the new fields of waving grain spreading ever westward across Minnesota and the Dakotas.

As the fortunes of the city's flour mills rose, they slowly began to displace sawmills at Saint Anthony Falls, helping remake Minneapolis's civic identity from two competing "sawdust cities" into a single city famous for its flour production.[48] As late as the middle of the 1870s, despite the growing profitability of flour milling, any casual observer at the falls could conclude—after taking in the river full of logs and listening to the steady whine of the gang saws at work in the sawmills lining the falls—that the timber industry still dominated the newly unified city. Indeed, taken together in 1876, the nineteen sawmills at Saint Anthony Falls were still far and away the city's largest employer and collectively sawed 53 million board feet of lumber.[49] But 1876 proved to be a turning point.

That year, as the Army Corps of Engineers finished its massive concrete dike project, it warned the two power companies that sawmills threatened the future of the falls, partly because logs and other debris continued to damage the lip of the waterfall and partly because sawmill platforms constrained the river channel in ways that exacerbated flood damage. Soon after the Minneapolis Mill Company announced that it would not extend the leases of the west-side sawmills after they elapsed. Between 1876 and 1880 it purchased the remaining sawmills and began to phase out production. Flour mills pushed into the opening but suffered a dramatic setback on May 2, 1878, when the Washburn "A" flour mill exploded, killing eighteen and leveling much of the west side. The tragedy, however, only prompted leading millers to double down on their commitment to rebuilding the industrial district with even larger, more technologically advanced flour mills.[50] By 1880 twenty-two flour mills were operating on the west side alone, although just two companies—C. A. Pillsbury and Company on the east side, then in the process of building the world's largest flour mill, and Washburn, Crosby and Company on the west side—together produced over half the city's flour.[51] Milling a combined 2 million barrels of flour valued at over $20 million in 1880, flour for the first time surpassed timber as the city's leading industry, pushing Minneapolis past Saint Louis as the nation's leading flour producer—just as the city's population nosed past Saint Paul's by 5,500 residents.[52]

The rise of the flour mills at Saint Anthony Falls did not spell the end for sawmills in Minneapolis, however; instead, in a sign of broader changes then sweeping over American industry, the timber industry continued to prosper after relocating to North Minneapolis. Although the sawmills remained dependent on both the Mississippi and northern forests for a steady supply of logs, sawmill operators found it easy enough to replace the power once provided by Saint Anthony Falls with power from steam engines. Steam cost more per horsepower than the inexpensive energy provided by falling water, but the margins were smaller for the timber industry than for most others, since it could repurpose its chief by-product—sawdust—as fuel.[53] There were also economic advantages to being able to spread out unencumbered on the inexpensive land north of the falls, which the timber industry quickly filled with mills, warehouses, railroad yards, and other associated industries even as it filled the river north of the city with new booms.[54]

Although the timber industry cut its ties to Saint Anthony Falls, it retained its easy access to the ample supply of logs floating down the Mississippi from northern forests, and by adding new methods and technologies, the sawmills were able to remain a leading industry in the city. In fact, their output steadily *grew*, climbing from almost 180 million board feet in 1880 to over 325 million board feet in 1890, making Minneapolis the nation's leading lumber producer through 1899, when it milled 594 million board feet. Thereafter production began to fall—slowly, at first, but then rapidly after 1915 when the northern forests gave out.[55] Propelled by its embrace of steam power, the timber industry's successes in the last two decades of the century marked a new era for Minneapolis, in which population growth and expanding industrial operations became less reliant on the power of falling water—and thus less tightly bound to the city's industrial waterfront—even as the city's industries continued to focus on processing the surrounding region's natural wealth for national and international markets.

As the millers at Saint Anthony Falls focused first on saving the falls themselves and then on boosting their industrial output, James J. Hill, Saint Paul's railroad magnate, moved to strengthen the position of his railroad empire by establishing firmer control over the wheat flowing into Minneapolis. In 1880 Hill oversaw the purchase of the Saint Anthony Falls Water Power Company and became its president in 1882, not because he was interested in adding waterpower development or milling to his activities but to give his railroad operations a foothold on the Minneapolis waterfront. After acquiring control of the power company in 1880, Hill moved quickly to begin construction of the Stone Arch Bridge, which even today remains the signature feature of Minneapolis's waterfront, aside from the falls themselves. Opening in November 1883, the 2,100-foot bridge snakes elegantly across the river below the falls, crossing twenty-three enormous stonework arches to connect the east- and west-side mills. Hill also opened a new Union Depot in 1885, a few blocks away from the west-side milling district.[56] These facilities fully integrated the waterfront flour mills into Hill's extensive network of railroads and grain elevators, establishing indelible connections to the western wheat fields and ensuring that a steady supply of wheat would flow along the tracks of Hill's transportation empire into Minneapolis's mills (Figure 1.2).

As Hill competed with other railroads to bind western wheat fields to Minneapolis's mills, the millers at the falls developed a keen interest in gaining more reliable control over the other key resource flowing into their mills: the water itself. By the late 1870s the millers had moved beyond a simple understanding of the immediate economic damages wrought by extreme river conditions—flood and drought—and were moving toward a new view of fluctuating water levels as a sign of economic inefficiency. When water was higher than normal, even if it did not cause flood damage, the water overtopping the dams at Saint Anthony Falls represented potential production capacity gone to waste; when water was lower than normal, even at levels that did not amount to drought, mills could not operate at full capacity. To solve this problem, Minnesota representative William D. Washburn—who, along with his brother Cadwallader C. Washburn and cousin Dorilus Morrison, was a major stakeholder in the

Figure 1.2. In this postcard (ca. 1890), a larger-than-life version of Pillsbury's massive flour mill sits astride Saint Anthony Falls on the right, with the iconic Stone Arch Bridge in the foreground. The west-side milling district, including most prominently the facilities of Washburn, Crosby and Company, which had been rebuilt after they exploded spectacularly in 1878, are on the left. Courtesy of the Minnesota Historical Society.

Minneapolis Mill Company and various flour and timber mills—began to seek congressional funding for a reservoir system in the Mississippi River headwaters region. A handful of strategically located dams to create reservoirs, he argued, would allow the Army Corps of Engineers to even out the flow of the Mississippi, impounding water when the river ran high and releasing it when it was running low. This would benefit navigation interests along the river—a legitimate interest of Congress—but would also benefit Minneapolis's millers, including Washburn himself.

Yet building dams would also create problems. Congress appropriated money for its first dam in 1880, and over the next decade built dams on lakes Winnibigoshish, Leech, and Pokegama, as well as on the Pine River. The dams had their largest impact, however, not on boats or mills, but on the Ojibwe residents of the headwaters region, where fluctuating water levels decimated wild rice grounds, fishing areas, and hay supplies, while damaging sugar-making trees and inundating graveyards. "I am heartsick over this whole matter," wrote Henry Whipple, an Episcopalian bishop and advocate for Native American rights, in 1883. "It is one of the many instances where we have clearly violated principles of justice."[57] The injustice proved all the more piercing because even as the Army Corps built the new headwaters dams, Minneapolis's flour mills were beginning to install auxiliary steam engines that allowed them to maintain consistent production schedules when the river ran low. The backup power systems were expensive to install and operate, but managers calculated that the profits from operating at a consistent capacity year-round would more than offset the extra costs that came with providing supplemental steam power when necessary.[58]

The introduction of auxiliary steam power in the mills at Saint Anthony Falls reflected the degree to which the world of the 1890s differed from the period of ascendant growth in which Minneapolis and Saint Anthony grew around their waterfront industries into a unified city. In the decades after the Civil War, the city's industries had used the power of Saint Anthony Falls to cash in on the natural resources of the many far-flung environments that fed industrial growth in Minneapolis—including trees in the North Woods, the rich soils of western farms, and even the water flowing to the city from headwaters lakes. By the 1890s, however, the city no longer depended on Saint Anthony Falls for its industrial success. Sawmills had given way to flour mills at the falls, and both industries turned in varying degrees to steam power, either to replace or supplement hydropower from the falls. In another sign of a new era, in 1889 British investors formed the Pillsbury-Washburn Company, consolidating the Washburn Mill Company, C. A. Pillsbury and Company, the Minneapolis Mill Company, and the Saint Anthony Water Power Company into a single entity that controlled roughly a third of the city's flour production capacity, signaling the growing influence of international wheat markets over local and national interests.[59] In a final momentous change, between 1890 and 1908 Pillsbury-Washburn erected new hydroelectric dams that slowly converted the falls away from directly powering industry toward producing electricity. On completion of the new Lower Dam in 1897, Pillsbury-Washburn signed a lease with the Twin Cities Rapid Transit Company to provide electricity for its expanding streetcar network. Both literally and symbolically, repurposing the power of Saint Anthony Falls to disperse the city's population along a far-flung network of streetcar tracks underscored the degree to which the industrial waterfront had ceased to be the primary focus of everyday life in Minneapolis.

In 1850 both Saint Paul and Minneapolis were tiny new river towns on the western edge of a rapidly expanding United States, each focusing on a distinctive set of waterfront activities and just beginning to nurture dreams of an urban future. A half-century later both had grown into important, prosperous cities. Neither city existed until the United States took control of Dakota and Ojibwe land, but both exploded when settlers flooded into Minnesota to claim the land as their own, with each city using its river location in a different way to connect the region's natural resources to eastern markets. Saint Paul's waterfront reflected the burgeoning city's role as a commercial center, built first around express company routes and steamboat landings and later remade around railroads. As the railroad network grew larger and more connected, first to the well-developed East and then to the developing transcontinental West, Saint Paul's waterfront became progressively less important to the city's civic identity. Minneapolis's urban waterfront also propelled the city's rise, first as a sawmilling center built around a natural waterfall and a river full of logs, and later as a sophisticated flour milling empire built around a highly engineered waterfall with extensive rail links to a vast hinterland of western wheat farms. By 1900, when Saint Paul's population reached 163,065 and Minneapolis's burgeoned to 202,718, both were well-established cities in the middle of a wealthy region, with maturing economies and sizable populations, and both had grown beyond a singular focus on the waterfront activities that had fueled their rise.

James J. Hill again helps encapsulate the nature and degree of these changes. By 1900 Hill was in his early sixties and was just beginning to achieve the peak of his power and influence. In 1893 he finished the transformation of the Manitoba into a fully realized transcontinental line, the Great Northern Railway, and in 1900 finally gained control over the Northern Pacific Railroad as well.[60] His pastimes included long late-night conversations with his friend and neighbor, Frederick K. Weyerhaeuser, a lumber baron who had built his empire from the forests of Wisconsin and Minnesota and whose mansion was just up from Hill's on Summit Avenue. As Hill's interests moved westward, his conversations with Weyerhaeuser began to include frequent, impassioned soliloquies about the forests of Washington State, which Weyerhaeuser regarded as too removed from existing markets to be of serious interest. By 1900, as Hill's influence over the Northern Pacific grew stronger, he proposed a deal that Weyerhaeuser could not refuse: nine hundred thousand acres of prime timberland in Washington for just $6 per acre. This was an expensive proposition, but also an incredible deal—accountants ultimately concluded that it worked out to just ten cents per thousand board feet of lumber. Weyerhaeuser said yes, and in doing so set in motion the American timber industry's shift from the Upper Midwest to the Pacific Northwest. To secure the deal, Hill offered preferential rates on hauling timber from Washington to the Twin Cities, ensuring Weyerhaeuser a competitive way to market his lumber and filling the Northern Pacific's eastward-bound cars with steady, rate-paying traffic that it would not otherwise have had.[61] In the twentieth century even Saint Paul–based enterprises like Hill's railroads and Weyerhaeuser's timber operations, both of which had been fundamental to the rise of the Twin Cities and the development of Minnesota, now focused much of their attention outside the region. After just a half-century of rapid growth, during which settlers built the Twin Cities by transforming Dakota and Ojibwe land into the thirty-second US state, taking advantage of two very different waterfront sites to link the region's abundant natural wealth to national markets, the first era in the history of the Twin Cities drew to a close.

DOWN TO THE FARM

Wheat Ecology and International Markets in Minnesota, 1850–1900

Thomas Finger

 THIS CHAPTER DESCRIBES MINNESOTA'S EARLIEST participation in the global economy. Its participation came in response to regional environmental processes and international market conditions in the wheat trade between the 1850s and 1900. For much of the nineteenth century Minnesota was an isolated market, producing raw materials at inflated prices for merchants in Milwaukee, Saint Louis, and Chicago. But after the 1870s Minnesota produced bushel upon bushel of wheat that would feed people in England, Ireland, and much of industrial Europe. The story of wheat's ascendency in Minnesota is not a story that can stay confined within the state's borders. Rather, the rise of Minnesota's wheat agriculture is an international story that connects farmers and businessmen in the state with natural processes around the globe and the growth of industrial cities in Europe.

The explosion of wheat agriculture in Minnesota during the 1870s and 1880s was the result of, first, millers and farmers engineering plants and machines in order to come to grips with the region's climate, hydrology, and pest ecology; second, the efforts of farmers and millers to streamline connections to steady consumptive markets; and third, the collective decision in response to widespread crop failure throughout Great Britain, that England would be a primary market for Minnesota wheat and flour. Contrary to the arguments of other historians, the development of the Minneapolis and Minnesota wheat economy came as the region tried to move *away* from Chicago, as wheat merchants and flour millers of the Twin Cities tried to move around the grasp on price exerted by established merchants to seek out the stable and growing markets of Europe.[1] The harvest and price variability that characterized early Minnesota connections to Milwaukee, Saint Louis, and Chicago had to be mitigated if wheat production in the Upper Midwest could serve as the bedrock of the region's economy. The only way that this variability could be engineered around was by connecting to the most stable wheat market in the world: Great Britain. Local environmental factors in Minnesota and Great Britain guided how this connection was forged. River flooding, grasshopper infestations, and wheat diseases in Minnesota, and weather in Great Britain during the late 1870s, served to subtly drive the markets of Minneapolis and Great Britain closer together. Thus the Minneapolis and British wheat markets developed in relationship with one another, connected both by a stream of investment *and* a series of environmental processes that guided the flow of wheat from Minnesota to Liverpool and London.

In telling the story of Minnesota's first participation in the global economy, I will venture from pioneer farms on river bottomlands in southeastern Minnesota, to burgeoning mills next to Saint Anthony Falls, to the great bonanza farms of the Red River Valley, to the churning railroad cars and grain elevators of James J. Hill's transportation empire, and finally to the greatest wheat markets of the nineteenth century: Liverpool and London. I will connect these places through the careers of Minneapolis flour millers C. C. Washburn and Charles Pillsbury, shipping magnate James J. Hill, and London grain merchant Sidney Klein, as each revolved around the use of European capital to grow, mill, and transport Minnesota's wheat directly to British markets. It is a story that requires me to switch scales at a moment's notice: from the cellular structure of a wheat grain to the superstructure of large-scale trade networks that connected production in Minnesota to bread consumption in the United Kingdom. This is a story that weaves together individual decision making and environmental processes to explain how global economic systems are built and gain coherence over time. Finally, this is a story of Minnesota, how settlers in the state first found their economic footing, how the Twin Cities grew from a frontier outpost into the world's milling center, and how the state contributed to the rise of a global food system by fueling industrial labor in Great Britain.

NATURAL EXTREMES, SPRING WHEAT, AND FARMING IN MINNESOTA

Before Minnesota farmers and merchants could grow wheat in any volume, they had to gain a level of control over the environmental systems of wheat production that made regional agriculture and marketing variable at best. Farmers in early Minnesota encountered extreme variability in nature and a market which forced them to adopt new breeds of wheat that could better withstand the extremes of temperature and precipitation characteristic of the Upper Mississippi watershed. While new breeds of hard spring wheat allowed new farmers to continue growing their main cash crop, their answer created a set of problems for merchants and millers, who found the new wheat difficult to process and sell.

Wheat is, first and foremost, a product of specific soil types, weather patterns, latitudes, and genetic compositions. How wheat responds to environmental stimuli has much to do with its cellular structure, growing habits, and ultimate quality as food for humans. Wheat is among the most versatile plants and can be grown in a wide variety of environments, but this range is human-engineered. Humans have created a vast number of wheat subspecies that respond to different temperature and precipitation conditions all over the globe. In general this variety can be simplified into two categories: the strength of the bran and the primary growing season. The strength of the bran, rendered as "hard" or "soft," determines a plant's ability to withstand variation in temperature and precipitation. This difference also dictates the comparative ability of milling wheat into flour. Soft wheat has a thinner bran, making it more susceptible to extremes of temperature and precipitation but easier to mill.[2] Hard wheat has thicker bran to protect the berry, but is more difficult to mill: the same bran that insulates the plant from temperature fluctuation also makes it difficult to mill into desirable flour.[3] An additional factor in planting and processing wheat is whether a particular breed is winter habit wheat, meaning most of its growth cycle takes place in the winter, or spring habit, meaning the plant matures throughout the spring and summer.[4] Winter and spring wheat are differentiated by the climatic conditions necessary to induce reproduction and germination. As a general rule, winter wheat is sown in the fall and harvested in late spring, while spring wheat is sown in the spring and harvested in late summer.[5] Winter wheat requires a prolonged period of cold temperatures to shift into its reproductive phase, but this fact leaves it particularly susceptible to a freeze later in the growing season—a common occurrence in March and April in Minnesota. These freezes often leave grain kernels flattened, shrunken, and shriveled and of little use to humans.[6] Spring wheat, because it grows during the warmer months, is more susceptible to pests.

REAPING WHEAT.

Figure 2.1. When early settlers began planting wheat in Minnesota, they used traditional methods and hand tools to try to gain a level of control over the environmental systems of wheat production, including the extremes of temperature and precipitation that made regional agriculture and marketing variable at best. *Source:* "Reaping Wheat—Wheat and Its Associations," *Harper's New Monthly Magazine*, June 1, 1857, 307.

The decisive factor in settling the Upper Midwest between 1850 and 1870 was the development of hard spring wheat varieties that could withstand the region's climate.[7] Settlers who arrived in Minnesota prior to the 1870s came from places such as New York, Ohio, Indiana, and Illinois, places that fostered the growth of soft winter wheat. These settlers found out quickly that winter wheat could not withstand the deep cold and snow accumulation of Upper Midwest winters.[8] Following a series of winter wheat harvest failures in Wisconsin and Minnesota from 1847 to 1853, farmers across the Upper Midwest slowly switched to spring wheat.[9] By the 1880s spring wheat was dominant in Minnesota.[10] This shift is important because, though winter wheat in general produces greater yields, spring wheat leaches the soil of its fertility in order to produce its protective outer casing.[11] This loss of soil fertility would come to plague spring wheat farmers and created a host of problems for the region's budding milling industry.

CONSTRUCTING TECHNOLOGICAL CONTROL OVER NATURE AT MINNEAPOLIS

While farmers wrestled with the region's variability in temperature and precipitation, millers in the Twin Cities had to gain a level of control over the various ecosystems associated with wheat agriculture and flour production before they could successfully create a market. For men like Charles Pillsbury and C. C. Washburn, flour milling was about constructing a stable production environment, maintaining a steady stream of wheat into their mills, and establishing reliable connections to potential consumers. All of these objectives involved dealing with nature. First, millers banded together to change the flow of the Mississippi River to prevent damaging floods and to channel water into their mills. Second, they borrowed or developed machines that could more effectively mill hard spring wheat into desirable flour. Finally, Minneapolis millers established reliable connections between new sources of spring wheat supply and potential markets in Europe. Thus, like wheat farmers, Minneapolis millers had to calibrate their actions at specific places in response to variable environments and unstable markets.

The Mississippi River and Saint Anthony Falls were at once the source of the city's great milling potential and its biggest obstacle. The flow of the Mississippi was seasonally variable, with spring floods giving way to a more modest flow in the summer. Second, and most crippling, the shifting geologic structure of the riverbed in the 1860s was not conducive to stationary industrial power. Simply put, the river and the waterfall both moved. Throughout the 1860s millers in Minneapolis struggled to dig tunnels, canals, and holding ponds that would deliver water when they needed it. These water systems, however, were literally built upon shifting sand. A series of devastating floods in the 1860s, culminating in an 1869 deluge that destroyed much of the milling district and threatened the waterfall itself with collapse, convinced millers that they would have to impose a level of stability on the river if they were to grow their milling operations. When local efforts proved insufficient, they managed to enlist the aid of the federal government to rescue the falls, resulting in the decadelong efforts of the Army Corps of Engineers to construct a new dike, dams, and apron.[12]

With the river's flow under control, millers, including C. C. Washburn and Charles Pillsbury, could shift their attention to the wheat coursing through their buildings. The main problem was that there was little market demand for hard spring wheat flour due to its hard outer casing. Washburn and Pillsbury simultaneously engineered around this problem by adopting the middlings purifier and the roller mill. Together these innovations comprised the milling revolution of the 1870s.

Prior to the mid-1870s there was no market for hard spring wheat because there was no way to effectively mill it. Until the late 1870s most millers in the United States still used large millstones that pulverized grain into flour. When this process was applied to hard spring wheat, the result produced a universally disfavored flour. Stones had to be set closer together to pound the hard bran, which invariably shattered it. The action made flour produced from hard spring wheat more grainy and discolored compared to softer winter varieties. This, coupled with the fact that most hard springs were of the red variety and would be slightly tinted even if the bran was completely removed, meant that until the 1870s most flour produced from hard spring wheat had to be sold at a marked discount.[13] Both Pillsbury and Washburn were quick to recognize that the future of the Minneapolis milling community was dependent on their ability to process desirable spring wheat–based flour.

The answers to the hard spring problem came to Washburn and Pillsbury in the form of the middlings purifier and roller mill.[14] Essentially, the middling purifier allowed the "middlings," or undesirable grainy matter from the hard bran, to be sifted out of the flour. Both millers searched for an acceptable purifier that would literally separate the wheat from the chaff and produce finer, whiter flour. The answer came from the work of a French emigrant named Edmund LaCroix, who came to Minnesota in 1860 and began to experiment with a purifying system. LaCriox would prove the prototype of the purifying systems adopted by Minneapolis millers a decade later. Middlings purifiers sent a blast of air through the flour mixture as it was being sieved, causing the undesirable bran to separate from the berries. LaCroix was soon employed by C. C. Washburn, and a series of modifications gave those milling spring wheat a competitive edge: flour that retained its nutrition and enjoyed fine bread-making qualities without the discoloration or grainy texture that had plagued earlier spring wheat flours.[15]

The second problem that Washburn and Pillsbury faced was more difficult to solve. How could they better grind the spring wheat to eliminate even more of the bran? They found the answer half a world away, in the hard wheat districts of Hungary. Like the Upper Midwest, grain and flour coming out of this region was subject to dramatic fluctuations in environmental conditions, producing a hard variety of wheat. Sitting at the convergence of the mild, dry maritime Mediterranean and cooler continental climates, Hungary had long produced a hard red variety prized in other European markets.[16] The roller mills used in these Hungarian mills substituted heavy millstones with a series of porcelain rollers which eased the grain out of its shell rather than pulverizing it. These millers tried to keep their process a strict secret.[17] In 1877 C. C. Washburn himself visited Budapest to investigate the Hungarian roller mills, and established correspondence with a Hungarian milling engineer centered on devising ways to adapt the technology in his Minneapolis mills. Never to be outdone by his chief competitor, Charles Pillsbury also traveled Europe that same year to view new milling technology. By this time a few experimental rollers were in use in Liverpool and scattered throughout the United States. Most were porcelain or marble, but Washburn, Fillsbury, and their engineers were convinced they would need rollers made of iron or steel if they were to adequately mill hard Minnesota wheat. Both the Washburn and Pillsbury mills set up experimental rollers in 1878 and by 1880 were rolling virtually all of their flour. When Washburn submitted his "new process" flour in a competition at the International Milling Conference in Cincinnati in 1880, his flour won the Gold Medal—and the name stuck.[18] Pillsbury named his new brand "Pillsbury's Best." Both companies quickly rose to the height of world fame by engineering around variability imposed by the main source of their milling power, the Mississippi River, and the source of their supply, hard spring wheat. Now all that remained was finding a new source of supply closed off to merchants in Milwaukee, Saint Louis, and Chicago. As fate would have it, at precisely the same time Washburn and Pillsbury were grappling for a way around the hard spring wheat problem, the world's greatest spring wheat district opened to their west on the flat plains of the Red River Valley.

BONANZA

In order for the Twin Cities milling revolution to gain momentum, millers needed access to a new source of supply where they did not have to compete directly with the established grain merchants and millers of Milwaukee, Saint Louis, and Chicago. This break was achieved by the opening of the Red River wheat bonanza in the Dakota Territory and western Minnesota after 1875. The Red River bonanza—like the development of spring wheat and the milling revolution—was in turn a product of overlaid environmental and economic concerns. First, the failure of the Northern Pacific Railroad in 1873 left much of the land in the valley available for investors who wished to plant wheat in order to turn a quick profit. Second, a series of grasshopper infestations in the early 1870s left much of this land unclaimed and wheat prices high. Third, as the Northern Pacific sold off its lands and grasshopper infestations dwindled, the unprecedented yields of early bonanza wheat farms led to a flood of settlement and the establishment of close business connections between Minneapolis millers and bonanza farmers. Finally, a series of stem rust outbreaks in the traditional center of wheat production in southeast Minnesota caused Minneapolis merchants and millers to turn their attention fully to the growing agricultural potential of the Red River Valley.

The Red River bonanza would not have been possible without the failure of the Northern Pacific Railroad in 1873, and the bonanza would not have exploded when it did in the late 1870s without endemic grasshopper infestations throughout the mid-1870s. Since the failure of the Northern Pacific and the ensuing Panic of 1873 have been well described elsewhere, I will focus here on insects.[19] Grasshopper infestations are cyclical, operating at epidemic levels for two to three years before disease and overconsumption of food cause a major decline in population. Grasshoppers often plagued portions of the prairie west recently opened by cultivation.[20] Breaking the sod for agriculture created temporary opportunities for the grasshoppers to lay their eggs in the easily opened ground, but as cultivation became more consistent and intensive, it became increasingly difficult for large swarms to find the soil conditions necessary for mass reproduction. This is what happened along the Red River Valley in the mid-1870s. The Northern Pacific failed simply because nobody wanted to settle in the table-flat Red River Valley that was notorious for floods, harsh winters, grasshopper infestations, and was in the frontier that marked white settlement to the east and Lakota-dominated lands to the west. The biggest threat was grasshoppers. James B. Powers, land commissioner of the Northern Pacific commented to railroad president Fredrick Billings in 1877 that "fear of grasshoppers is the great obstacle" to settlement.[21]

"Hoppers" had plagued the region since at least the 1810s, and a series of infestations beginning in 1864 kept away settlers who had been driven away by the Sioux Uprising two years earlier. A farmer near Yankton, Dakota Territory, recorded the spectacle of the 1864 swarm to a local newspaper. "The grasshoppers invaded the fields like a living river pouring upon it," the farmer wrote, "the stream stretched away to the south and west as far as one could see in either direction and the flutter of their wings created a roaring noise that was almost deafening."[22] The insects covered all crops, ate, and laid their eggs in the open ground—giving rise to subsequent infestations in 1865 and 1866. Then, they went away—until 1873. Small grasshopper outbreaks were reported throughout the northern plains in 1871 and 1872. But in 1873 and 1874 grasshoppers appeared in such numbers as to clog farm machinery. Grasshoppers flowed into Minnesota in a swarm large enough that insects were reported simultaneously at all telegraph stations between Moorhead and Mankato—a distance of 225 miles. Grasshoppers returned in even greater numbers in 1876.[23]

The main effect of the grasshopper plagues was to retard settlement in the Red River Valley and western Minnesota until the late 1870s, leaving the price of wheat high throughout the region. The infestations temporarily raised the price of wheat to more than a dollar a bushel in the mid-1870s, a price that would incentivize the settlement and monocropping of the bonanza regions. The failure of the Northern Pacific left large amounts of land up for grabs, land the railroad hoped to sell to investors. One harvest changed everything. In late 1874 one of the farmers who had bought a forty-acre tract from the railroad arrived in Fargo with 1,600 bushels of wheat which he sold at the then-astronomic price of $1.25 per bushel. On hearing this news settlers poured into the valley, though much of the choice Northern Pacific grant was bought by eastern capitalists hoping to turn their worthless Northern Pacific stock into hard cash by growing wheat. By 1875 large swaths of the Red River Valley in Dakota and western Minnesota were in private hands.[24]

The first bonanza farms were wild successes, known the world over for their vast size and unprecedented yields. The farms were largely investment strategies designed to recoup investors' losses from the failure of the Northern Pacific Railroad. Farmers and farm managers were thus incentivized to plant as much land as possible both to raise the value of that land and to sell the wheat to the waiting millers of the Twin Cities. This extensive plowing also served to reduce grasshopper infestations, who found it difficult to lay their eggs in constantly turned soil.[25] But this was an expensive venture. Breaking the virgin soil and turning it to agricultural production could cost thousands of dollars. During the first year of planting on a bonanza plot, farmers broke ground after the first thaw of spring. The compacted soil around the Red River made this a particularly arduous task. The deep roots of native grasses allowed vegetation to quickly rebound from the initial soil turning, and thus the same lands would have to be backset (re-plowed) during the fall of the first few years. An account from the William F. Dalrymple bonanza farm shows that this process of breaking and back-setting cost on average $4.50 more per acre on new ground than old. This doesn't seem like much until we account for the full scale of the bonanza farm. Spread out over ten thousand acres, this $4.50 difference per acre translates into $45,000.[26] The costs of getting the land into first production were enough to lead some investors to walk away after one or two poor harvests.

It took another natural phenomenon to fully establish the Dakota-Minneapolis nexus: a disastrous stem rust outbreak in southeastern Minnesota, which until then had comprised the main source of supply for Minnesota millers. In 1878 wheat in Minnesota was reportedly on its way to a record crop. As late as July 10 the estimated statewide yield was nineteen to twenty bushels an acre.[27] Then rust appeared. Stem rust occurs on all varieties of wheat plants in all regions of the world. It is a fungus that attaches itself to the aboveground portion of the plant, creating dusty or reddish pustules that turn black as the plant matures. As this happens, the grain dries and shrivels until there is nothing worth harvesting.[28] In 1878 rust hit with the greatest ferocity in the southernmost counties of Minnesota. As a result 1877, the year before the stem rust outbreak, was to be the high-water mark of wheat production in southeastern Minnesota.[29] Just as their new milling process was coming online, Minneapolis millers were forced to look elsewhere for their wheat. Free from grasshopper infestations and already practicing their innovative form of expansive agriculture, bonanza farms sat ready and waiting.

During the 1880s and 1890s the bonanza farms of Dakota and Minnesota produced wheat in unprecedented volumes. In 1877 the Cass-Cheney bonanza farm in Dakota had 3,692 acres in crops, of which 3,415 were in wheat.[30] This decision was based partly on the knowledge that Minneapolis mills sat ready to accept their new surplus, but also stemmed from a belief that extensive farms could absorb some of the losses of soil depletion after successive monocropping. In this vein the Dalrymple farm in 1880 consisted of a whopping 75,000 acres, of which only 25,000 were planted with wheat. The remaining acres were unbroken. Dalrymple's intention was to eventually put all 75,000 acres under cultivation, leaving 3,000 acres of the most depleted soil fallow each season.[31] In 1876 the Northern Pacific Railroad reported to the state of Minnesota that it carried 6,007 tons of grain within the state. By 1896 that number had ballooned to 99,218 tons.[32]

The size of bonanza farms dramatically altered the ecology of the Red River Valley. Bonanza yields soon grew smaller following successive cropping. The unprecedented yields of the mid-1870s meant that much of the region's dormant soil fertility was exported in the form of grains of wheat to Minneapolis to be used for human consumption. The sheer amount of wheat harvested from the bonanzas and agricultural practices that did not favor long-term husbandry contributed to the depletion of the soil and the eventual failure of many bonanza farms. Fred Rutledge, who worked on various bonanza farms in northeastern Minnesota, described the results of monocropping on soil and vegetation: "First year, bumper crop. Second year, less and the last I saw of it, 7 years later, nothing but rose bushes about 6 inches high was growing on it."[33] This depletion in soil fertility was tied directly to the imperatives of bonanza farms: produce as much wheat as possible one year to guard against crop loss by pest or drought the next. By the 1890s yields and wheat prices dipped throughout the bonanza farms.[34] Additionally, a series of severe droughts in the 1890s cut yields further, and fostered the breakup of huge farms into smaller, diversified units that could better handle changes in weather and dips in the wheat market.[35] However, by

THRESHING.

Figure 2.2. In the late 1870s the end of grasshopper infestations coupled with the unprecedented yields of early bonanza wheat farms in the Red River Valley led to a flood of settlement. As farmers embraced a combination of new technologies and massive scale to bring unprecedented volumes of wheat to market, James J. Hill tightened his control over transportation in the region, ensuring that the crop flowed first to Minneapolis—and then to the markets of Liverpool. *Source:* "Threshing Wheat—Dakota Wheat Fields," *Harper's New Monthly Magazine*, December 1, 1879, 60.

the late 1870s, just as Washburn and Pillsbury adopted the new process milling, there was a huge supply of wheat waiting for an outlet along the Red River, with the ability for it to be processed effectively at the Falls of Saint Anthony. Now all that was needed was an effective method to bring it to markets—both in Minnesota and beyond.

JAMES J. HILL AND AN EMPIRE OF WHEAT

The milling revolution of Minneapolis and the Dakota bonanza hit their stride in the late 1870s and early 1880s as James J. Hill tightened his control over wheat transportation in the region, as millers began to market their wheat in England, and as a series of poor crops in Europe meant that merchants there were looking for new supply sources. The problem for James J. Hill was not so much in gaining control over nature; much of that had already been achieved by Minneapolis millers and Dakota bonanza farmers. His problem was how to connect and integrate sources of supply, production, and consumption. This he did by establishing business connections to the money market of London and the wheat market of Liverpool. Hill then sat at the center of these emerging international connections as he used his connection to European capital to stitch together a network of railroads, elevators, and flour mills that connected the Dakota bonanza to Minneapolis, and Minneapolis to London and Liverpool. By the 1880s Minnesota wheat and flour production sat amid an emerging international trade in wheat that saw money flow westward to fund the development of transportation, and wheat flow eastward to feed the growing populations of Great Britain and Europe.

While the Dakota Territory's bonanza farms came of age within the Northern Pacific system, by the late 1880s James J. Hill and his empire were expanding their operations to siphon off much of the wheat heading east from the Red River Valley. Hill's system redirected some of the Red River's grain away from Milwaukee and Chicago, across the Great Lakes to New York City and the Atlantic. This was a conscious business tactic to connect the bonanza farms to English markets; Hill himself was a bonanza farmer largely backed by the major British banking houses, a business connection forged through his association with Gaspard Farrer, a partner in the vast London banking firm Baring Brothers & Company.[36] His activities in the 1880s and early 1890s provide an example of the increasing centralization and integration of the grain trade in Minnesota.[37]

Hill began his business career as a steamboat operator on the Red River, shipping wheat from farms to railroad depots up and down the river. By the mid-1870s he had diversified his portfolio to include stock in the St. Paul & Pacific Railroad—a line that would eventually be the keystone of his Great Northern Railway system. To ensure that his steamboats and railroad cars carried as much freight as possible, Hill also began to invest directly in agriculture, milling, and storage. During the 1880s Hill came to own all or portions of the Saint Paul, Minneapolis, and Manitoba Railroad; Red River Rolling Mill in Fergus Falls, Minnesota; two bonanza wheat farms (one in western Minnesota and one near Saint Paul); a chain of grain elevators from the Red River through Duluth and on to Buffalo, New York; a steamship line on the Great Lakes; and the St. Anthony Falls Water Power Company, which controlled all the waterpower required for milling on the eastern shore of the Mississippi River at Minneapolis.[38] So by the late 1880s and early 1890s Hill had come to own much of the transportation and storage capacity that connected wheat farms in western Minnesota to the Twin Cities and beyond.

Until now this story has been one of regional development. But what really differentiates the developments in wheat agriculture in the 1870s and 1880s in Minnesota is the extent to which it was dependent on British environments, capital, and markets. By the 1830s the industrial cities of Europe, particularly Great Britain, were in need of large wheat imports in order to feed their growing populations. A concurrent growth in domestic wheat output provided much of England's caloric need until the late 1870s. England's wheat output decreased throughout the 1870s as a result of international competition, migration to cities, and a series of poor harvests prompted by record-breaking cold and wet seasons from 1878 to 1883.[39] As this was happening, Britain experienced a continued demographic boom that multiplied its population from 8 million in 1801 to 32 million in 1901.[40] The majority of that boom occurred within cities. From the 1810s a major concern within England was provisioning these cities with cheap food in an attempt to reduce working-class volatility resulting from the incongruity of low wages and the high cost of bread, the primary caloric base of the working poor.[41]

The milling revolution and wheat bonanza of Minnesota and Dakota coincided with a dramatic increase in demand for and interest in American wheat on the English market.[42] As economic historian Henrietta Larson notes, "The production of a growing surplus [in Minnesota] was contemporaneous with an increase in the demand for food in the industrial areas of western Europe—especially England."[43] Prior to the milling revolution, hard spring wheat grown in America had a poor reputation in England. Until the repeal of the Corn Laws in 1846, the majority of bread consumed in England came from the wheat-growing districts surrounding the Thames and Ouse Rivers.[44] These regions produced a soft winter wheat, which in turn made a fine flour that by the early 1800s had become the universally favored bread by English millers and consumers alike.[45] The Minneapolis milling revolution changed this equation, however, and enabled British millers and bakers to mix their soft wheat with imports from the American interior. Larson notes that by the 1870s British millers had bought increasing amounts of Minnesota wheat.[46] Because the new-process rolled flour was strong without being overly coarse, it mixed well with other wheat types, providing added nutrition, structure, and stability to bread. The result was a universally favored bread that combined the taste and appearance of soft wheat with the strength of hard wheat needed to produce bread via a new mechanized processes.[47] British millers were willing to pay a premium for it.[48]

From the mid-1870s Minnesota millers and wheat merchants began to reach out to the now-ready English market. C. C. Washburn was the first Minneapolis miller to envision England as a market for his wheat. In 1877 Washburn sent erstwhile Philadelphia flour merchant William Dunwoody to England in an attempt to sell flour directly to Liverpool merchants, bypassing middlemen in New York City. Partnering with the Liverpool grain firm Horne Brothers, Dunwoody consistently wrote and cabled Washburn with news about the British market at the same time that he forged new accounts with other London and Liverpool merchants.[49] By 1881 Dunwoody was coordinating complex wheat shipments and credit accounts between the Bank of Liverpool, London grain merchant J. J. Walker, the Bank of Montreal in New York, and Washburn, Crosby & Company in Minneapolis.[50] William Edgar, writing a history of the Washburn companies in 1925, noted that "many of the firms to whom Mr. Dunwoody sold flour on this trip in 1877, nearly fifty years ago, remain customers of the Washburn Crosby Company to this very day."[51]

Capitalizing on his investment in transportation, production, and his connection to the British market, James J. Hill opened an account with Liverpool wheat merchant Stolterfoht, Sons & Company in the late 1870s. The two firms corresponded regularly about the state of the flour market in both England and the United States. Stolterfoht noted that English merchants particularly favored one of the mill's new-process brands—"Our Best"—and would be willing to pay slightly higher prices to obtain it.[52]

While the business plans of Washburn and Hill led them to seek out foreign markets, a series of failed harvests throughout Great Britain in the late 1870s and early 1880s produced unprecedented convergence between Minnesota wheat and the markets of Liverpool and London. Poor harvests in England began with the wet autumn of 1875. This was followed by abnormally heavy rainfall in the winter of 1876–1877. Spring of 1878 opened very wet, which continued well into the fall harvest season. Following 1878 were two and a half years of exceptionally cold and wet weather. There was significant flooding throughout England in March 1878, and livestock numbers were dramatically reduced due to the cold and duration of the winter of 1878–1879. For the rest of 1879 the average temperature was below the thirty-eight-year average *in every month.* Throughout these wet years, the rainfall was greatest in the south and east of England, precisely in the same regions dominated by wheat production. The Midlands experienced another long and severe winter in 1880, followed closely by another wet summer that brought more flooding to Lancashire. Soils were further saturated by heavy snows in January and February 1881, and the summer of 1882 was as cold and wet as the notorious summer of 1879.[53] The period from 1875 to 1883, in short, was devastating to English agriculture.

Minnesota markets were primed to respond. Bonanza farms were exploding, Twin City mills stood ready to process that wheat, and English markets looked far and wide for relief. In between was a network of railroad men and grain merchants known variously as "the elevator system"—or, more harshly, the "railroad monopoly"—by successive agrarian movements. As exemplified by James J. Hill's career, by the late 1870s wheat became a middleman's game. Grain merchants like David Dows of New York or William Rathbone of Liverpool became rich as they invested in American railroad securities to pro-

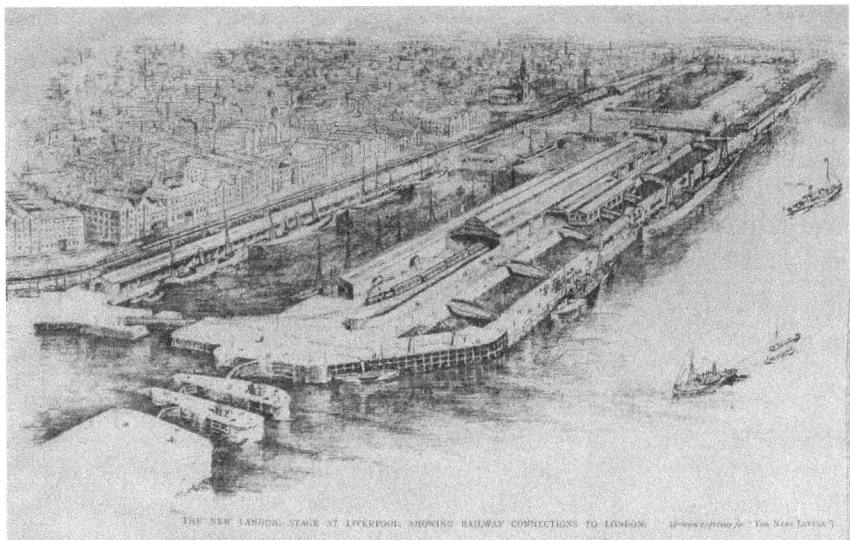

THE NEW LANDING STAGE AT LIVERPOOL, SHOWING RAILWAY CONNECTIONS TO LONDON. [*drawn expressly for "The News Letter."*]

Figure 2.3. The milling revolution and wheat bonanza of Minnesota and Dakota coincided with a dramatic increase in demand and interest in American wheat on the English market. The Liverpool Docks, pictured here in 1895, became a primary destination for Minnesota flour in the 1880s, helping make the American and British wheat markets virtually indistinguishable from each other and attracting direct British investment in Minneapolis's flour mills. *Source:* "Liverpool Docks," *American-European Newsletter*, May–June 1895, 70, Liverpool Record Office.

tect their investment in wheat. Hill himself used the opportunity afforded by the wheat bonanza, milling revolution, and failed English harvests to build his empire. In summer 1884 Hill reported, "Last year we carried over one-fifth of the entire spring wheat crop of the United States and this year I think we will carry one-fourth."[54] In doing so, Hill consolidated his hold over railroad lines in the Upper Midwest—what would become the Great Northern Railroad—based primarily on his shipment of grain and his growing connection to the London money market.

By the late 1880s the Anglo-American grain trade radiated outward from Liverpool and London, following the stream of investment through New York City, along the main transportation corridors of New York and Pennsylvania, to the branch lines, country elevators, and family farms of the American Midwest. It is the reason that by the late 1870s the United States exported between 25 and 40 percent of its crop directly to England.[55] So complete was the new connection between American wheat and British markets that a New York State committee on railroads and grain prices in 1879 found "Liverpool, as a rule, governs New York, and New York, as a rule, governs Chicago . . . and many other places . . . the European market is the basis; our exportations for the last year have been very large, and we expect they will be very large this year."[56] These connections were forged by businessmen like C. C. Washburn, Charles Pillsbury, and James J. Hill as they sought to contend with environmental processes and market variability at their own place within the trade. Between the 1860s and 1880s the American and British wheat markets became virtually indistinguishable from each other. This convergence became even more complete in the late 1880s as English merchants sought greater access to American, and in particular Minnesotan, wheat.

CONVERGENCE

The ecological and business events that produced the Minneapolis milling revolution, the Red River bonanza, and the increasing convergence of the Minnesota and British markets formed the basis for the final climactic event in the development of the Upper Midwest's wheat industry: the 1889 purchase of the Pillsbury and Washburn mills by a British company. At first glance, these purchases seem a solely business move that could be explained by its remaining within the board rooms of American and British companies. But we have seen how this purchase was in fact built on the measures that various merchants in both countries took to build a steady wheat trade upon the shifting sands of the natural world. The final triggering event, the systemic failure of wheat agriculture in southeastern England during the early 1880s, prompted British merchants to attempt a greater control over the business ties that had previously been dominated by American merchant middlemen.

Following the widespread failure of British wheat crops in the 1880s, British companies were particularly active in purchasing American grain—and the required investments in railroads, farm mortgages, and trade infrastructure that went with it.[57] Eighteen eighty-nine was a particularly active year. In that year a British company with the seemingly American name of Chicago and Northwest Granaries Company, Ltd. bought Star Elevator and Van Dusen and Company which together held a string of granaries that connected the Dakota bonanza to the milling centers of Minneapolis and southeastern Minnesota. Another British company bought the City of Chicago Grain Elevators, Ltd. With these purchases, British grain merchants sought to gain greater control over the flow of grain by removing the intermediaries between American farms and British stomachs or, in the words of one contemporary, "to free the British of the grip Americans had over the grain trade."[58]

The most significant deal came in Minneapolis. In 1889 Pillsbury and part of the Washburn company were bought by a group of British investors led by London grain merchant Sydney T. Klein. The British corporation that resulted, named the Pillsbury-Washburn Flour Mills Company, Ltd., came into existence with a board headed by Richard Glyn, director of the British Bank of North

America, and J. Flower Jackson and Sydney Klein, both London flour merchants.[59] It is clear that James J. Hill's connection to the London money market helped the effort. Several of his associates, including Lord Mount Stephen,
were among the company's original investors, and when the company experienced severe difficulties in 1908, Baring Brothers' then-director and longtime
Hill associate, Gaspard Farrer, took a close look at investing in or buying out
the enterprise in full. Klein's London wheat headquarters then became the chief
agent for Pillsbury-Washburn in Great Britain.[60]

With this one transaction the British company assumed control of the majority of wheat traveling from country elevators in Minnesota and Dakota to
the terminal elevators in Minneapolis, the facilities to process the wheat, and
the water to power the processing. The new Pillsbury-Washburn Flour Mills
Company held three Pillsbury mills and two Washburn mills, two elevator
companies, and the capital stock of the Minneapolis Mill Company. Knowing
they had to control the natural systems which surrounded milling, the British
also purchased the St. Anthony Water Power Company directly from James J.
Hill, and thus came to own most of the milling power taken from the Falls of
Saint Anthony. These holdings meant that this single British-owned company
controlled more than half of the flour output of Minneapolis, and 170 grain elevators in Minnesota and the Dakotas. The system could produce 14,500 barrels
of flour a day, making it by far the single largest milling outfit in Minneapolis,
and very likely the world.[61] Pillsbury and Washburn were British-owned until
the families bought back the operating rights in 1908 and the parent holding
company in 1928. This means that for almost forty years, the two largest American milling companies were owned by the British.

The 1889 purchase of Washburn and Pillsbury illustrates the convergence
between the Minnesota and British markets by the 1880s. By 1900 the US Bureau of Statistics reported that "the influence of the foreign market upon the
internal grain trade is becoming constantly greater."[62] While the English agricultural depression years of the late 1870s and early 1880s represented the
high-water mark of grain shipments from the United States to Great Britain
until World War I, exports throughout the 1880s and 1890s remained high relative to the 1860s and early 1870s. From 1879 to 1883 the United States exported
35, 40, 37, 31, and 29 percent of its *total* wheat crop, respectively. This percentage spiked again when the United States exported on average 34 percent of its
total wheat crop between 1894 and 1898.[63] During this period, 67 percent of
those exports, on average, went directly to the United Kingdom.[64] This means
that by 1900 one-quarter to one-third of a given farmer's crop could end up in
the stomach of an English man or woman, providing them with the energy to
work. While it is difficult to follow Minnesota wheat directly through to the
Liverpool market in exact numbers (due to the practice of mixing grain in ele-

vators, train cars, and ship holds), remember that the rise in American exports to the British market occurred simultaneously with the rise of Minnesota as the top wheat-producing state in the nation. Between 1883 and 1900 Minnesota was the nation's leading grain producer in every year but 1888, when California momentarily took the lead.[65]

To demonstrate how this rise in production was connected to the British market, let's follow the wheat as it flowed out of Minnesota during the 1880s and 1890s. By this time the wheat produced in Minnesota flowed out of the state not to the once-dominant markets of Saint Louis and Chicago but through Duluth and eastern Great Lakes ports. As the spring wheat district exploded in the 1870s, however, new methods of agriculture, production, and transportation combined to move the routes of grain to Minneapolis and Duluth. By the end of the nineteenth century James J. Hills's Great Northern system—a gigantic combination of bonanza farms, country and central elevators, rolling mills, railroads, and steamships—carried nearly 26 percent of all wheat arriving in Minneapolis (20 million of a total 77 million bushels).[66] By 1880 both Minneapolis and Duluth rivaled Chicago as wheat centers. In 1898 wheat receipts in Minneapolis and Duluth were both larger than those at Chicago. Compared to Chicago's 35,741,556 bushels, Minneapolis received over double—77,159,980 bushels—and Duluth received 62,286,712.[67]

What does this all mean? First, we can see that these numbers highlight the center of production within the American grain trade at the end of the nineteenth century as the Upper Midwest, particularly Minnesota and the Dakotas. Second, the ultimate destination of much of the wheat was the world's primary consumptive markets: Liverpool and London. Between Minnesota and England were the forwarding markets—Duluth, Chicago, Buffalo, and New York City. Duluth and Buffalo were transshipment points, as were Chicago and New York, but the latter also acted with London and Liverpool as the financial nerve centers of the trade. To get a sense of this network, envision a flow of grain moving from Minnesota, across the Great Lakes, loaded onto rail or a canal boat at Buffalo, onto an ocean vessel at New York, unloaded at London or Liverpool, and distributed to millers and bakers throughout England via that country's rail network, finally ending up in the stomach of an industrial laborer in Great Britain.

This chapter has offered broad strokes to illustrate the wide ecological and economic relationship forged between Minnesota farms and British factories in the last quarter of the nineteenth century. The energy and nutrients collected in wheat plants in Minnesota, carried through the system by milling innovations and transportation investments, powered industrial labor in Great Britain and the growth of the world's largest economy.

The explosion of wheat agriculture in Minnesota during the 1870s and 1880s was the result of (1) millers and farmers engineering plants and machines in order to come to grips with the region's climate, hydrology, and pest ecology; (2) the efforts of farmers and millers to streamline connections to steady consumptive markets; and (3) the collective decision, made in response to widespread crop failures there, that Great Britain would be a primary market for Minnesota wheat and flour. As sketched through the careers of C. C. Washburn, Charles Pillsbury, James J. Hill, and Sydney Klein, the growth of Minnesota agriculture and its connection to the global wheat trade were equal parts environmental processes and individual decision making. Their daily business decisions were made in the context of an ever-changing environmental setting, and their collective desire to build stasis into that change drove the increasing stability of wheat and flour production in Minnesota and its growing convergence to the industrial nations of Europe. These individuals were thus forced to synthesize a complex network of natural and market forces in order to maximize their own profits and reduce the risk they shouldered.

The movement of the Minnesota wheat frontier—the very development that made the state the nation's top wheat producer—would be unthinkable without the ecological forces of stem rust and grasshopper infestation over which humans had little to no control. But, crucially, humans could control their response to these environmental forces. They did so by reinventing the market. In this way the vast application of capital to build railroads, to innovate new milling techniques, and to expand the scale of agriculture was all meant to deal with the inherent environmental variability of wheat ecosystems. But this money could not originate in the Twin Cities. Rather, it came from the center of the nineteenth-century world economic system: England. The need for stability and the means to achieve it thus originated in the industrial cities of England's North, where large populations sat hungrily waiting for their daily bread and merchants employed their vast fortunes to bring that bread forth.

Environmental historians are quick to note the vast environmental changes associated with the rise of commercial agriculture on the plains and prairies of the American interior.[68] They have been less detailed, however, in following how the energy and nutrients drained from American farms cycled through trade networks powering work and wealth creation elsewhere. The decline of wheat yields on bonanza farms during the 1880s and 1890s translated through a series of cultural and physical conversions into wealth and power for people like Washburn, Pillsbury, Hill, and Klein as the energy and nutrients flowed away from those farms and toward Great Britain. All of these individuals forged their own economic growth directly from this stream of wheat.[69] Conceptualizing the flow of wheat this way thus preserves the importance of place, process, and individual action at the same time and allows historians to follow the full implications of environmental change through human society.

Places are defined by their outside connections as much as their internal structure, by vision as much as material process.[70] As anthropologist Anna Tsing has noted, specific ecological relationships at particular places—and the tension between environmental realities and human goals—do much to steer the evolution of global economic systems.[71] Historians would have an incomplete story of wheat agriculture in Minnesota if we focused on either grasshoppers or the export of British capital at the expense of the other. Floods and pests would mean little to this story if they hadn't contributed to the economic vision of farmers and Minneapolis millers and its eventual realization as a physical trade.

Ideas and environmental processes guide the shape of developing trade networks, but the financial and technological means humans adopt to impose stasis determine their long-term structure. In the response to local environmental and market patterns, individuals forge large-scale processes that connect far-flung regions into direct and increasingly stable relationships. Wheat came to dominate Minnesota's economy and identity for more than a century following the 1870s. Its implications are still felt as one wanders through Minneapolis's Mill District, and the traces of Washburn, Pillsbury, Hill, and Klein are woven into the very fabric of the city, its region, and its state. Even today, when a Minnesotan goes "down to the farm" to visit family, they return to the structures of the 1870s milling revolution, when the Twin Cities became the economic center of the region by collecting wheat from family farms, processing it, and sending it abroad. The actions of individuals like Washburn, Pillsbury, Hill, and Klein meant that this was not a simple flow of wheat, but a series of relationships made stable by railroad tracks, grain elevators, and banks—technologies that may have vanished from the landscape long ago, but gave rise to the shape of Minnesota today. Traveling down to the farm, then, replays an unspoken connection to Minnesota's first participation in the global economy, built from the ecology of wheat.

COMPETING HINTERLANDS

Saint Paul, Madison, and the Landscape of Burnett County, Wisconsin

David A. Lanegran

CITIES MAKE THEIR HINTERLANDS and in turn hinterlands make their cities. This urban–rural interdependence has been described and analyzed in many contexts, especially by the geographer John R. Borchert and the environmental historian William Cronon. In *Nature's Metropolis*, Cronon analyzes the intimate relationship between Chicago and its hinterland, "the Great West."[1] In *America's Northern Heartland*, Borchert discusses the relationship between the Twin Cities' hinterland and the developing hierarchy of urban places within it.[2] Both authors write with a broad brush as they discuss large-scale changes in land use, demography, transportation, industrial change, and commodity flows. Cronon focuses on one city and its hinterland, while Borchert uses an approach more clearly based in Central Place Theory, analyzing the interplay between cities and their hinterlands at differing levels in the hierarchy. While both authors reference the role of politics and the effect of political capitals, neither draws out the continuing tension that emerges when the hinterland of an economic metropolis (highest-order city) is also governed by the political jurisdiction and public policy realm of another city.

For many reasons, the territories controlled by state capitals are not coterminous with the primary economic hinterlands of large industrial cities in the

United States. This historical interplay between political and urban/economic geographies has produced fascinating settlement patterns and economic landscapes. The industrial and environmental history of western Wisconsin offers an unusual variation on the dynamic relationships among cities discussed in Borchert's and Cronon's seminal books since it is a place that has simultaneously been part of two different hinterlands: Madison's political and administrative hinterland and the economic hinterland centered on Saint Paul.

In western Wisconsin's early phases of economic development, Saint Paul dominated. The area's fur traders, timber businesses, and agriculturalists were linked to their markets by waterways and simple roads that extended from river landings. In the three decades after the American Revolution, transportation in western Wisconsin focused on the Mississippi and Saint Croix Rivers. Fort Snelling and the traders at the confluence of the Mississippi and Minnesota Rivers dominated the political economy of the region's watersheds. Gradually the timber companies expanded operations up the rivers that flow into the Mississippi from the east, including the Saint Croix. Although the area was nominally part of Michigan Territory and then Wisconsin Territory, the soldiers at Fort Snelling were the law north of Prairie du Chien.

When the politicians in southeast Wisconsin applied for statehood, a struggle for control of a rich hinterland exploded. The territorial government in Madison wanted to set the western border of the new state at the Mississippi River, the western boundary of the Northwest Territory. This border would not only have given the wealth of western Wisconsin to the new state, but the industrial potential of the Falls of Saint Anthony and the head of navigation on the Mississippi River as well. The residents of the Saint Croix Valley settlements and the Mendota community were unwilling to join Wisconsin, however. The lack of roads across Wisconsin and the annual freezing of the rivers isolated them from the main population centers, and they argued that this would cause their needs to be ignored. Instead, they wanted to create a Minnesota Territory that would include the Saint Croix watershed settlements and have an economic and political capital at Stillwater or Saint Paul. After a great deal of maneuvering, Congress drew a compromise border, south from Fond du Lac to the Saint Croix River, then along the midline of the Saint Croix to its confluence with the Mississippi. This compromise set up a rivalry between Madison and Saint Paul for the domination of northwestern Wisconsin.[3]

Although rivers are frequently political borders, they also unite the natural environment and provide conduits for economic activity. Because of these dynamics, the Saint Croix border between Minnesota and Wisconsin provides an excellent case study of the political and economic struggles between an industrial center and political capital for hinterland control.

As the Wisconsin government organized the area's new political landscape, the Twin Cities continued to dominate the economic life of western Wisconsin. The gradual extension of Madison's control over its northwestern Wisconsin hinterland was abruptly reversed in 1912 when the Crex Carpet Company purchased twenty-three thousand acres of marshlands in Burnett County to harvest sedge (*Crex stricta*). The now-defunct Crex Carpet Company, once the largest industrial employer in Saint Paul, brought a new scale of commercial agriculture to the area. After the company's financial failure, Madison exerted its authority over the company's former holdings in Wisconsin. The marshes have since become public property, managed for wildlife production and leisure activity for urban dwellers.

This essay will document the ebbs and flows of political and economic forces that emanated from the two urban centers by combining the well-documented history of the Crex Carpet Company with geographical analyses of the development of the marshlands in western Wisconsin. Ultimately, decisions made in Madison stimulated settlement processes and land-use changes in the area over the past 150 years and interfaced with a variety of economic and geographic processes that emanated from Saint Paul. The essay will trace the early settlement patterns of western Wisconsin, development of the "Carpet Swamps," the abandonment of the agricultural production landscape, and the subsequent planning process that is attempting to create a landscape of wetlands and pine barrens for the enjoyment of the urban populations. It will argue that the ability of either city to dominate this part of western Wisconsin has waxed and waned with changing technologies and political agendas.

To gain a full understanding of the process, we will focus on an examination of the landscapes at critical times in their evolution, and we will examine the various images of northern Wisconsin or narratives about the region that seem to have guided its development. We will begin with the presettlement landscape and the early efforts at wresting a living from the land through agriculture. During the agricultural frontier era, Madison's political policies promoting forestry and then agricultural settlement made possible and complemented the emerging economic and transportation infrastructure of the Saint Paul hinterland. Then we will examine the activities of the Crex Carpet Company in Saint Paul and Burnett County. During this burst of economic activity, Saint Paul dominated nearly every aspect of local life except the election of political officials. Finally, we will discuss how the political and cultural establishment in Madison regained control of the area and has created its contemporary recreational landscape.

PRESETTLEMENT LANDSCAPE

At the end of the Wisconsin Glacial period (10,000–15,000 years ago), a layer of glacial till ranging from three to three hundred feet deep covered what is now Burnett County. In addition, Glacial Lake Grantsburg covered much of the county. The deepest part of the lake evolved into shallow sedge marshes and low ridges. The uplands are sandy while the lowlands have several feet of organic matter over a layer of sand that varies from one hundred to three hundred feet. The area known as the Pine Barrens extends from Burnett County in western Wisconsin north to Bayfield County on Lake Superior. The original vegetation was apparently a pine savannah or "bush prairie" with a few large pine trees per acre as well as grasses and flowers. Fires also maintained this community. After agricultural settlement the suppression of fires resulted in the growth of an oak and jack pine forest.[4]

MADISON ORGANIZES A TERRITORY DOMINATED BY SAINT PAUL

The earliest maps of Wisconsin and Burnett County convey a clear impression of the agricultural frontier in the nineteenth century. The first maps show only the Saint Croix River and its tributaries. This is to be expected because rivers were the primary transportation routes for Native Americans, fur traders, and lumbermen.

Although there were two small fur trading posts on the Yellow River in operation from 1802 to 1805, the first permanent European settlement occurred in 1852. The first place-names on maps indicating settlements are Anderson, Burnettville, and Grantsburg in the southwestern portion of the county. Located on the Saint Croix River, Burnettville was listed as having a post office in the 1873 *Wisconsin Legislative Manual*, but there is no evidence that anything more than a single house was ever built. For many years a ferry was located at the site. Today it is a canoe landing.

In 1854 Canute Anderson established himself at what is now Grantsburg on the Wood River just south of the large marshes. There he built a dam, a sawmill, hotel, and store. Population growth was slow until a group of Norwegian immigrants took up homesteads in 1863. Even though Burnett County split off from Polk County in 1864, the entire county was governed as one township between 1865 and 1875, with Grantsburg as the seat of local government. Grantsburg soon eclipsed the Saint Croix River towns and Hickerson, which was located where the tote road from Saint Paul to Bayfield crossed the Wood River about one mile south of Grantsburg.[5] Lumbermen privately built the tote road to provide connections between the river towns in Minnesota and the pineries in northern Wisconsin.

The first settlers were interested in harvesting timber. The sawmills bustled during the 1860s, and farmers put the cutover land into crops. By 1875 the settlement at the current location of Grantsburg had become a trade center for the surrounding frontier, boasting two sawmills, a shingle mill, a grist mill, a

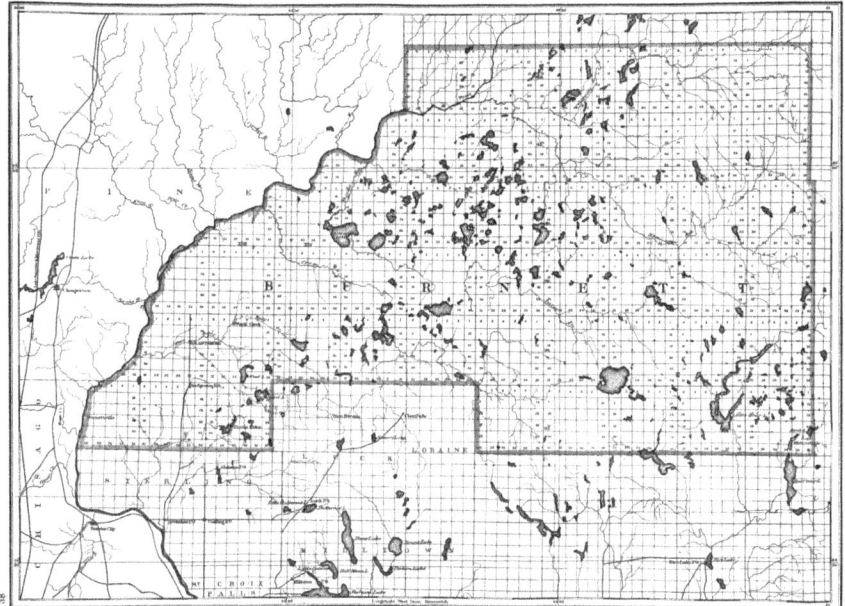

Figure 3.1. The early settlement pattern and transportation system in the western portions of the county were developed in the frame of the rectangular land survey. The map also shows the system of rivers and lakes that attracted early farmers. *Source:* Untitled (Burnett County Wisconsin), *Historical Atlas of Wisconsin* (Milwaukee: Snyder, Van Vechten, 1878), author's collection.

blacksmith shop, a saloon, two churches, and a school. In 1886 the settlement incorporated as a village with a population of 311 people.[6]

Early maps of the area show Grantsburg literally at the end of the road (Figure 3.1). This part of Wisconsin was very isolated from the state's population centers in the south, and Grantsburg's primary link was the road that ran due south via the Anderson post office to Saint Croix Falls and on to Saint Paul. Saint Croix Falls had a land office in the 1870s and was the gateway to northwestern Wisconsin. A grid of township roads gradually replaced the dendritic pattern of early trails and traces.

In 1884 a branch line of the St. Paul and Pacific Railroad connected Grantsburg to Rush City and Saint Paul. This firmly established Grantsburg as part of Saint Paul's economic hinterland and also directed traffic away from Saint Croix. A series of ferries crossed the river at several points. Finally, a toll bridge was built across the Saint Croix due west of Grantsburg. This provided western Wisconsin with a connection to the Minnesota territorial road that connected Saint Paul with Duluth and the lumber camps of eastern Minnesota. The completion of the railroad and bridge firmly embedded the area in Saint Paul's economic hinterland.

MADISON'S POLITICAL AGENDA SPURS AGRICULTURAL SETTLE-
MENT AND DEVELOPMENT

Despite Saint Paul's economic influence, the policies and politics of Wisconsin's state capital guided many forms of development in the area. After the fertile lands in southern Wisconsin were occupied, the railroad companies, timbermen, and political leaders began to intensely promote the agricultural potential of cutover lands in northern Wisconsin to Northern European immigrants. From 1880 to 1920 a small number of farmers moved into the area, lured by cheap land and promises of a bright future. Intense pressure from the state legislature and the dean of agriculture at the University of Wisconsin fanned the migration. On direction from the legislature, Dean William A. Henry produced the highly illustrated, two-hundred-page *Northern Wisconsin: A Handbook for the Homesteader* in 1895. About fifty thousand copies of the full report were published, and sixty thousand copies of a highly illustrated abridged version were published in English, German, and Norwegian. The photos were probably contrived to show the best possible aspects of the region. Although limitations are mentioned, the overwhelming tone is one of golden opportunities for the hard-working farmer.

> If this Hand Book shall be an instrument in removing the great ignorance and even prejudice which prevails in the southern half of our own state concerning the agricultural possibilities of northern Wisconsin, and if it shall convey to our people and those of other states and countries a true acknowledge [*sic*] of this region, much good will have been accomplished. . . . With farms supplanting the forest, northern Wisconsin will not revert to a wilderness with the passing of the lumbering industry, but will be occupied by a thrifty class of farmers whose well directed, intelligent efforts bring substantial, satisfactory returns from fields, flocks and herds.[7]

According to the authors, the great marshes of Burnett County and northern Wisconsin presented special opportunities for agricultural entrepreneurs. Rainfall amounts had been low during the years prior to the pamphlet's publication, which prompted the writers to predict great changes in the marsh environment.

> Many of these swamp lands have natural permanent and decided advantages over the higher lands which surround them . . . when swamp lands have been adequately drained and where the organic matter is not too thick and is sufficiently decayed, they are far superior for many purposes to the higher and naturally drier lands. The superior value of these lands when once reclaimed grows out of the large supply of organic nitrogen in the soil and the natural

system of sub-irrigation to which they are perennially subjected. The sub-irrigation has its advantages not simply in bringing an abundant supply of water but in bringing with that water and dissolved in it ready for the use of the plant, a large amount of plant food of the in-organic types which has wasted from the surrounding higher lands into their drainage waters, and which are steadily traveling towards outlets in the drains of the swamp areas.[8]

The scholars expected a thrifty class of farmer to tame the landscape.

Burnett County promoters picked up this theme and focused it on their county in a 1902 pamphlet titled *Official Report of the Burnett County Board of Immigration*, written by Edward L. Peet, a local newspaperman and land dealer. Their goal was to encourage increased immigration to the county. Peet told his readers that he knew about all available land in the county. His publication not only extolled the virtue of the entire county, but he wrote about each township and produced maps of the available land. He knew he was in competition with land dealers selling the rich prairie lands in western Minnesota and the Dakotas, so he faced the issues of the cutover land directly: "It may seem hard clearing up a timber country, but patient industry here is always well rewarded. In the new north is a diversity of labor for each season of the year and an abundance of building material not found in prairie regions. The result is a healthy vigorous people with good homes and farms rapidly increasing in value conditions which must produce sturdy, industrious and liberty loving citizens."[9] He was quick to point out that land was cheap because it was distant from railroads. The railroads were distant, he contended, because the population was small. He argued that once the land was taken up, the railroads would follow. He made the usual comments about the climate not being as bad as people believed. Furthermore, he wrote that although most people considered the entire county to be sand barren, that was not actually the case. In his view, this incorrect image resulted from travelers' accounts of the landscape along the timbermen's tote road and the railroad. Both the road and the railroad ran through the sandy part of the county. He also had plenty of ideas about how to get profitable crops out of that sandy soil by growing potatoes and other vegetables.

Despite all the potential for crops Peet saw in the county, he believed that the future would revolve around the dairy farm. "The intending settler should not lose sight of the fact that the cow is the thing one must tie to for the most practical results. All kinds of stock do well but there is nothing so reliable as the cow. Whether the cow farmer works to sell beef, butter or cheese the point is the same. He must plan his crops that his profit comes through the cow. Should Burnett County always remain isolated from the railroads the cow will furnish a condensed product that can be taken a long distance to market at little cost."[10] Peet foresaw an agricultural landscape that would produce a variety of crops and livestock products. He did not anticipate how specialized farmers would become to compete with farmers occupying the more fertile prairie lands to the southwest.

CREATING THE LANDSCAPE OF AGRICULTURAL PRODUCTION

The farmers that were lured to Burnett County bought up the sandy moraines and rolling hills in the watershed of the Wood River to the south and east of Grantsburg. Never a highly productive lumber region, the major lumber companies soon abandoned the area. The land transfer from the federal government followed familiar patterns. Speculators bought up larger sections of government land for resale to farmers and some agriculturists bought directly from the government. The result was a patchwork of ownership and land use. Large parcels were held by wealthy individuals, syndicates, and land companies. Smaller parcels, some only forty acres, were taken up by families who practiced a combination of subsistence and commercial agriculture. Most farmers grew a combination of livestock and crops. The northern Europeans were used to growing small grains and consequently planted the fresh fields with wheat, oats, and rye. The Wood River dam provided ideal locations for milling, and the Hickerson Roller Mill produced high-quality flour from local wheat. In addition, it produced and sold a variety of animal feeds. The region may have been abnormally fertile during the first years of farming because the wood ash left by widespread fires added nutrients to the soil. The wheat crops of the early years must have been profitable because in 1879 the local paper carried notices of local farmers buying steam threshing machines. The farmers also organized the first county fair in 1878 to show off the fruits of their labor. The Hickerson Roller Mill prospered into the 1920s; the local paper contains regular notices of expansions and improvements in the mill and its machinery.[11]

Most farmers continued to practice the mixed-crop and livestock farming prevalent throughout the Midwest. The short growing season and poor soil limited the profitability of grain farming, and eventually dairying became a specialty of the area. The major stores in town accepted eggs and butter produced on local farms as currency for many years. As the cooperative movement grew in Wisconsin, many farmers realized they could make more money by shifting from on-farm butter making to providing milk to cooperatives. By 1900 local farmers created the Branstad, Grantsburg, Trade River, Trade Lake, and Wood River Creameries to process and market their liquid milk and butter. Although the Hickerson Roller Mill continued to mill grain for farmers after World War I, it was heavily engaged in retailing animal feed produced by national corporations to local farmers. Grantsburg was also home to two other feed mills that did not use waterpower.[12]

The economic and land-use progression from fur trade to timbering to mixed-crop and livestock farming with an emphasis on dairying was typical of the northern edge of the great Midwestern farm belt. Although agricultural experts at the University of Wisconsin thought the peat lands could be productive, they did not foresee the two crops that would, for short times, yield a profit from the swamps: cranberries and wild sedges. At a time when geographers and geologists were just developing theories about glaciations and erosion cycles, the local farmers developed their own views about geomorphology and what we now know as Glacial Lake Grantsburg:

> It is easy to believe that nearly all of Burnett County was at one time a great lake and the swamps, meadows and small lakes that now exist were the bottoms of the "deep holes" and the present streams were the channels of great currents flowing through the great lake. This lake, if it ever existed must have been drying up for ages by natural processes. Within the memory of present residents of the county, lakes have changed to marshes, marshes to meadow lands and meadows to good dry plow lands. . . . There are few swamps in the county that will not at some time in the future be turned into profitable farmland.[13]

The bogs had potential for specialized monocropping. Cranberry production had enjoyed success in New England and southern Wisconsin. Undoubtedly encouraged by these developments farmers began to produce cranberries in the late 1880s. At first they harvested wild cranberries. The local paper carried a notice of two hundred barrels of cranberries shipped out in October 1887, and in 1890 berries were being shipped to Saint Paul and Seattle.[14] Much of the marsh was acquired by the Baker Land and Title Company in 1880. In 1892 the Marshland Farming Company of Boston, Massachusetts, and Hudson, Wisconsin, began to acquire the marshland from Baker, and in 1893 that company owned about twelve thousand acres of wetland known as the "Big Meadow." The company apparently intended to harvest both hay and cranberries from the area, and Marshland Township would become the most valuable property in the county. In 1899 a carriage maker turned land developer from Iowa named L. C. Erbes bought out Marshland Farms. He continued to drain the lands and sold some of the property to small farmers, but he began to sell large tracts of land to the Crex Carpet Company in 1911. By 1913 he had sold essentially all of his holdings in the Big Meadow to Crex Carpet. The local paper reported that the company paid about $450,000 for twenty thousand acres.

SAINT PAUL TAKES CHARGE

The agriculture developments that shaped the area before Crex arrived were the result of various policies of the leaders and administrators in Madison. The changes during the next phase were not. After the best timber in the

Saint Croix watershed was harvested, commercial interests in Saint Paul were not particularity interested in northwestern Wisconsin. That attitude would change dramatically as the railroads controlled by James J. Hill and the Minnesota state legislature began to promote settlement on the prairies and cutover lands in Minnesota.

As the mechanization of agriculture progressed, wheat farmers worked with capitalists to develop machines that would replace the cutting and threshing of small grains by hand. The first set of integrated machines, called harvesters, would both cut the grain and use twine to tie up the grain stems into bundles or sheaves. These were then later picked up and carried to a threshing machine. The harvesters and threshing machines enabled the rapid expansion of very large grain farms in North America. Of course, the machines needed twine. In the early years twine was expensive because it was made with fibers imported from Mexico and Asia. American industrialists wanted a local and cheaper source of fiber, and for a few years it looked like wire grass was the answer.[15]

The American Grass Company, a Delaware corporation with national headquarters in New York City, set up operations in Oshkosh, Wisconsin, during the mid-1890s with the intention of solving the "twine gap." The American Grass Company was itself the product of a merger of Wisconsin Grass Twine Company and Northwestern Grass Company. American Grass Company became the Crex Carpet Company in the early 1900s as a reflection of its changing business plan. The Crex Carpet Company was vertically integrated in a distinctive manner. Its employees made carpets and other furnishings from naturally occurring sedge harvested by temporary workers on extensive marshlands owned by the company in the vicinity of Saint Paul. Before opening operations in Burnett County, the company that formed for the purpose of supplying the market for binder twine had moved on to new uses for the sedge, including first wicker furniture and then carpets. Its business plan was to harvest the wild sedge (*Crex stricta*), or wiregrass, from the swamps. Then, in its Saint Paul factory, workers wove the dried fibers into lightweight, attractive carpets for indoor and outdoor use (Figure 3.2).

Harvesting the sedge not only changed northwestern Wisconsin, but it also had a great impact on Saint Paul. At its peak, the Crex Carpet Company was the largest industrial employer in the city. American Grass and its successor, Crex Carpet, dominated the wiregrass industry from the mid-1890s until 1933, when competition from Japan and the economic travails of the Great Depression forced the company into bankruptcy.[16]

The root of the Crex Carpet Company lies in Oshkosh, Wisconsin, where the company sought to take a free product, wiregrass, and turn it into a product that could be sold at great profit. By using wiregrass, the industrialists could

Figure 3.2. Advertisements like this showcased Crex carpets in the comfortable domestic settings of middle-class households. *Source:* Crex De Luxe Rugs advertisement, *Independent*, March 2, 1918, author's collection.

simply harvest the wild-growing plant. No planting or tending was needed. By modifying the technology used in other fiber industries, an inventor named George Lowery, who was in the employ of a group of investors led by James F. O'Shaughnessy, developed machines that processed the tall wiregrass stems into twine. After proving the workability of the machine in a factory in Oshkosh, the managers of the American Twine Company moved their company to Saint Paul to be closer to the center of the railroad net that served the booming grain belt, and to be close to a sizable area of marshland that produced a very large supply of wiregrass. The company's prospects looked bright. Salesmen traveled the farming districts of the region to demonstrate the product, and soon profits grew.

The company had a significant presence in Saint Paul where, at the height of production, it operated two separate manufacturing plants. One, located at Front and Como Avenues in Saint Paul's North End, first made binder twine and later carpets. The other, located in Hazel Park, on the eastern edge of Saint Paul, was the home of the Minnie Harvester, a machine for harvesting wheat. The Minnie was designed to use the company's twine, because competitors' machinery frequently had trouble with the grass twine. In 1903 the two plants together employed about 860 people, divided roughly equally between the two operations. In that same year the company sold the harvester plant to International Harvester, got out of the binder twine business, and focused on other uses for the grass. For several years the twine produced in Saint Paul was shipped to New York where it was used to manufacture wicker furniture. Because carpet making proved to be the most profitable, management restructured and renamed the company Crex Carpet Company in 1908, phasing out its twine and wicker operations.[17]

Carpets woven from natural products were very popular in this era. The rugs were lightweight and easy to keep clean. They were marketed as high-quality products for use in middle-class families to decorate formal dining rooms; many of the ads featured scenes in households that employed domestic help. They were particularly popular as porch floor coverings. The national demand for carpets was so great that the company expanded its factory and built a warehouse. The carpet factory employed around three hundred workers, over half of which were women and girls, who did difficult work for low wages. Research indicates that a sizable fraction of the workers lived within walking distance of the plant, which was typical of that time.[18] Business was profitable for several years during the 1910s. The profits invited competition, most significantly from Japan. In the early 1920s cheaper imported carpets made from rice straw destroyed Crex's profitability. By the mid-1920s the company was losing money. After creating great wealth for its owners and providing steady work for men and women in Saint Paul, where it essentially created its own company neighborhood, the company went bankrupt in 1935 and was largely forgotten. Operations in Crex Meadows ended in 1933.

SAINT PAUL'S IMPRINT ON WISCONSIN

In 1915 Crex Carpet Company owned the majority of Marshland Township, along with a large holding to the east in Lincoln Township and another to the south in Anderson Township (Figure 3.3).

These were not the only marshy areas in the county, but they were the largest swamps. South of Grantsburg and east of the Anderson Township holdings, most of the land was divided into parcels sized for family farms. There were many small holdings of 40 acres and a few as large as 260 to 300 acres. The state and the land speculator L. C. Erbes also owned scattered parcels. The land along the Saint Croix was owned by the Minneapolis General Electric Company (later Northern States Power Company) and R. J. Graff, an attorney from Manitowoc, Wisconsin, who apparently speculated in land. Other parcels of land were held by the Baker Land and Title Company of Cushing, Wisconsin, the largest land dealer in the county. Several parcels did not indicate any buildings and may also have been held by investors. There were a few farms that were given names, but most of the holdings have just their owners listed. Most of these farms were likely still owned by their original homesteading families.

The local transportation system was not highly developed prior to World War I. Here as elsewhere roads were a local responsibility. State highway systems would not emerge until after the war and long-distance motorists were few. We can clearly see the direct connections between Saint Paul and western Wisconsin with ferries across the Saint Croix at several points, as well as the Northern Pacific rail line from Rush City to Grantsburg. This gave the town a connection to the national markets and opportunities to travel and solidified its role as part of Saint Paul's economic and cultural hinterland. The local newspaper contains several notices of the comings and goings of residents and their relatives on the train to Saint Paul.

WORKING THE CREX MARSHES

The writers of the *Book of Minnesota*, a folio-sized vanity promotional piece published by Saint Paul's Pioneer Press Company in 1903, waxed eloquently about the transformation of the marshlands of Minnesota harvested by Crex. The description is equally appropriate for the Big Meadow lands in Wisconsin.

> One of the greatest difficulties in the early stages of the industry was the gathering of the grass and its transportation over the marshes. To subdue the wilderness was a seemingly impossible task. The trackless marshes, much of them underwater, presented difficulties to appall the stoutest heart and to

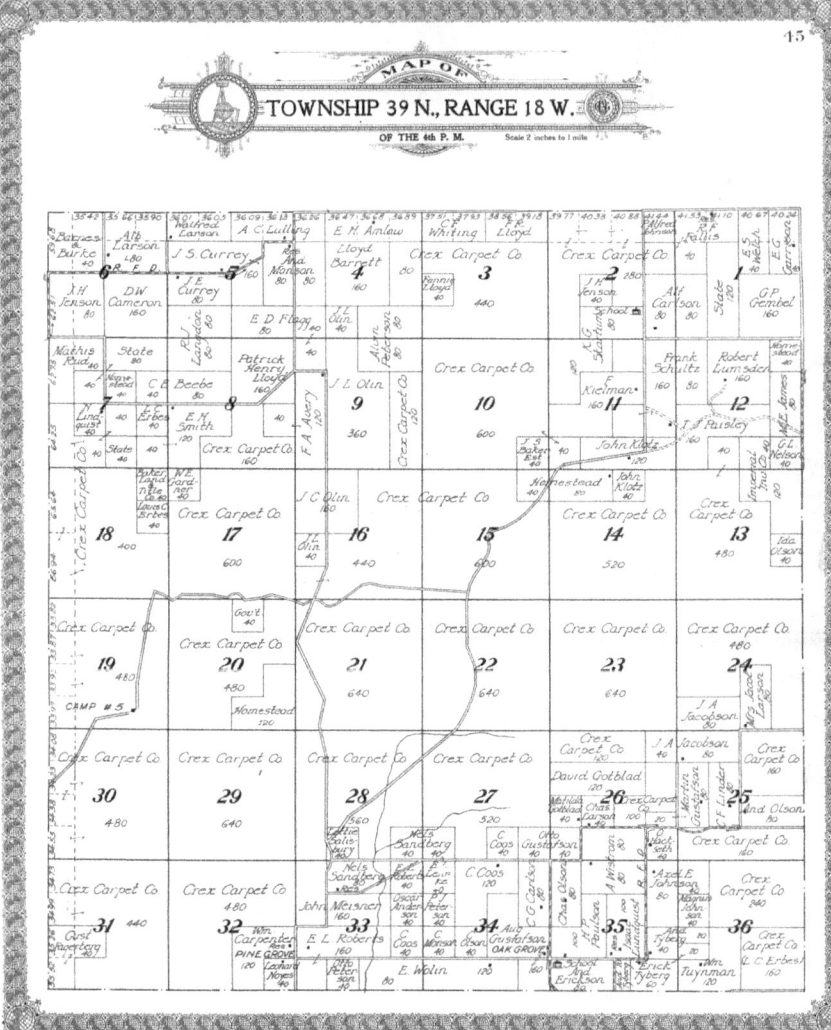

Figure 3.3. The patterns of landholding shown on this map illustrate the sharp contrast between the large-scale commercial agriculture of Crex Co. and the smaller subsistence farms of early settlers. *Source:* George Ogle, Map of Township 39 N, Range 18 W, West Marshland Township, *Standard Atlas of Burnett County* (Chicago: George Ogle, 1915), author's collection.

discourage the longest purse. Wagon roads were constructed where the foot of man had never trod. Tracks from the nearest railroad had to be laid to the warehouses in which the grass was stored. Sheds hundreds of feet long, warehouses of enormous proportions, barns, camps and other accommodations for an army engaged in the harvest were provided in the very heart of apparently useless marshes. This subjugation of these wild marshes is one of the most fascinating victories of modern industrial methods over nature's apparently insuperable obstacles.[19]

The process of harvesting the grass was like hay making. The work season began when the marshes began to dry out in midsummer and lasted until the frost. Horse-drawn mowers cut the grass and laid it into windrows. Workers followed the mower, stacking the sedge into piles for drying. The grass stacks had to be turned several times to ensure they were properly cured. All this work required back-straining labor.

In the larger Marshland Township fields, the piles were compressed into bales and hauled to Grantsburg where the company had two warehouses and an office on Madison Avenue. These facilities were located near the end of the Northern Pacific rail line that ran between Grantsburg and Rush City and then on to Saint Paul.[20] The large bundles of dried sedges were shipped from the Grantsburg warehouse to the carpet factory in Saint Paul on an "as needed" basis.

The workforce in the marshes was largely seasonal and focused on the harvest. Because the marshlands were in lightly settled areas without a large supply of laborers, most of the work was done by migrant workers; but there were also jobs for local men, women, and children during the harvest season. The migrant workers lived in two rather large camps in the midst of the marshes in much the same fashion that lumber camps housed lumberjacks in the pineries. One curious feature of the harvesting process was the need for a special attachment for the horses called a bog-shoe. These worked on the principle of snowshoes and helped prevent the horses from sinking into the muck. It is now not possible to determine exactly how much of the Crex property was harvested. Not all the company's land was marsh, and the 1915 plat book shows several isolated forty-acre parcels of company land.

There were no buyers for the company's land during the 1930s. The Grantsburg area was on the northern edge of Wisconsin's agricultural zone. Settlement continued to expand until the early the 1920s and then stopped in the face of bad weather, poor yields, and declining agricultural prices. Agricultural settlement was always less important than the mining and timber industries of northern Wisconsin, and many farmers worked as temporary laborers off the farm to make ends meet. Only 6 percent of the land in the northern counties of the state was ever cleared for farming. In 1927, 25 percent of the land in the northern counties was offered for sale as tax delinquent, and of that only 20 percent was sold.[21] Those farmers who stayed in business tightened their belts and tried to avoid foreclosure. Many were reduced to near subsistence levels of living and dependence on their reliable cows. Though the marshland owned by Crex Carpet Company could be bought cheap during the Depression and drought years of the 1930s, there were no buyers at any price.

MADISON RESUMES CONTROL

As the forces of the Saint Paul economic hinterland weakened, the influence of the Madison political hinterland grew. In addition to foreclosures, surrendering land for nonpayment of taxes, and having no cash, powerful people rebranded the settlers of the cutover in unflattering terms. They were no longer considered hardy pioneer yeoman farmers by the state and national governments. Instead, President Franklin Roosevelt and his New Deal administrators determined that the "white trash" farmers did not belong in the cutover. They were a drain on resources and troublesome. They lived independent lives, and many were notorious bootleggers. Conservationists of the Progressive Era believed that agriculture should be replaced by forestry in most northern Wisconsin counties. In 1929 legislation was passed by the state that enabled counties to designate portions of their territory as forests and closed to farming. Advocates argued that although forestland yielded profits only after several years, the poor farmers in the area were a drain on the economy. Those farmers who did show some potential were encouraged to enlarge their holdings and specialize in dairying. The newly planted forests would take a generation to mature, and so the tourist industry was promoted as an economic base for the area.[22] National forests, state parks, and a large public relations program were created to turn the cutover into a playground for the growing number of families with cars and the leisure time to take vacations. The basic model was the fishing camps or lodges that were found on the many lakes of north-central Wisconsin.

The 1932 map of Wisconsin produced by the state's Highway Commission welcomed visitors to the region as "the Playground of the Midwest." The map was accompanied by a letter from Governor La Follett encouraging automobile tourists to seek "refreshment in the countryside" and read a chapter in the westward march of "our people and the creation of a new Commonwealth." The accompanying text and photos extolled the amenities of all parts of the state, including the northern forest. There was, however, no mention of Burnett County. According to the map, there were no state highways in either the northern or western sections of the county except the east–west gravel-surfaced Highway 70, which led to the toll bridge over the Saint Croix. Only a thin scratching of sandy county roads led north from Grantsburg, skirting the southern margins of the great marshes. There was no need for county or township roads in this empty quarter of the county. This road pattern had changed little since turn of the century; it would take special efforts to turn this landscape into a tourist attraction during the automobile era.

Following World War II, policies and politics in Madison created a new image and narrative for the marshlands. The landscape was to be managed for wildlife, hunting, and tourism as part of the conservation ethic that developed during the Dust Bowl years. Conservationists such as Aldo Leopold eloquently argued that agriculture should retreat from the frontier and the land should be returned to nature. In his view the ecology was more important than tourism, but he compromised, allowing tourism for the sake of ecology. One of the people greatly influenced by Leopold was Dr. Jake A. Riegel of Saint Croix Falls. Dr. Riegel was a member of the first Game Committee, a volunteer organization that Leopold formed in 1934. It later became the Conservation Congress. Riegel also served on the Wisconsin Conservation Commission from 1947 to 1953. In all these organizations he was a tireless advocate for the preservation of wildlife and wild landscapes in northern Wisconsin. He is generally regarded as the most important voice advocating for the creation of Glacial Lake Grantsburg Wildlife Management Complex (GLGC.)

In 1946 the state purchased twelve thousand acres of tax-delinquent land from Burnett County for public hunting and a wildlife refuge, forming what has become the GLGC, made up of the Crex Meadows, Fish Lake Wildlife, Amsterdam Sloughs, and Danbury Wildlife Area. Most of this land had been formerly owned by Crex Carpet Company. There were many small farms in tax forfeiture that were taken into the reserve. A considerable but unknown number of houses were moved off the refuge and other buildings were scavenged for building materials.

Today little evidence remains of the farmers who once eked out a living from the area's sandy soils. Crex Meadows is a 30,000-acre property of wetlands, brush prairies, and forests. Another 14,000 acres are in the Fish Lake Wildlife Area. The Danbury Wildlife Area originated in 1956 when 1,233 acres of tax-delinquent land was leased from Burnett County. In 1968 the state purchased 1,202 acres from the county at $8.04 per acre. The original project acreage goal was 2,039 acres, but a boundary expansion in 1973 increased that to 2,866 acres. The wildlife area currently has 2,244 acres under state ownership. Acquisition for Amsterdam Sloughs began in 1956. Currently the project boundaries include 7,233 acres (over 80 percent in state ownership).[23]

Although neither Crex Meadows nor the other three areas are producing food or fiber for human use, they are intensively managed and more highly regulated than at any time in history. In Crex Meadows, twenty-two miles of dikes have been constructed to re-flood the drained marshes, and to regulate the water levels in the 6,000 acres that comprise its twenty-nine flowages or reservoirs to provide maximum food for migratory waterfowl. The water levels in flowages are raised and lowered with a system of thirty-four "waster control" buildings, eight miles of ditches, and a pump. Land outside the flowages has been managed with clearing and prescribed burning to restore 7,000 acres of brush prairie. Each year about 3,500 acres are burned to maintain the brush prairie and sedge marsh. During a normal year there are about 6,000 acres of open water in the flowages, four natural lakes, scores of smaller openings in the marsh, and hundreds of pothole ponds dug out of the marsh. Fish Lake and Amsterdam Sloughs are managed in similar ways.

The landscapes are modified and maintained in ways that will maximize the recreational harvest of game birds and animals. It is estimated that about 20 percent of visitors come to hunt or trap. Hunting is critical to the areas' survival because most of the cost of the landscape maintenance is provided by income from hunting and fishing licenses.[24]

The Wisconsin Department of Natural Resources (WI-DNR) would like to own all of the property and manage it for hunting, wildlife observation, and preservation of rare and endangered species, environmental education, and nonmotorized outdoor recreation. This image or narrative is essentially the landscape of recreation. It produces no goods for sale, and it depends on a transfer of funds for its continuance. This image of a landscape preserved for wildlife and recreation has been developing for several decades. Formerly the Department of Conversation, the WI-DNR has legal responsibility for the land owned by the state and the wildlife occupying it. The control of hunting, fishing, and trapping licenses, together with burning permits, gives the WI-DNR a great deal of power. In addition to its direct management activities, the WI-DNR maintains a public information program of publications and electronic media which provides details about its mission and the landscapes it controls.

Although policies in the state capital created the landscape, the project has been successful because the GLGC Management complex is located in the economic hinterland of Saint Paul and Minneapolis. The new Madison-based vision of conservation and recreation policies created a landscape that draws people and money to Wisconsin from the greater Twin Cities. This stands in stark contrast to the economic influence of the Twin Cities in the previous era, when urban industry created a manufacturing landscape in Saint Paul that depended on the marshes and working population of Burnett County. Now Madison-based policy makers have created a recreational landscape that depends in large measure on the spending of people from the Twin Cities.

The WI-DNR estimates that Crex Meadows attracts more than one hundred thousand visitors each year. The combination of a large area with diverse plant and animal communities with a relatively extensive and well-maintained set of roads, observation areas, and a rest area and hiking trails attract people from across the Midwest, but the vast majority of the visitors are from the Twin Cities. Although initially intended for hunters, the landscape has become increasingly popular with city dwellers seeking other forms of outdoor experiences. Most of the visitors come to view wildlife. Two hundred and seventy species of birds have been recorded in the area and nearly every animal native to Wisconsin can be found here. The fall bird migration is spectacular, involving seven thousand sand hill cranes, twelve thousand geese, and thousands of ducks. Large numbers of tour groups visit Crex Meadows during these months. The spring migration is less massive, and the summer bird population is more modest and draws fewer visitors.

THE PAST AND FUTURE OF BURNETT COUNTY'S POLITICAL ECONOMY

The establishment of Crex Meadows is a case study that illustrates the environmental impact of the waxing and waning of influences of the overlapping hinterlands of Madison and Saint Paul. The retreat of the agriculture frontier during the twentieth century has left Burnett County a landscape dominated by uncultivated land. The limited agriculture that remains viable in the county is focused on dairying. The Burnett Dairy Cooperative Cheese Plant is the destination for the milk produced here. The number of farms has declined slightly during the past decade, as has the production of most crops. Over eighteen thousand acres (or 22%) of agricultural land in the county shifted to forest and residential land use between 1998 and 2007.

Although not all the townships in the county have adopted zoning ordinances, most of the private land is now zoned for recreational housing, hobby farms, forest, and agricultural residential. The extensive areas zoned for low-density housing indicate the demise of the landscape of production and the expansion of the retired population and households with one or more members commuting to employment outside the county. Housing outside the villages is concentrated along lakeshores or in small-acreage properties zoned either as recreational residential or hobby farms.

Although the recent recession has slowed the outward expansion of the eastern and northern edges of the Twin Cities metropolitan area, there is no evidence that the urbanization of western Wisconsin will cease. While the economic development policies of the villages and towns in the area promote the establishment of small manufacturing firms, most of the land will be used for residential and recreational purposes. The rural area is being made over into a low-density image of the city.

There are unresolved questions involving property taxes and local government affairs. None of the owners of recreational property can hold office or vote in local elections, but all must pay property taxes and follow the dictates of the WI-DNR. The newcomers' thirst for land drives up land values and assessments, to the chagrin of locals. At present relationships are harmonious, but as prices and taxes rise that may change, and a new balance of powers in the hinterland may emerge.

UPSTREAM, DOWNSTREAM

The Flooding of Anishinaabe Lands by Upper Mississippi Dams

Michael D. McNally

OF SIGNAL IMPORTANCE TO THE COMMERCIAL DEVELOPMENT of the Twin Cities was a network of dams built in the 1880s on Mississippi headwaters lakes far to the north. The US Army Corps of Engineers built the dams to secure reliable flows of the river over Saint Anthony Falls to power the mills and to render the river reliably navigable from Lake Pepin to the falls, connecting the Twin Cities to markets and the nation in a manner that could compete with the railroads. At the behest of milling magnate and Congressman (later Senator) William Washburn, the US Army Corps of Engineers began in 1880 to build a series of dams at the outlets of lakes on the Upper Mississippi drainage: first at Lake Winnibigoshish (or Winnie), Leech Lake, and Lake Pokegama, and later at Pine River, Gull Lake, and Sandy Lake.

This chapter concerns the upstream displacement and loss that underwrote downstream development. The dams took a dramatic ecological toll on the lakes they stopped up, but also on the delicate ecosystems surrounding those lakes, given the flat topography of the Upper Mississippi watershed. As a result of the first two dams alone, Leech Lake rose four feet and Winnibigoshish rose fourteen feet above benchmark levels, flooding 46,920 acres of land around

Winnie, by a conservative government estimate. Later estimates that formed the basis of a federal settlement more than a century later identified 178,000 acres of lands submerged by those two dams alone.[1]

Crucially, the flooded lands were all on reservations, cherished homelands not ceded in treaties by Anishinaabe (also known as Ojibwe) communities who continued to live in profound relationship with the ecosystems on and around the lakes. As if losing almost two hundred thousand acres of their reservation when the ink on the treaties was barely dry were not enough, the flooding devastated the resource base on which their subsistence relied. It inundated the vast beds at water's edge of wild rice (*manoomin*), a staple, and sacred, food. Rising waters drowned other lakeside plants and grasses with which the Anishinaabeg made medicine and hay to feed horses, and displaced fish populations, threatening their ability to reproduce. Flooding lay waste to homes, gardens, villages, graves, and, according to the chorus of Native voices, the social and spiritual bonds of an aspirational land-based community that made life not only economically possible in the region, but ethically grounded and aesthetically beautiful.

Because it was felt in these multiple registers, the damming was among the more vexing issues of the 1880s and 1890s for Anishinaabe leaders and the federal officials and missionaries trying to settle the matter. Historian Jane Lamm Carroll has ably examined the extensive government document record to narrate the technical history of the dams and the ensuing controversy over the monetization of the lost economic resources in what might be called the managerial discourse of "damages," a controversy that took more than a century to legally settle, and which were never entirely resolved, as contemplated federal drawdowns during the drought of 1988 made clear.[2] As David Treuer has pointed out in a critique of Carroll's article, the readily available government documents, focused as they are on the settlement of damages, obscure the full scale on which the dispute over flooded homelands played out. The dams were contemplated, built, and their ill effects managed by a federal government that was unilaterally rewriting the premises on which relations with tribes were based, away from the government-to-government recognition of collective Native rights based on treaties to one of cultural assimilation of Indian individuals by breaking up powerful social, economic, and spiritual bonds to kin and land. Treuer argues that the very concept of damages was shaped by a rhetoric that coded economic and cultural attachments to land as savage and obsolete and that framed the discussion of compensation in terms of a program of cultural and economic assimilation. "Damages did not mean damages to the health, well-being, culture and traditions of the Ojibway," Treuer writes; rather, "damages were any occurrences or effects that prohibited or limited attempts to civilize them."[3] Environmental effects of the dams were tangible signs for Ojibwe leaders that the rug was being pulled out from under them and their imagined future.

Building on the insights of Treuer and the work of Carroll, this chapter tries to understand the dam controversy, and the difficulty of resolving it, in broader terms that consider these multiple registers—economic and ecological, but also political, cultural, and spiritual—in which Anishinaabeg were living in profound relationship to those lakes. If Anishinaabe leaders were loath to settle on a monetized figure for their losses, it was not simply because they were holding out for more cash. The length of the controversy and the words of Anishinaabe leaders suggest instead that while they were strategically trying to contain losses and to build a life for themselves on the shifting grounds of US Indian policy, their concerns were rooted more deeply than the discourse of monetizable resources could access: a relationship to those lands and lakes, and a sense of peoplehood tied to those lands and lakes that were surely economic and ecological, but also political, cultural, and spiritual.

DAMS AND CONTROVERSY

The Army Corps of Engineers began as early as 1850 to explore the feasibility of headwaters dams to promote the downstream interests of milling companies, but it was not until the late 1870s that Congress asked the Corps to estimate the potential effect on reliable steamship navigation to Saint Paul and Minneapolis, deemed a broader public interest in part because it could compete with the railroads.[4] Clearly the driver behind the dams was William Washburn, the grain milling magnate elected to the US House of Representatives in 1878 and to the Senate in 1888. As a senator, Washburn championed the project, with the full support of elected officials in Minneapolis, from his position as chair of the Committee on the Improvement of the Mississippi River and Its Tributaries.[5]

Convinced of these beneficial downstream effects, and through Representative Washburn's behind-the-scenes efforts, in June 1880 Congress allocated $75,000 to build an initial dam at Winnie.[6] Effects of the proposed dam on the several hundred members of the Winnibigoshish Band of Ojibwe, or on the remains of their many ancestors buried along those shores, clearly were of little concern to Congress, given the cap it set on damages in the authorizing legislation: "All injuries occasioned to individuals by overflow of their lands shall be ascertained and determined by agreement, or in accordance with the laws of Minnesota, and shall not exceed in the aggregate five thousand dollars."[7]

If the $5,000 cap seems small, even by the standards of the day, for a project sure to flood tens if not hundreds of thousands of acres, an even deeper issue lay in Congress's language of "injuries occasioned to individuals." The United States viewed the ill effects of the dams as a cost of business according to the logic of eminent domain, without acknowledging the more fundamental question about the unilateral building of a dam that would flood treaty-secured reservation lands without consent. Reservation lands were not at the time matters of individual property; they were collectively held and specifically reserved in treaties by the Ojibwe along with a bundle of usufruct and other reserved rights. In short, the Ojibwe weren't "given" reservations to live on; they reserved them as cherished homelands.

There was knowledge enough in Washington of the legal contradiction to prompt an opinion by the attorney general as to the lawfulness of the dam's authorization. He found that the provision for "the injury to individuals" could not apply to the tribal lands involved, and thus that further legislation and Ojibwe consent were necessary before the construction could proceed.[8] Congress raised the cap for damages to $22,500 and the Interior Department's Indian Bureau, in collaboration with the War Department, dispatched the first of what would become three commissions charged with ascertaining damages caused by the dams and obtaining Ojibwe agreement on those damages.[9]

The 1881 commission worked from maps estimating the flooded acreage at 101,940 acres and set an award of $15,466.90 for the Winnie and Leech Lake dams, including individual property damage ($1,936) and tribal damages to "fisheries" ($750), "sugar trees" ($100), "hay meadows" ($1,000), "cranberry marshes" ($75), and "injury to Indian graves" ($240). To these enumerated losses was added an astonishing penny per acre of submerged tribal lands.[10] Ojibwe leaders, increasingly mindful of the devastating impact of the dams and insulted by the penny per acre for their most productive and cherished lakeside land, flatly refused to consent and continued to do so despite further impoverishment caused by the flooding by dams that were built anyway.

Carroll has related the details of this protracted controversy over damages, and the various sums proposed are shown in Table 4.1, but the differences between various commissions and Native leaders amounted to more than differences on the numbers. Wise to the gulf between commissioners deputized to enumerate single-payment "damages" for unilateral federal actions and the fuller diplomacy of nation-to-nation treaty making, at each stage of the story Carroll tells, Ojibwe leaders asked to be received formally in Washington as a treaty delegation. Also at each stage, federal commissioners expressed their frustration that Ojibwe leaders were not complying with their attempts to ascertain the dollar value of enumerated losses. Indeed, it was all advocates could do to persuade Ojibwe leaders from simply destroying the dams outright, which many clearly wanted. Ojibwe leaders were united in their 1883 insistence on payments of $250,000 every six months for "as long as the dam stands."[11]

While Ojibwe outrage continued apace, two other commissions in the late

Table 4.1. Summary of Actions over Dam-Related Damages

Year	Government Action	Recommended Award for Dam-Related Damages	Outcome
1880	Authorizing Legislation	Up to $5,000 for injuries to individuals.	Opinion by Attorney General Devens prompts new legislation, requires Ojibwe consent.
1881	Revised Legislation	Up to $22,500 for injuries to individuals.	Simpson Commission to settle on damages with Ojibwe leaders.
1881	Simpson Commission	$15,466.9, including individual property damage ($1,936) and tribal damages to fisheries ($750), sugar trees ($100), hay meadows ($1,000), cranberry marshes ($75) and "injury to Indian graves" ($240), plus $.01/acre of submerged lands.	Rejected by Waubunoquod.
1883	Sibley/Blakely Commission	$10,038 for damage to property and $26,800 annually for ongoing economic losses (wild rice, fish, berries, maple sugar; hay).	Rejected: Ojibwe leaders ask for $250,000 every six months "for as long as the dam stands," and in any event Congress never appropriated the funds for the ongoing economic damages, so neither did Commissioner of Indian Affairs release the property damage.
1887	Northwest Commission	One-time payment of $150,000.	Recommended, but Congress never appropriated.
1889	Rice Commission	Fulfillment of $150,000 payment recommended by Northwest Commission.	Purpose to secure agreement with "Nelson Act" for land allotment. No settling on damages from dams.
1985	Out-of-Court Settlement	Pillager Band: $700,000; Mississippi Band: $27 million, including 178,090 acres of submerged reservation land and 5% accumulated interest since 1884.	Rejected by Ojibwe leaders.

Sources: Jane Lamm Carroll, "Dams and Damages: The Ojibway, the United States and the Mississippi Headwaters Reservoirs," *Minnesota History* 52 (Spring 1990): 2–15; "Damages to Chippewa Indians: Letter from the Secretary of the Interior, transmitting a copy of a communication from the commissioner of the Department of Indian Affairs, together with papers and report of Messrs. Marshall, Gilfillan, and Sibley, appointed December 22, 1882, to re-examine and ascertain the damages accruing to the Chippewa Indians residing upon Lake Winnibigoshish and Leech Lake Indian Reservations in the State of Minnesota," 48th Cong.: 1st Session, House of Representatives, Executive Document 76, February 1884; *Minnesota Chippewa Tribe et al. v. the United States of America,* 29 Indian Claims Commission 211 (1972); *Statutes at Large,* 21:1:193; 47th Congr., 1st sess., 1881, *House of Representatives Executive Document,* no. 1; U.S Department of the Interior, "Message from the President of the United States," "Chippewa Indians in Minnesota," 1890.

1880s attempted to resolve the matter. First, the 1886 Northwest Indian Commission recommended a onetime settlement of $150,000, to which some Ojibwe leaders agreed but which Congress ultimately failed to approve.[12] Then in 1889 the so-called Rice Commission, as part of its effort to promote voluntary removal of Ojibwe from all over Minnesota to White Earth, called again for a $150,000 onetime payment, together with 5 percent interest, and added to that payment of $1.25 for each acre of flooded tribal lands.[13] But Ojibwe people would receive no monetary damages for the dams until the 1980s.

BEFORE THE DAMS

To more fully understand Anishinaabe approaches to the damages controversy over its *longue durée*, we must begin with Indigenous relations to the lands that the dams flooded. That story begins with an Anishinaabe story of peopling the land, and of the sacred and reciprocal bonds and obligations extending between Anishinaabe people and the other-than-human people belonging to that land established in and through this narrative.

Citing early documents in the Euro-American written record, ethnohistorians place the movement of Anishinaabe peoples into the region in question in the late seventeenth and early eighteenth centuries.[14] Native oral traditions don't dispute this migration to the region, but they place it in two important contexts. First, Anishinaabe people understand this as a return after a long sojourn to original lands. Second, and crucial to the concerns of this essay, the migration story plots the movement/return to the region not as a historical happenstance but as the fulfillment of a moral purpose, a sacred calling. From the Atlantic seaboard, the Anishinaabeg followed prophetic traditions that they should move west, following a visionary shell through a seven-staged migration, until they reached the place where "food grows on water."[15]

Especially when one considers that this was perhaps the central narrative of the Anishinaabe, the significance of ancestral lands, plants, and animals can hardly be overstated. The *is* becomes seen as an *ought*: it is not that Anishinaable people merely *are* living along these lakes, but that they *ought to be* there according to core religious narratives. Indeed, the Ojibwe language term for their territory, *anishinaabe akiing*, is a possessive construction that works both ways, translatable either as the land of the people or the people of the land. For this and other reasons, it is no mere poetic license at work when Ojibwe people like Marvin Manypenny, an activist at White Earth, proclaims, "We and the land are the same. What happens to the land, happens to us."[16]

Bimaadiziwin is the Ojibwe term that describes the integration of the economic and the ecological, the moral, and the spiritual. *Bimaadiziwin* is a substantive formed from the Ojibwe verb which refers to motion that passes by continuously, and can be rather flatly translated as "life" or "nature," but is imbued in Ojibwe culture with a sense of profound relatedness, beauty, even ultimacy, tied to lives lived in correspondence with the seasons. The Roman Catholic missionary Frederic Baraga translated *anishinaabe-bimaadiziwin* as "the savage life," giving voice to a view that Ojibwe people were incapable of separating themselves from the rhythms of the northern forests and largely unconcerned with any eternal consequences beyond living well in this world.[17] From Anishinaabe perspectives, living close to the rhythms of the land, moving with the seasons, was no mere economic necessity or well-honed cultural ecology; it was also a fulfillment of sacred obligation. This was observed by A. I. Hallowell, the ethnologist who spent years among Anishinaabe east of Lake Winnipeg. Hallowell translated *bimaadiziwin* as "the Good Life," making reference to the ultimate concern of Ojibwe religion and culture. "The central goal of life for the Ojibwa," Hallowell wrote, "is expressed by the term *bimaadiziwin*, life in the fullest sense, life in the sense of longevity, health, and freedom from misfortune. This goal cannot be achieved without the effective help and cooperation of both human and other-than-human persons, as well as by one's own personal efforts."[18]

In the lake region of the Upper Mississippi watershed, what was known as "the seasonal round" characterizing anishinaabe-bimaadiziwin involved, and has continued to involve, what Frances Densmore described as a "systematic" tradition of local movement to seasonal resource locations.[19] Anishinaabeg could be found in late March and early April at their "sugarbushes" when maples offered their sweet sap for sugar making. Shortly thereafter, streams and the graveled shallows of lakes offered opportunities to spearing spawning suckers and other fish. Then, moving to summer residences, Anishinaabeg planted gardens of corn, potatoes, and squash in clearings near lakes. In succession then came the picking of strawberries, blueberries, raspberries, gooseberries, cranberries, and the gathering of "swamp tea" and medicines in their seasons. Summer village life also occasioned setting nets for a range of fish. At summer's end was the celebrated harvest of the year's wild rice, together with the drying, "dancing," hulling, and roasting of the rice for winter storage, hunting migrating waterfowl on the lakes and near the rice beds, and spreading out for big-game hunting, snaring small game, and ice fishing through the winter months. Specific details of a seasonal round of Ojibwe groups varied considerably over the range of microenvironments of their expansive territory throughout the Great Lakes region, but for the Ojibwe who called the Upper Mississippi lakes their home, clearly much of the seasonal round took place in the zone on or near the shores of the two lakes.

The seasonal round was more than a mere resource strategy; to call it "a way of life" on Native terms is not to overlook its economic or ecological facets but to recognize the crucial ways in which plants and animals on which such a life depended were more than resources. The Ojibwe language structures experience in gendered terms, not of male and female, but in terms of animate and inanimate. All plants and animals, including manoomin, fish, and sugar maples, are gendered animate in the Ojibwe language. That is, they pertain to an entirely different set of transitory verbs and are referenced by the pronoun "him/her," not "it." While one can argue that such linguistic structures are ultimately arbitrary, many contemporary Ojibwe language speakers speak of how the animate gender of plants and animals grounds and reflects a distinctive valuation of the natural world not as so many "natural resources" but as a society of nonhuman persons with whom human people live in morally regulated relationships.

Beyond even the notion common to resource management discourse as a "cultural resource" these plants, animals, landforms, weather phenomena, and even bodies of water, have been moral persons, spiritual others; and Anishinaabe peoplehood, not to mention livelihood, has depended on reciprocal relationships with them. Respect for these other-than-human persons has not meant purely sentimental attachment; Anishinaabe narratives and ritual practices have supported the moral regulation and ritual obligations with which hunts and harvests properly proceeded. This would surely be true of the trees, berries, medicines, and the lakes themselves, but the effect of the dams was perhaps most injurious in terms of fish and wild rice. Making sense of these time-honored traditions, the Anishinaabe intellectual Basil Johnson writes, "In essence each plant being of whatever species was a composite being, possessing an incorporeal substance, its own unique soul-spirit. It was the vitalizing substance that gave to its physical form growth, and self-healing."[20] This didn't mean that such foods as wild rice or fish were not resources available for the furtherance of human community, only that they were not reducibly or *only* resources of an economy. "Wild rice is consequently a very special gift," the ethnomusicologist Thomas Vennum agrees, "with medicinal as well as nutritional values—a belief reflected in the Ojibway use of wild rice as a food to promote recovery from sickness as well as for ceremonial feasts."[21]

Grounded in sacred narrative, this special relationship with manoomin has been reiterated and renewed in ritual. As White Earth elder and activist Joe LaGarde puts it, wild rice and water are the only two things required at every ceremony. Manoomin accompanies Anishinaabe celebrations, mourning, initiations, and feasts, as a food and as a spiritual presence. It holds special significance in traditional stories, which are told only during ricing time or when the ground is frozen. "In these stories, wild rice is a crucial element in the realm of the supernaturals and in their interactions with animals and humans; these legends explain the origin of wild rice and recount its discovery" by Nanabozho, the principal spirit in sacred narratives.[22] In the words of White Earth's tribal historian, "wild rice is part of our prophecy, our process of being human, our process of being Anishinaabe . . . we are here because of the wild rice. We are living a prophecy fulfilled."[23]

Importantly, the integrity of these relationships did not depend on some aboriginal balance incapable of change. Exchanges of Old World diseases brought significant disruptions to aboriginal populations in the region, and the fur trade involved the exchange of surplus wild rice, fish, and game for valued goods that clearly had environmental consequences, but fur trade–era encounters with Euro-Americans from the early seventeenth to the early nineteenth centuries did not alter appreciably the highly regulated ecological, moral, and spiritual balances on which Anishinaabe sustenance from the land depended.[24] It was the westward expansion of settlers and of those seeking timber and other resources to support that expansion that changed everything in the Upper Mississippi in the second half of the nineteenth century. In the Treaty of 1855 Anishinaabe leaders ceded lands to the United States but reserved rights (hence "reservation") to "a sufficient quantity of land for the permanent homes of the said Indians" in areas surrounding Gull, Sandy, and Rabbit Lakes for the Mississippi Bands, and nearly seven hundred thousand acres surrounding Leech Lake and Lake Winnie for the Pillager and Winnie Bands. The Gull, Sandy, and Rabbit Lake reservations were later ceded in an 1863 treaty to access coveted timberlands, on the one hand, and to consolidate Ojibwe populations toward their assimilation into American economy, culture, and society, on the other.[25] The Treaty of 1867, the last treaty as such with the Minnesota Ojibwe, recognized a new reservation at White Earth, a location that missionaries and reformers intended to absorb Ojibwe from elsewhere in the state into a consolidation designed to equip a transition from the seasonal round along the wooded lakes of the new reservation's eastern parts to agrarian living on fertile glacial Lake Agassiz prairie in its western part.[26]

It is ultimately no coincidence that serious talk of dams on the headwaters began in the same decade, the 1870s, that saw the elaboration the new formal assimilation policies of the United States implied in the White Earth removal scheme. Indeed, by the time of the Northwest and Rice Commissions in the late 1880s, the justice Anishinaabe leaders sought for the damming of their lands had been thoroughly folded into the fully elaborated US policy of land allotment and cultural assimilation.

TRYING TO LIVE WITH DAMS

Even without Ojibwe consent to the recommended monetary settlement of the 1881 commission, and thus compliance with the attorney general's legal opinion, construction resumed on Winnie dam in 1881. To make matters worse, Congress authorized further appropriations to build contemplated

Figure 4.1. Canoeists paddle across the reservoir behind the dam on Lake Winnibigoshish, in Cass County, Minnesota, 1902. The United States viewed the ill effects of the various headwaters dams as a cost of business according to the logic of eminent domain, without acknowledging the more fundamental question about the unilateral building of a dam that would flood treaty-secured reservation lands without consent. Courtesy of the Minnesota Historical Society.

dams at Leech Lake (1882) and Lake Pokegema (1884), and construction there too went on despite the failure of various commissions to reach or deliver a settlement.

The Winnie Dam was completed in 1884. When its gates first closed, the lake rose a reported fourteen feet, drowning rice fields, gardens, meadows, villages, and cemeteries while denaturing the growth around it into an eerie landscape. "Just look around this lake!" Sho-kah-ge-shig told the 1889 Rice Commission gathered on the shores of the reconfigured reservoir, reminding the commissioners that the Ojibwe had lived up to their word that they would not attack the dams. "There are no persons who have been so badly damaged. Look around here. It is not fire that makes it look so barren around the lake. It is the effects of the water caused by the overflow."[27] "When that [dam] was erected here, they told me they would never shut the gates for good," added May-quom-me-woub, "that they would leave the water to run pretty freely until the whole thing had been settled for."[28]

At Leech Lake the initial rise in lake level by a reported four feet was less visually dramatic, but perhaps even more impactful given that more than a thousand Anishinaabe were making a living, and trying to live the Ojibwe good

life, from a seasonal round along its shores. "It cannot be put plainer than this,"
Kay-me-wun-oush told Rice: "there is something grows there by nature which
is put in my hand. The white man takes that morsel put there for me, from my
mouth. The reservoir built there has taken away my subsistence. It has created
not only hard feeling, but hardships."[29]

Clearly the ecological impacts of most concern to Anishinaabe leaders were
the dams' immediate impacts on manoomin beds and on fish populations and
access to them. Of particular concern was wild rice, the staple food which was
easily stored and rich enough in nutrients to see the people through the bitter cold winter months if it was harvested in sufficient amounts. In fact, Winnibigoshish leaders were several days late for an early September meeting with
the Rice Commission because they were traveling from alternative rice camps
at some remove from the flooded lake, a point that they wished the commissioners not to mistake for insolence. May-quom-me-woub told Rice, "This is
the only subsistence and the only chance we have of making a living for the
winter. . . . Almighty God has provided for all his creatures. He has provided
that kind of subsistence for the Indians. If they do not take it in time they must
go without it."[30] Wild rice (*Zizania palustris* or *Zizania aquatica*) is an annual
wild grass that roots in the underwater alluvial mud on the shallow edges, up
to three feet, of the region's lakes and slow-moving rivers. It grows to the surface in its floating leaf stage and eventually extends above the water, where its
grain matures. Stands of rice provided crucial habitat and food for the migrating waterfowl that played another key role in the seasonal round. But for all the
"vastness of its natural stands," writes Thomas Vennum Jr., "wild rice is a relatively delicate plant."[31] The plant requires good water quality and circulation of
mineral-rich water. It is also quite sensitive to water levels. While wave action
and the natural cycles of spring flooding aerates the mud to the benefit of the
germinating seeds, sustained high water above a seeded stand can prevent the
plant from reaching the surface and falling water levels can cause the mature
plant to collapse on itself. Above all, manoomin's shallow root system does not
tolerate rapid fluctuations in water levels.

Vennum draws on fur trade archives to note considerable differences from
year to year in the yield of the manoomin harvest depending on natural fluctuations in water levels, and to be sure beaver dams could dramatically change
local water levels.[32] But the devastation of initial flooding from the Upper Mississippi dams was something else altogether. Subsequent water level fluctuation, especially when not managed with manoomin stands in mind, hindered
the restoration of the plant in new conditions.

Even the more modest rise of Leech Lake had a dramatic impact on Anishinaabe livelihood there. In the five years after the dam was built, No-din-
ah-quah-um of Leech Lake told the Rice Commission in 1889, "We have been

unable to get two crops of rice."[33] Federal officials admitted, before the dams were operational, that they would obliterate wild rice beds. Ojibwe leaders remembered how a logging dam built at Mille Lacs in the 1850s had destroyed wild rice beds there, and therefore knew what was coming. Members of the 1881 and 1883 commissions were clear that stands of wild rice would be initially destroyed.[34]

But when it came to fish, those officials were unconvinced dams would have a serious impact. Captain Russell Blakeley, the head of the 1883 commission, argued with Leech Lake leaders that government engineers had a very reliable list of all the lands to be flooded, and while they were prepared to come to terms about the wild rice stands and hay marshes that would surely be flooded, and about timber cut to build the dams, "we think you can catch just as many fish every day in the year after the dams are built as you do now."[35] He reasoned that the way the dams would be managed, the lake levels would rise five to six feet during winter and spring, and fall back to their usual levels "by the time fall fishing commences."[36] Noting that they took different fish at different times of the year, Anishinaabe leaders disputed whether officials had sufficient local knowledge of their seasonal round or the ecological systems on which it depended to judge the extent of the impact.[37]

While government documents show surprising awareness on the part of federal officials on the ground of the more immediate impact of the rising water levels, it was clear that the discourse of damages involved only the economic value of the initial loss, not the long-term losses that would be borne by Anishinaabeg. Those documents also show that those officials could not concretely persuade decision makers in Washington of the extent of the ecological impact, much less the implications, economic and cultural, for Anishinaabeg. We turn now to the cultural registers in which the violence of flooding was felt.

WHAT HAPPENS TO THE LAND, HAPPENS TO THE PEOPLE

William Marshall wrestled more explicitly than Blakeley with the enormity of their shared task on the 1883 commission: "As a question of material damage it is not easy to get at a just estimate. I doubt if any commission could arrive at it. The possessions of the Indians, the fishing privileges, rice marshes, sugar-making and canoe-making grounds, etc. have not a marketable and commercial value, such as the possessions and privileges of white men."[38] Marshall continued to describe what he called "a large sentimental damage, not material, but not less real" that was incurred because of dams: "Their accustomed haunts are broken up, their paths, roads submerged, they will feel compelled to relocate their villages, will have to adapt themselves to new surroundings, a thing a white man could readily do, but not an Indian."[39]

Marshall had good instincts about what fueled Anishinaabe resolve in the matter, for their concerns about dams and flooding stemmed from attachments to place that were not purely economic, though neither were they vaguely sentimental. To put it bluntly, the dams destroyed the ecosystems that made the entire Anishinaabe economy and way of life possible. "I love my land and country," an Ojibwe leader told Rev. John Enmegabowh in response to the paltry offer of the 1881 commission:

> Its beautiful scenery and the beautiful shore where my children played for years: can this be true that the palefaces are coming to destroy my beloved land and country its beautiful scenery and the abundant resources from which I have derived my daily food. They say what is fifteen thousand dollars can do—how many years will that give me food and clothing for three or four generations my store houses have always been well filled I never go hungry have always plenty of wild rice, sugar, cranberries, rabbits, and plenty of fish.[40]

Recalling that the Ojibwe language grammatically engenders as "animate" all animals and plants, we should imagine utterances by Native leaders not in the broken English of their dictation, but in the idiom of their original language, speaking of manoomin not in terms of "it was destroyed" but "he/she was destroyed," and invoking the fuller semantic range of its associations with peoplehood and purpose. Eyeing construction of the Winnie dam, an Anishinaabe leader told Enmegabowh: "Now it would be the greatest event that shall and will happen to us since our eyes can see. For generations past I have cook and ate my food—my food is my dish—such food as the Great Spirit gave me for many generations."[41] Namewinini (Sturgeon Man) gave voice to the ways that the losses were more than merely economic in his words to the 1883 commission: "No white man knows of the damage that will be done to us. As long as the sun shall pass over our heads we would have been able to live here if this dam had not been commenced. Every year what supports us grows on this place. If this dam is built we will all be scattered, we will have nothing to live on."[42]

Native leaders were particularly incensed about the obliteration of hundreds, perhaps thousands, of human remains from cemeteries along Leech Lake and especially Winnibigoshish. The Winnibigoshish leader I-aub-aince addressed the Rice Commission with notable precision about the loss of dead relatives. Reminding the commission that two years earlier the Northwest Commission had themselves seen "bones sticking out of the ground where they had been washed out by the waves":

> The covering of the graves and the coffins have in large numbers drifted out onto the waters of the lake. The cemeteries are now just as level as this ground here—nothing to show what was there. This was at the mouth of Third River

and at the mouth of Cutfoot Sioux River, and also right alongside of the Mississippi River here. We bring this to your notice, as we would like to have action taken in the matter, although I don't think it would be of any use now, as the bones are all scattered, the skulls are here and there all along the shore.

If the scattered remains of relatives, including children, epitomized for Anishinaabeg the fundamental chaos and violence caused by the dams, it also suggested how unnecessary—and irrevocable—was the damage. The 1881 commission had valued the "injury to graves" at Winnie at $240, the anticipated cost of moving the remains to higher ground at government expense, which pleased those "who had fathers and mothers and children buried there," according to I-aub-aince, "but the promise was all we ever saw of it."[43]

THE DAMS AND ANISHINAABE PEOPLEHOOD

Just as the story of the dams and the ensuing controversy cannot be understood by monetizing the value of "natural resources," neither can its intractability be understood merely by reference to what Marshall called the "sentimental damages" of cultural attachment to the land understood by bimaadiziwin.[44] Here, the environmental history of the dams was also caught up in the history of US Indian policy; dams built explicitly to promote the commercial development of the Twin Cities were also seen implicitly to support the emerging US policy of assimilation.[45] This policy, operationalized in Minnesota by the 1889 Nelson Act, meant that the Anishinaabeg were no longer viewed as the nations with whom the United States had made treaties reserving collective rights to land, culture, and religion but as groups of individuals to be assimilated within American economy, culture, and society. Allotment of collective reservation lands into individual plots of land was the key feature of the Nelson Act, along with both carrots and sticks to secure their removal to White Earth from ancestral lands and the seasonal rounds tied to them, but it also came with a host of administrative policies that compelled English-only education in boarding schools, rewarded Christianization, and criminalized traditional ceremonies.[46] With allotment came Euro-American settlement of "surplus" reservation lands, timber operations, fences, and a range of environmental impacts that hindered anishinaabe-bimaadiziwin.

In the years before the Nelson Act, it was perhaps the controversy of the dams through which Anishinaabe leaders came to realize the sea change that had undermined not only the Treaty of 1855 but the nation-to-nation premise of treaty making altogether. When Waabaanakwad (White Cloud), the key Mississippi Band leader, responded to the 1881 commission's offer of a penny per acre for flooded reservation lands, by saying "as poor as I am . . . I would throw it away," his was no mere negotiating position; it was a comment on the framework of the negotiation: "At Washington is an understanding, a strong one, in which a mention is made of our reservations, also that a white man should take nothing from those reservations or little on them. That is what the Great Father and ourselves understood each other. . . . We could and did not give assent to the damming of the river."[47] In August 1883 a general council of Pillager and Mississippi Band leaders convened at Leech Lake, demanding a stop to the construction of dams until "we have settled with you our rights," and petitioning for a diplomatic delegation to Washington, DC, to settle those questions.[48] Later that month Marshall asked the head of the Indian Bureau to dismiss the 1883 commission on which he was serving, suggesting that the question was "one of treaty or agreement, rather than of valuation, and of the necessity, finally, of visitation from these Indians who are to be damaged by these reservoirs."[49]

But the leader of the commission, Captain Blakeley, argued the United States would no longer agree to a diplomatic approach and urged Native leaders to be "more reasonable" than he viewed their unified position of $250,000 every six months for the lifetime of the dams: "The Great Father never asks his white children their permission to build dams or cut timber. He always appoints a commission to say how much the Great Father shall pay the white children when he takes their property for dams or anything else, and that they have to take, and no complaint."[50] Even Bishop Whipple, in whom White Cloud and Flatmouth (the leader at Leech Lake) placed considerable confidence and who early on had joined the call for a diplomatic approach, knew the direction in which Indian policies were going—indeed he had been a key voice in their design. From his view the tragedies of the flooding could be redeemed by hastening what he and others deemed a necessary transition from the seasonal round on ancestral lands to an agrarian life imagined for them at White Earth. From his perspective, maximizing the monetized losses was advocacy, for it meant further federal support to capitalize that agrarian transition. It was in this context that the Northwest Indian Commission of 1886 and the Rice Commission of 1889 tried to steer unease about unsettled claims for the dams into their stated purpose to persuade Anishinaabe leaders to remove to White Earth: "Now, White Earth is a beautiful place. If any of you wish to leave here, there is no finer place to go to than White Earth. . . . With your game gone, nothing is left for you to subsist upon but the fish in the lakes, and no nation, no people, can live and raise their children successfully on fish alone."[51]

So even as Anishinaabe leaders were coming to terms with the swollen reservoirs that constrained their Good Life, they were also coming to terms with a dramatically altered legal landscape where treaty agreements were no longer recognized and survival required adjusting to that reality. As Treuer points out, historians who take their principal cues from the readily available government documents wind up naturalizing the assumptions on which those government decisions were based, consequently obscuring much of what was at stake behind the claims of Anishinaabe leaders.[52]

AFTER THE DAMS

Resourceful Anishinaabeg did their best to make do, paying close attention to the new rhythms of the reservoirs and striking new balances with their environment. But adapting the seasonal round by moving to other rice beds and resource areas was made harder still by tightened enforcement of the perimeter of the reservations. In what historian Bruce White aptly calls the "criminalization of the seasonal round," state fish and game laws coming online in the 1880s and 1890s to protect sport and tourism interests fueled further arrests of off-reservation Anishinaabeg either returning to accustomed hunting and gathering grounds or finding new ones to make up for the ecological impacts of the reservoirs and timber companies on the reservation.[53]

The privations of the 1880s and 1890s led to deep challenges to Anishinaabe health and well-being. As Melissa Meyer documents, the environmental stresses on the seasonal round, together with missionization, land allotment, and unequal disbursement of treaty annuities, led to the social stresses of factionalism and violence on reservations.[54] Tuberculosis, alcohol abuse, and other illnesses exacerbated by poverty swept through reservation villages. Indeed, an 1882 smallpox outbreak was attributed by many Anishinaabeg to the Army Corps of Engineers' construction camp at Winnie. The scourge of untimely death and a despair associated with the dams were clear in the broken English of a leader dictated by Enmegabowh: "I wish the government would let us alone to let me die peacefully where I have buried my fathers and dear children—cannot the government have pity upon my poor dying people being few years more your red brothers shall all disappear very soon our loving homes shall be made vacant. Then you can come and build all the dams on the Mississippi River."[55] But constrained as they were, as Brenda Child has shown, Anishinaabe people also proved remarkably resourceful at adapting their practices and making do in the context of new ecological and economic conditions.[56]

By the 1930s the system of locks downstream from the Twin Cities settled navigability questions and allowed more local concerns to inform federal management of Upper Mississippi dams. Indeed, after the disavowal of formal assimilation policies and the formation of federally acknowledged tribal governments under the 1934 Indian Reorganization Act, the Army Corps of Engineers could work with tribal leadership to stabilize lake levels and promote stronger wild rice yields despite natural fluctuations in wet and dry years.[57]

This newer potential for cooperative management with tribal governments, combined with final settlements of the damages questions in the mid-1980s, made it seem that the controversy about the dams had come to a close. A 1985 out-of-court settlement with provisions related to the dams required a $3.39 million federal payment to the Leech Lake tribal council, and a 1985 ruling by the Indian Claims Commission involved $700,000 to the Pillagers and $27 million to the Mississippi Band for 178,090 acres of reservation land taken by the reservoirs, loss of wild rice marshes, and 5 percent accumulated interest since 1884.[58]

But a major drought in the summer of 1988 showed how environmental stresses could create conditions under which downstream benefits to the Twin Cities could still potentially trump Ojibwe concerns about effects of the dams on their reservation. With water levels so low they threatened Twin Cities public works and sanitation systems, Governor Perpich made a formal request in late July asking the Army Corps of Engineers to release water from the headwaters reservoirs, drawing the ire of resort owners and Ojibwe communities.[59] Warren Tibbets, a Leech Lake activist, said "We believe this water belongs to us and our treaties say so and the Constitution of the United States says that our treaties are the supreme law, it's that simple."[60] Tibbetts told the *Star Tribune* that he and other activists encamped and keeping vigil by the dam would stand on it in an attempt to prevent a drawdown.[61]

At the last minute, the heaviest rains in nearly a year drenched central Minnesota, restoring healthy river flows by August 4, and the Corps officially rejected Governor Perpich's request. Rains continued through August, and the encamped Ojibwe activists holding vigil at Winnie dam picked up their stakes and dried out their tents. But the tense summer revealed that, when push came to shove, the events of the 1880s and the 1980s were oddly alike. Despite all the changes in US Indian law in the century since—the rescinding of assimilation policy, the recognition of tribal governments, and government-to-government relationships rooted in affirmed treaty rights—the federal government maintained ultimate control of the dams, and acknowledged treaty rights were derogated by extraordinary environmental circumstances. Challenges by tribal members were folded into the managerial discourse of monetary damages.[62]

The environmental history of the Upper Mississippi dams shows that the commercial vitality of the Twin Cities has come at considerable expense to Anishinaabe communities upstream. These costs were surely economic, but they have also been political, cultural, moral, and spiritual. This essay has not only tried to countenance the fuller story that consideration of government documents recording the monetization of natural, even cultural, resources alone does not access. It has aimed to do so first by challenging what the discourse of resource and resource management obscures: the integrative nature of the relationship of Anishinaabe livelihood and peoplehood to what the Ojibwe language grammatically construes as nonhuman persons—the fish, wild rice, and game—whose livelihoods and peoplehood also rely on the ecological integrity of those lakes. Second, this essay has considered Anishinaabe challenges to government attempts to monetize and manage the damage as a thorough critique of a federal Indian policy where nothing is sacred, neither the affected environment, the treaty commitments, or the first peoples of the land.

MAKING STUMPS AND FIELDS

Working Environments in the Woods and on the Cutover, 1890s–1930s

Kevin C. Brown

"WOOD SERVES SO MANY PURPOSES," University of Minnesota professor of horticulture and forestry Samuel B. Green wrote in 1902, "that it may indeed claim to be the most useful of all natural products, excepting only food."[1] Indeed, wood formed an essential part of both industrial and agricultural landscapes in Minnesota during the nineteenth century, in the structures of factories and houses as well as in ubiquitous wood products like fences, railroad ties, and furniture. This strong demand for wood had substantial implications for the state's heavily forested northern counties. Starting in earnest after the Civil War, wage workers (often hired and directed by Twin Cities–based lumber companies) cut vast swaths of the region's old growth white pine (*Pinus strobus*), red pine (*Pinus resinosa*), and other marketable tree species, setting logs on a course for distant mills and markets. By some estimates loggers removed about 75 percent of all "standing timber" in the state during the thirty-five years between 1880 and 1915.[2]

Such logging did not simply destroy the forest; it also created a new landscape: "the cutover." In this ecologically degraded space a second environmental transformation began after 1890. Lumber companies, having already sold the land's most marketable asset (lumber), sought to turn a second profit from their acreage by selling it to prospective farmers. These new owners in turn planned to convert the land—alternatively rocky and swampy, and littered with stumps, brush, and rocks—into a stable agricultural environment. Unlike the rapid deforestation that swept across northern Minnesota, however, the transformation of the cutover to a farming landscape never took hold on a wide scale. "What little agricultural development there is," a state committee appointed by Minnesota governor Floyd Olson reported in 1935, "faces serious difficulty."[3] Conditions were so dire that for much of the northern part of the state the committee recommended reforestation and public ownership rather than continued efforts to farm.

Historians and environmentalists are familiar with this broad story: forest became wasteland became farmland became (new) forest. But too often this narrative overlooks the fact that as timber workers and farmers turned forests into logs and stumpland into farmland, they also constructed and inhabited temporary environmental spaces, what environmental historians call "working environments." Forests partially cleared of trees and farm fields partially cleared of stumps were each defined by the social organization of production and the variable environmental conditions and risks that timber workers and farmers encountered, endured, and sought to mitigate. Such "working environments" blurred the line between "made" and "unmade," the human and the natural.[4] As the geographer Neil Smith put it, "We do not live, act, and work 'in' space so much as by living, acting, and working we produce space."[5] Indeed, loggers working in the woods and farmers working on the cutover each created distinctive working environments that were neither wholly old growth forest nor farm.

Recovering these working environments from obscurity allows us both to document the everyday acts that fueled the hasty environmental change of the region and to paint a portrait of the northern part of the state during two key historical moments. Even more important, exploring the working environments of logging and land clearing helps explain why logging produced such widespread and rapid changes in the land, while farming never took hold, despite the fact that both activities occurred in the region's same harsh environments. In short, working environments provide the key to understanding why logging "succeeded" and farming "failed" in northern Minnesota.

TIMBER WORKERS AND CUTOVER FARMERS

Developing such an understanding begins with a simple observation: not all working environments are created equal. In addition to comprising unique environmental conditions, working environments are conditioned by their particular relationship to *capital*. Used most often as shorthand for invested money, *capital* also denotes (1) the technology and machinery used in labor (such as a sawmill or a hoe), (2) a system of power relations defining how and by whom that technology is deployed, and (3) the ability to connect the products of labor to markets. Cutover farmers' working environments, for example, were starved for capital—not just in a basic monetary sense, but also in terms of access to technology, manpower, and markets. This dearth shaped both how cutover farmers went about transforming the land and how successful their labors ultimately were. We can say that "capital shaped specific working environments," but only if we recall the several meanings conferred by the term and understand that this occurred in relation to certain material conditions.

In the late nineteenth century Minnesota's working environments were mediated both by the environment of the northern forests and by their connections to lumber firms and distant cities. Though the first commercial logging operation began on the Saint Croix River before Minnesota's statehood, logging and sawmilling exploded after the Civil War. Between 1869 and 1899 lumber production in Minnesota grew dramatically, from 242 million board feet to over 2.3 billion board feet per year, making the state the third largest producer in the country.[6] The industry took advantage of the state's extensive river and lake system to channel logs into the Mississippi, Saint Croix, and Rum Rivers, which carried them to downstream mills near the Twin Cities, Winona, and even farther south to Iowa.

The presence of wheat farms and railroads in the region served as both cause and consequence for the increased demand for wood products and the deepening of the connections between the Twin Cities and the northern forests.[7] "Wheat and lumber," the historian Agnes Larson noted in 1949, "were the warp and the woof in the weave of the pattern that is the Northwest. And the railroad was the shuttle."[8] Rivers provided the lumber industry with its first great transportation system, but the gradual expansion of railroads in the state after the 1870s allowed it to stretch its connections farther into the north and farther from major river systems. Frederick Weyerhaeuser, the preeminent "lumber baron" in Minnesota, for example, started his lumber career in Rock Island, Illinois, and gradually expanded his operations into Wisconsin. Realizing the potential of northern Minnesota as a source of much needed wood, Weyerhaeuser and his partners made several large land purchases in the state. He relocated his operations to Saint Paul by 1890, and effectively used both rivers and railroads to integrate the state's northern reaches into the economic orbit of the Twin Cities.[9]

The relationship, however, was unstable. Few industry leaders imagined the lumber economy would be a permanent feature of northern Minnesota; the structure of the production process itself ensured that the industry would move on.[10] With widespread clear-cutting and limited reforestation efforts, according to the third chief of the US Forest Service, William Greeley, the industry was a "great nomad" that moved steadily from place to place in search of more "merchantable" trees.[11] Frederick Weyerhaeuser's operations exemplify the trend: one year after the highest recorded lumber production in the state in 1900, Weyerhaeuser and his partners bought nine hundred thousand acres in western Washington State and established the Weyerhaeuser Timber Company there.[12] By 1920, with its supply of timber exhausted, Minnesota was a net importer of lumber.[13]

A different type of nomad followed. Migrants to northern Minnesota settled the cutover as part of a broader turn-of-the-century "back-to-the-land" movement. Born from immigrant and native dissatisfaction with the problems of industrialization and urbanization, mixed with a cultural heritage that romanticized the life of yeoman farmers, and fueled by the intense boosterism of northern Minnesota landholders (mostly lumber firms) who had few other uses for their now denuded land, people moved onto former logging lands to establish farms.[14]

By following the labor process of logging and then the labor of turning the cutover into farmland, the role that each working environment played in the state's environmental transformation comes into focus. More broadly, this approach explains why one environmental transformation succeeded and the other attempted transformation failed. In northern Minnesota loggers participated in working environments comprised of a hybrid of the forest ecosystem, the climatic realities of winter in northern Minnesota, and especially the technologies and labor practices instituted and maintained by the Twin Cities–based timber companies that reaped the profits of workers' labor. Though conditions were dangerous for workers, timber company investments of money, technology, and labor assured that logging occurred on a wide scale. As a result, the state's northern forests became the cutover. By contrast, farmers laboring in working environments on the cutover—comprising a mixture of the stumps, rocky soils, and the seasonal cycles of northern Minnesota—were characterized by an extreme *lack* of capital. Their attempts to transform the cutover into productive farmland were part of a slow, labor-intensive, and backbreaking endeavor. Especially in the face of other changes in American capitalism, however, they proved to be impossible to sustain.

WORKING ENVIRONMENTS IN THE WOODS

Logging during the late nineteenth century in the Great Lakes region was highly dependent on the seasons as part of a process that bridged northern forests and urban markets. Lumber firms relied on and manipulated frozen winter conditions by cutting, hauling, and storing logs on snow and ice, before driving logs through a network of temporary dams on swollen spring streams and rivers. Sawmilling then peaked in the summer and fall.

The working environment of logging—the first scene in the industry's annual cycle—comprised a mix of harsh winter conditions, old growth forests,
and a technologically sophisticated, if still labor intensive, production process
in order to remove logs from the forest. For workers this combination meant
both the need to develop considerable knowledge about their workspace and
the reality of persistent danger. For the northern Minnesota landscape itself,
this working environment produced significant and rapid change. This section
explores this working environment, focusing on key steps in the process of removing logs from the forest and identifying how technology, labor, and danger
defined the working environment and resulted in widespread change.[15]

Herman Haupt Chapman was a Yale University Forest School professor
from 1903 to 1946. During the winter of 1903–1904, he wrote a detailed analysis of the logging process in Minnesota based on his observations of a company
then logging north of Duluth. Chapman began by describing the ideal process
of felling trees. He explained, "The tree is notched close to the ground, 6 or 8 in.
deep in large trees," by the "undercutter." Using an axe to make this first incision
in the tree, the undercutter's placement of the notch determined the way that
the tree would fall. The other two members of the felling crew, "the sawyers,"
then cut "about 6 in. higher than [the] bottom of [the] notch and on the opposite side" until the tree was nearly cut through and fell.[16]

Chapman explained how the direction of the tree's fall could be altered:
"If necessary a tree can be felled to the right or the left of the direction of the
greatest lean. . . . Besides starting the cut right, [small metal or wooden] wedges
can be used to over-come a slight lean, or aid in diverting the direction of fall."
Since trees do not grow in a perfect grid in a forest or stand on perfectly even
ground, altering the direction of a tree's fall was necessary to avoid other standing trees and in order to help the tree land in the area where removing branches
and sawing it into logs would be easiest.[17] Workers' safety and ease of labor, as
well as the firm's profitability (which depended on unbroken logs) required that
laborers know how to drop a tree in a specific place.

For sawyers the working environment combined human knowledge and
technology with the uncertainty and variability of the forest. Logging and
sawmilling—the two sides of the lumber industry—were, in fact, among the
most dangerous occupations in Minnesota during the early twentieth century, trailing only mining and railroad operations. The Minnesota Bureau of
Labor, Industries, and Commerce calculated that during 1909 "fifty-five fatal
accidents and 1,115 non-fatal ones were reported by the lumbering and wood-
working industries of Minnesota."[18] The state's statistics did not separate the
accidents occurring in logging from those in sawmilling, but it seems that the
two principal segments of the industry were almost equally dangerous. Though
in sawmills gruesome injuries resulted from unguarded, rapidly spinning saws

and fast-moving logs, in logging the risk of simply being crushed by a tree or log were much higher. Of the fifty-five fatal accidents in Minnesota in 1909, roughly half seem to be attributable to logging operations, and of those sixteen were the result of being crushed. Felling trees in particular offered plenty of opportunity for this to occur. That same year the Bureau of Labor described accidents in which "two woodsmen were killed by the trees they were felling because they ran the wrong way [after making their cut]. Another woodsman became excited when two men yelled to him to get out of the way of a tree about to fall and ran right under it."[19]

Leonard Costley, who worked in the Minnesota woods during the early twentieth century as an undercutter, told an oral history interviewer that the sawyers could also be exposed to danger when a tree fell into other trees. "When a tree fell, there was sometimes what was called a 'widow-maker,'" he said. "The widow-maker was a limb that was broken off the tree when it was falling and that caught on another tree. That tree would bend over and send the limb back like an arrow. If you weren't watching you'd get hit with it. I've seen some serious accidents from that."[20]

Louis Heinzer, also an undercutter in the Minnesota woods during the early twentieth century, explained another type of risk to a feller if the tree was leaning. The notch in the tree helped the tree fall in the desired direction, but there was always a possibility that the weight of the leaning tree could begin to snap the tree before the sawyers completed their cut. This would create a situation, according to Heinzer, where "you'd have what you call a barber chair. Maybe pretty near half the butt be standing there. The rest would kick back and maybe hit you."[21] Essentially, the tree would fall with part of it still connected to the stump, splitting the tree along its length and forcing its butt back toward the feller. Even in 2012 the Occupational Safety and Health Administration explains that the "barber chair" is one of the main risks in tree felling.[22]

Once a felling crew brought a tree down and cut it into logs (usually between 12 and 16 feet long), a separate group of workers inhabited another part of the working environment, blending snow and ice, sleds and tools, and domesticated animal power to bring the logs out of the forest. The process of skidding is exactly what it sounds like: dragging logs across the ground toward a railcar or sleigh. In camps where sleighs were the primary means for transporting logs across the frozen ground to mills, H. H. Chapman explained, "The character of the country makes it possible to put in branch [ice] sleighroads within 20 or 30 rods [110 to 165 yds.] of all bodies of pine." This meant that skidding would be done by a two-horse team with the aid of a small sled placed under the front end of the log, usually called a "dray." In camps where railroad lines funneled the logs out of the woods, the high capital costs of that infrastructure made it more difficult to put in as many spur roads as with sleigh ice roads. This fact made skidding longer distances unavoidable. In these camps, according to Chapman, "skidding distances . . . increased occasionally to 70 or 80 rods [385 to 440 yds.]; and for all distances above 20 or 25 rods, drays are used."[23] These drays eased the labor of the horses needed to pull logs. In a logging camp consisting of ninety men, Chapman calculated that forty-five—a full half—of all workers would be laboring in skidding operations.

Skidding required crews of "seven men—one cant-hook man on [the] skidway, two teams and teamsters, and four swampers."[24] The "cant-hook man" used a tool, unsurprisingly called a cant-hook, to roll and manipulate the log while it was connected to the horse team and being dragged. A cant-hook is a long wooden pole with a simple tongs-like mechanism attached to the end, which gave the user leverage in rolling the log. After being attached to the log, two teams of horses and teamsters drove the logs to the landing, while the swampers used axes to remove branches from trees and cleared paths for logs to be skidded. Swamping was one of the lowest-paid jobs in the woods, often filled by workers unfamiliar with lumbering. This inexperience may have increased their risk for self-inflicted axe wounds. Leonard Costley, the former undercutter, recalled, "I've also seen some severe cuts from axes by swampers [because they] . . . were a class of men that went broke in the cities and had never worked in the woods. They'd come up there without experience on account of an employment agency shipping them out, and they really got cut pretty bad sometimes."[25] With no clear training systems in place, workers learned on the job, sometimes suffering self-inflicted injuries before they learned less risky ways to manipulate an axe.

After the skidding crew moved logs to the landing, the loading crew transferred them to sleds and railcars, incorporating yet another group of workers and forms of capital to move logs through the forest. Loading crews included about ten men out of the ninety-man camp Chapman described in 1904. Though during the early twentieth century a variety of technological innovations slowly mechanized the loading process, the main way that logs were moved onto sleds or railcars for longer-distance transport was through the use of a horse team and a device called a jammer. Essentially, a jammer was a twenty- to thirty-foot-high wood frame with a pulley at the top through which a rope was passed. On one side of the jammer a "cross-haul" horse team with the rope attached stood facing away from the jammer. On the other side, sleigh or railcars pulled up in between the jammer and log pile. Logs would be secured to one end of the rope with a set of heavy cast-iron tongs, and the horse team would walk away from the jammer, pulling the rope through the jammer into the woods, and thus lifting logs on top of the sleigh or railcar.

The "top loader" directed this process while standing on the logs already piled on the car. Frank Werthner, who worked on a loading crew in the early twentieth century recalled how loading logs required that all of the workers involved pay close attention to the process and the commands of the top loader. "And this guy up on top loading, he gives the signals, you know. When he say 'go' or 'go ahead,' whatever they have, that [horse] team is already going and when he says 'ho' that team will stop and hold that log right there. And when he hollers 'come back,' you want to be ready to come back because they gonna trample all over you."[26] Especially the worker driving the horses had to be ready, because the horses took commands from the top loader not the driver. "The man that drives those . . . horses," said Werthner, has "got be on the jump because when that, when that top man, on top there loading logs hollers 'come back,' you want to be already to start back because them horses turn around just like that and they're coming and get right up to the road and they back right up."[27]

Such a process, though it efficiently moved logs out of the woods, also produced danger for workers who participated in this working environment. The state's 1910 report showed that "twenty-five of the 45 men severely injured by logs in 1909-10 were hurt while loading or unloading logs and 6 were teamsters in the woods hauling logs to the loaders." Several things could go wrong in this process. As the report recounted, "One man was injured because a sleigh load of logs tipped over; another because a broken stake allowed a log to roll from a car; several were squeezed between logs on the skidways . . . several were struck by falling logs while loosening chains around loads of logs."[28]

The final task for workers in logging camps involved using horse teams to drive the sleighs to a river or rail spur for further transport to a sawmill. Herman H. Chapman calculated that "a team can travel on a logging sled, about 24 miles per day," a distance that would ideally represent several round trips to and from the rail or river landing.[29] The construction, maintenance, and use of logging roads for hauling required a significant amount of labor and specific environmental conditions which blurred the human and the natural. These roads could be undermined by swings in the weather, events like midwinter blizzards or unseasonal warm spells. Jacob Pete, a lumberjack from Ely, Minnesota, explained how a thaw could begin to affect the transport of logs via sleigh roads: "Water tanks were used to freeze the roads, so they could haul big loads. And then, you know, every once in awhile they'd get a thaw in this country. And if you don't ice your roads, a thaw would break your roads up and stop your operation. But if you have a good layer of . . . heavy ice, even in thaw you can still use that road."[30] Keeping a "good layer" of ice on a road required careful construction and continual upkeep. Before logging began, a crew would clear a road and use horse teams to drag ruts into the ground eight feet apart. These ruts would be filled with water gradually until, as Pete explained, "they had ruts built into the ice then so that sleighs would follow these ruts. And they'd ride very smoothly and a team of horses here, if they once get the load moving, they'd move a very large load."[31]

After ice roads were built, their maintenance was critical for making these sleigh trips successful. Leonard Costley worked as a "road monkey" sustaining these highways: "A road monkey is a fellow that goes out to keep the roads clean, and takes out all the bark, and watches the ruts to see that there's no slipping or tendency to slew off sideways. In those days when horses were used entirely for pulling, you had to keep the roads clear of the leavings of the horses."[32] James Reid, who ran a small logging operation in northern Minnesota during the 1910s and 1920s, explained that in addition to keeping the roads clean, a crew was need to "repair them [the ice roads] every night."[33]

Each phase in the logging of northern Minnesota during the late nineteenth and early twentieth centuries—felling, skidding, loading, and hauling—entailed the creation and maintenance of a working environment that was not solely defined by cold conditions or by migrant workers, but the many forms of capital and specific environmental conditions that met and blended in the woods. The lumber extracted from this working environment built the infrastructure of industrial capitalism in the Midwest, but it also transformed forests into "the cutover": the piles of branches ("slash") removed from logs, the tree stumps, and the newly exposed rocky or swampy land.

WORKING ENVIRONMENTS ON THE CUTOVER

In the early 1900s the *Bemidji Daily Pioneer*, Beltrami County's first daily newspaper, wrote confidently that the future of the region rested in the conversion of abandoned cutover land into a sea of prosperous smallholder farms. Reviewing this newspaper from the period leaves the reader with the impression that agricultural prosperity had either already arrived in the region or was just around the corner. In one 1907 editorial it even dismissed those who called for reforesting the region—something that lumber firms were not required to do after cutting their land—and joked that such "misguided enthusiasts" who sought to "reforest Northern Minnesota will need to hurry or face eternal failure. It is impossible to graft the pine tree on the cow, and equally difficult to cross the pine cone and the potato vine."[34] Agriculture was on the march, the *Pioneer* believed, and any reforestation would be a moot point after potatoes and cows dotted the land.

By the early 1930s, though, the landscape across northern Minnesota suggested neither established nor impending prosperity, and increasingly residents had come to doubt that it would even be desirable. Land-use statistics tell part of the tale: by the early 1930s, 74 percent of the land area in the fourteen northeastern "cutover" counties remained classified as "forest," while just 17 percent lay in agriculture, according to data from the Census Bureau and the US Forest Service compiled by researchers at the University of Minnesota.[35]

Such numbers, however, conceal as much as they reveal of the environmental changes and social problems that characterized the preceding decades. Just 7 percent of total land area was classified as "saw timber." The rest fell into lower-value designations, including primarily "cordwood" land, suitable only for use as firewood (33%), and "restocking" land (29%), which was simply cutover logging ground slowly growing new trees—mainly low-value aspen and birch—without any active forest management. Although 80 percent of land lay in private hands in 1931, 47.6 percent of that land was tax delinquent: lumber firms would rather forfeit the low-value land than pay taxes on it, and marginal farmers rarely earned profits sufficient to pay their taxes.[36]

The farmland that did exist in the region, of course, was cutover land sold by lumber companies. And though it accounted for 17 percent of land in northern counties, little of that area had actually been "improved" and put into crops.[37] In Beltrami County, the home of the *Daily Pioneer*, for example, the census of 1920 showed that just 19.5 percent of farmland was improved. By contrast, among farms across the state (including the expansive wheat farms of the south and west), improved farmland as a percentage of total farmland never dipped below 70 percent during the first three censuses of the twentieth century.[38] These low rates of improvement were the direct result of a difficult working environment, and the figures of "improvement" may themselves have been generously granted. A visit by government investigators in 1931 to one Lake County township, for example, exposed the fact that, though "the census for 1930 shows 2,043 acres in farms in the township . . . interviews with settlers revealed that only 892 acres were owned by operating farmers and only 138 acres were cleared."[39]

If figures of land-use statistics in the post-logging environment reveal the variable and uneven development of the cutover, it was still true from the first decade of the twentieth century that agriculture represented an important potential future for this space. In fact, the low rates of farming, through the rose-tinted glasses of cutover boosters, suggested a future of nearly unlimited agricultural growth. The same forces that had marshaled capital to remove logs from northern Minnesota used their privileged position, vis-à-vis landownership, to sell cutover land to farmers. They did so by promoting a vision of northern Minnesota's future in which smallholder farms on cutover lands made sense. In the press and through advertising, lumber companies' marketing promised to solve two structural problems, one environmental and one political, for the firms: the risk of fire on abandoned cutover land and the tax burden owed to the state.[40]

To explain why boosters' vision did not come to pass, it is necessary to turn to the working environments that new farmers faced on the cutover. Transforming the cutover into a productive agricultural hinterland required new relationships between labor, environmental conditions, and forms of capital, particularly technology and money capital, to complete a variety of tasks, especially where stumps could be removed and soil tilled. But in general the farmers who heeded the call of land firms and moved into the region lacked the resources to build out the farms they now owned. In essence, new farmers' working environments were shaped by a lack of access to capital.

Stump pulling was a prime example of the difficulties facing farmers. It was a labor-intensive task that was crucial for transforming the cutover into "improved" arable farmland. Practical only when the ground was not frozen, it was often performed with a crew of two people splitting the stump into segments, then using a "stump puller" that employed pulleys and a rope connected to a horse to rip the stump and roots from the ground. Removing stumps this way was slow going. As one cutover farmer from Minnesota later related in an oral history interview, "Well, you do well, you know, a couple of men do well to pull a couple stumps a day."[41] Without a horse, the process could be even slower. Another Minnesotan raised on a cutover farm recalled clearing the land by explaining, perhaps with some hyperbole, "Everything was done by hand—we didn't have nothing; we didn't even have a horse—there wasn't a horse in the country."[42]

Other methods of stump removal were proposed at the time, but the widespread implementation of these alternatives would have demanded changes in the technology, money, and cutover farmers' relationship with the state government in Saint Paul. Charles Carter of Beltrami County explained in 1908, "We want the State of Minnesota to furnish, at the lowest possible cost, to actual settlers, dynamite enough to transform this country, in five years, from a wilderness of stumps and slashings into one of the richest farming and dairy countries on earth."[43] Despite Carter's plea, sending money into the region for this purpose was apparently not a priority; it failed to pass the legislature. Without state aid an editorial in the *Bemidji Daily Pioneer* called for a more creative approach. Instead of buying fireworks, it read, "supposing the entire north country would celebrate by blowing stumps out of the ground on each and every Fourth of July for the next ten years how many acres could be cleared in this manner!"[44] Though maybe more fanciful than state-subsidized dynamite, the editorial appealed to the same sentiment. Transforming the landscape into a productive agricultural hinterland would require a different working environment.

Stumps, meanwhile, were not the only feature of the cutover that made the working environments that farmers faced difficult. Clifford Alghren, the son of Finnish immigrants who grew up on a farm in Lake County in the early twen-

Figure 5.1. Pulling stumps on cutover land near Duluth, Minnesota. Stump removal was crucial for transforming cutover forests into "improved" arable farmland, but it was a slow and labor-intensive task, even under the best of conditions. Courtesy of the Minnesota Historical Society.

tieth century, recalled that his family would clear "about two or three acres, depending on how much time was available" each year, time that competed with the other needs of the farm. They would use a team of horses to remove stumps, but it was also the rocks that Alghren later recalled as a major impediment to remove and the most difficult work:

> When we plowed and harrowed we had to haul out the rocks. And that was a real job. If you ever traveled in that country you'll see the long, long fence rows of rocks. All of the rocks came out of these little fields. Hauling rock was heavy work. Some of the rocks which were too large to lift, had to be skidded with a chain and horses. If you removed the rock consistently for a few years, most of the field would eventually be free of rocks, and easy to work with. We would treasure these fields because of all the work that went into them.[45]

By the 1930s Alghren's family had cleared approximately forty acres of their ninety-acre farm, a relatively high percentage of cleared land by regional standards. As Alghren's story reveals, though, the difficulty in clearing land in the cutover explained the low percentage of "farmland" that had been "improved" relative to the rest of Minnesota.

The difficulty of clearing the land even formed part of the humor of the region. A 1909 *Princeton Union* article related the tale of a man selling his automobile, due to its impracticality. The car's brakes had failed during a drive with his father-in-law, forcing him to veer into a meadow to slow the car. The ride ended after the car hit a stump and pulled out its roots, to which a passenger remarked, "My boy, there is no reason whatever in your going so far from home to do stump pulling. You would do me a favor by clearing that piece of cut-over land that I own."[46]

The lack of money capital that farmers had to invest in clearing land reflected their low incomes, a fact that led many to "subsidize" their farms by participating in other jobs in the region. Some farmers intermixed working on the cutover with working in the (rapidly contracting) lumber industry. Alexander Carno, who also worked winter logging and spring log drives, recalled how homesteaders on the cutover in the Great Lakes region, like his father, used logging wages as supplemental income while expanding their farms: "That time, work in the woods in the wintertime, understand see, and spend the rest of his money trying to open that land."[47] Another son of a homesteader, George Eitel, explained that aside from the lumber industry "there was nothing else there for a man to do" during the winter in the region. Working in the woods represented a chance "to get a little cash money that was needed for these homesteaders."[48]

The difficult conditions for farmers on cutover forestland made cash income in the winter all the more important. Eitel recalled, "Sometimes these homesteaders cleared ground and tried to cultivate among the stumps and stones that looked like purgatory itself. So there was really nothing for them to do when winter came, but to head for the woods to make a living. My brother was four years older than I and so we both headed for the logging woods and earned a little money and pulled it out over the summer."[49] Ironically, Eitel and his brother's labor produced stumps during the winter, and then worked to remove them on their own land over the summer.

The development of roads and drainage ditches throughout the northern counties also became an important source of income for many cutover farm families. When one oral history interviewer asked William Kaukola, a Finnish immigrant who grew up on a cutover farm in northeastern Saint Louis County, where his family's income came from, Kaukola replied that in addition to his father's labor in logging during the winter, "well—there was a lot of road work, too."[50] Such work, though it occurred during the summer when time could have been spent expanding the farm, allowed workers to earn wages that could be used to subsidize their unremunerative and undercapitalized working environment on the farm.

Another Minnesota resident who grew up on a cutover farm, John Ollila, also remembered the importance of public employment for his father during the early twentieth century. He recalled how little money came in from the small amounts of milk and cream he was able to sell and then added, "He had to hire out because he didn't make any money on farming—no, at that time, you know, there was no money in that.... And when all those roads were made then in the summertime, too, he took contracts of making these roads—making the ditches on the roads."[51] Building public infrastructure therefore subsidized the incomes of some residents for whom the dream of a self-sufficient farm was unrealized.

The difficulty of building a farm on the cutover without adequate money or technology was captured in 1916 by a US Department of Agriculture (USDA) report, which suggested that in the cutover "the cost of clearing land ranges from $20 to $60 per acre, depending mainly on the kind, condition, number, and size of stumps; the acreage to be cleared; the amount of wood and brush that must be removed or burned; and the quality and condition of the soil."[52] Such costs dwarfed the purchase price of the land itself. In 1917, Immigration Land Company, the scion of the Weyerhaeuser-controlled Pine Tree Lumber Company, sold its lands in northern Minnesota for an average of $12.95 per acre, or between 40 and 500 percent less than the cost of clearing it.[53] Settlers "often overlooked" such costs, another USDA report concluded, "or are purposely discounted by the land agent who wants to make a sale."[54]

In 1902—well before the 1930s, when the long-standing problems of the cutover region became part of a broader discussion about Minnesota's economic and social future—the *Minneapolis Journal* captured the central issue that would plague farmers in the region over the next several decades:

> The *Journal* is in receipt of an interesting letter from a man who took the advice so freely proffered to the poor of the city to get out on the farms, to get back to the land.... After one year's experience he is like the man who said: "When I came to this country twenty years ago I hadn't a rag on my back; now it's all rags." His trouble is, of course, lack of capital. Not having horses or machinery he was able to cultivate this year only an acre-and-a-half of land, and an unusually early frost ruined most of the crop on that small patch.[55]

Though it of course did not use the phrase, this account of the lack of capital (horses, machines) and the particular environment (weather) described a *working environment*, the specific features of which made the region's transition into an important agricultural hinterland in Minnesota unlikely. Other forces (such as the market prices for any given crop) also always affect the success of a farm, be it in northern Minnesota or elsewhere, but the *Journal*'s editors excluded such considerations and focused on the people and the environment, concluding the venture would lead only to "all rags" for those who headed "back to the land."

Like the *Journal*'s brief description of the conditions that beset the anonymous letter writer at the point of production, this essay argues for the centrality of working environments in understanding the content of regional environmental change and offers three lessons to historians and citizens. First, even large-scale environmental change, such as the logging of northern Minnesota during the late nineteenth century, was made up of day-to-day human actions and environmental conditions. Logging in northern Minnesota occurred not only because of the demand for lumber existing in the Twin Cities and elsewhere, but also through the elaborate (and dangerous) working environments operating in the forests themselves. Second, not only could these working environments *produce* new environments at a regional scale, but they also *constituted* a temporary environment that blended nature and human work. As this chapter shows, a logging road required constant attention to keep the ice tracks set, and stumps anchored to the earth required removal before plowing could occur, with timber workers and farmers in each case acting in a space both "made" (human) and "unmade" (nonhuman). Finally, the particular construction of a working environment can help us to explain the success or failure of different industries and agricultural endeavors in the past, and they point to the importance of capital (in many forms) in shaping working environments and regional environmental change. Logging during the nineteenth century in northern Minnesota did not suffer from the same inability to transform the land as farming in the region would in the early twentieth century. Capital—appearing in the form of saws, horses, management, and a labor process—met and melded with the conditions of the northern forest to establish a working environment capable of rapid change in the landscape, even if it produced dangerous conditions for workers. Without analogous forms of capital suitable for farming and facing an equally harsh set of environmental conditions, the transformation of the cutover into a prosperous farming region did not occur.

"FOLLOW THE ARROWS TO THE ARROWHEAD"

The Environment of Tourism in the Interwar Years

Aaron Shapiro

ARRIVING IN NORTHERN MINNESOTA IN 1886, Joseph Ruttger worked in the woods, met his wife, and settled on an island in Deer Lake. Hearing reports of good food and fishing at the Ruttgers, railroad travelers arrived to inspect their veracity in the 1890s. As business grew, the Ruttgers added cottages, charging $5 a week for bed, board, and boat. Joseph Ruttger moved from hauling timber to enticing tourists.[1] Wilson Dunn arrived in Minnesota from Wisconsin in the 1880s, purchasing lakefront land north of Pelican Rapids and building a resort that attracted hunters, fishermen, and families from the Twin Cities. By 1929 Wilson's son, Roy, expanded the resort to include two dozen cabins and a main lodge. He also served in the Minnesota House in Saint Paul, rising to majority leader, and returned to operate the resort during summers.[2] Facing a declining timber industry and agricultural failures, Ruttger and Dunn were among the many early operators who joined tourists, residents, and the state to make the vacation landscape in Minnesota's North Woods, and in the process helped make "up at the lake" and "up north" common refrains for Twin Cities residents.

While geology and history mark time "up north," natural features took on new meaning in an expanding consumer society as vacationers from the Twin

Cities and beyond enjoyed northern Minnesota's tourist camps, cabins, resorts, state parks, and national forests during the interwar years.[3] The market's intrusion into the nineteenth-century countryside established urban-rural connections that exploited natural resources, altered lives, and transformed the landscape. By promoting vacations as a consumer item that improved health and productivity while providing an escape to a supposedly more natural environment, twentieth-century advocates hoped tourism could diversify the economy and address land-use concerns. Like earlier lumbermen, they also saw profit in nature. But unlike lumbermen who cut and ran after felling the forest, they relied on nature's regenerative forces to provide a new cash crop, a forested and lake-dotted countryside offering outdoor recreation for the masses.[4]

Experiencing deindustrialization earlier than the Rust Belt cities to which it sent natural resources, Minnesota's North Woods residents participated in transforming a landscape of production into one of consumption. Familiar with the boom-and-bust cycles of extractive industries directed by outside capital, they wanted a more secure future. A growing market of vacationers boosted such aspirations.[5] Promoted and experienced as a haven from the industrial world despite the North Woods' intimate connection to it, vacationing offered a respite in the place that had provided resources to build cities. But this time, the consumers—tourists—traveled to enjoy the landscape rather than having its resources extracted and shipped to them. The ascription of aesthetic values on natural places commodified nature in new ways, as Minnesotans realized consuming nature for pleasure demanded new approaches to managing natural resources. Vacationing with family at a rustic cabin, embarking on mental journeys by reading the Work Progress Administration (WPA) guide to Minnesota's Arrowhead Country, enjoying summer camp near Bemidji, and hiking trails constructed by the Civilian Conservation Corps (CCC) in the Superior National Forest, they redefined nature in consumer terms through leisure.[6] Although scenic beauty helped make the region a destination, tourism is not a natural product, but rather is managed and packaged by people and organizations with often competing interests.

Tourism in the North Woods had its own distinctive character. Historian Hal Rothman characterized western tourism as a devil's bargain because it provided economic benefits but vested power with outsiders and contributed to a loss of the uniqueness that made places worth visiting. North Woods tourism attracted nonresident interest and altered the landscape, but residents exerted greater control and experienced tourism as less of a devil's bargain than their western counterparts.[7] Local and state activities to develop tourism and manage land use presaged federal efforts, revealing the roots of New Deal conservation and land-use planning. Nationally, rural land-use policies emerged within an urban industrial society, and conservation efforts were increasingly connected to urban consumer needs, from early twentieth-century fights over water in California's Hetch Hetchy and Owens River valleys to growing urban desires to enjoy leisure in nature. Ultimately, the activity of local residents co-operating with and contesting conservation regulations proved crucial in making northern Minnesota a vacation destination.[8]

Government facilitated tourist development—through promotion, conservation, and construction—and thereby became a tourist industry stakeholder. The creation of the US Forest Service in 1905 and the National Park Service in 1916 led states to establish more parks and forests for conservation and recreation. Minnesota played a pioneering role, creating Itasca State Park at the Mississippi River's headwaters in 1891. While many rural residents expressed concerns about interference in local affairs, they participated in consumer culture on their own terms. As the preeminent symbol of that culture, the automobile gave a boost to tourism, democratizing travel and increasing access to the North Woods.[9] Working within the region's environmental benefits and constraints, state and private interests looked to tourism to diversify a region suffering from industrial decline and an inability to farm northern lands. Fueled by the region's engagement with a modern consumer culture in which changing technologies and expanding market relationships transformed land, labor, and leisure, creating the North Woods involved organizational leadership, new land-use policies, and effective development and promotion.

CULTIVATING A LANDSCAPE OF TOURISM

Depicting a farmer holding a hoe with "publicity" emblazoned on the bottom, looking at a bush full of moneybags denoting $450 million in tourist expenditures since World War I, a 1920s *Minneapolis Daily Star* cartoon (Figure 6.1) captured tourism's impact on Minnesota's economy, people, and landscape. Beside the farmer stood a watering can with "hospitality" in prominent letters and a scarecrow telling him, "Seems to Me This is One Crop Worth Cultivating." Government agencies, regional organizations, and residents, some avidly and others reluctantly, helped tourism replace farming as the crop to cultivate. The lingual link to agriculture was no mistake. In transforming a cutover landscape to one marked by lakes and reemerging forests, they looked to tourism to improve land and life.[10]

In the late nineteenth century the state recruited farmers to northern lands. Upon arrival, individuals discovered stump-filled lands and tilled rocky soil with hopes of farming. After decades of unsuccessful efforts, organizations like the Minnesota Arrowhead Association (MAA) emerged, working with state government to develop tourism. The area did not initially offer an aesthetically pleasing landscape, so advocates supported a program of reimagining it, conserving and even reconstituting its natural scenery in order to sell its charms to

Figure 6.1. As this *Minneapolis Daily Star* cartoon from 1929 attests, tourism emerged as big business in Minnesota. As residents of the Arrowhead worked to cultivate a reliable crop of tourists, however, they quickly realized that doing so demanded new approaches to managing natural resources. *Source: Minneapolis Daily Star.*

urban dwellers. Often promoted as untouched and natural, the reality was that the land had long experienced the imprint of human hands.[11]

Rather than rely solely on private efforts, in 1918 Minnesota appropriated funds for tourist promotion and development, a decade before Michigan and nearly two decades before Wisconsin. In language highlighting tourism as a cash crop, the *St. Paul Pioneer Press* suggested the state was now "officially interested in the exploitation of its natural scenic resources." It declared Minnesota had not fully capitalized on these assets, viewing tourism as a benign land use.[12] Inquiries poured into the state, including many echoing the sentiments of Des Moines resident Charles Parker, who labeled Minnesota "the most won-

derful summer resort state in America."[13] Initially hoping people would come
to Minnesota as tourists and then settle, by 1923, with tourists depositing money in state and local coffers, state agencies in Saint Paul viewed tourism as a
solid industry in its own right.[14]

Organizations like Mathias Koll's Northern Minnesota Development Association (NMDA) backed state efforts. Limited farming opportunities led
the NMDA to advance a conservation agenda emphasizing "preservation and
propagation of fish in our Ten Thousand Lakes; creating in the pure cool waters
of this state a continuing industry and one of growing recreational and commercial importance."[15] Like earlier agricultural advocates, Koll initially viewed
tourists as potential settlers. He ultimately encouraged the state to promote attractions, suggesting that tourists helped local businesses and provided a market for farm products.[16] The State Immigration Board also promoted northern
Minnesota, assuring vacationers of abundant fishing, hunting, and lodging options in Minnesota's lake-dotted north while obscuring mining's and tourism's
impacts on the land.[17]

The MAA, NMDA, and the state were not alone in believing tourism could
offer employment, return abandoned land to tax rolls, and diversify the economy. Federal agencies aided local efforts by providing recreation on public lands.
Minnesota benefited from having national forests earlier than Wisconsin and
Michigan, positioning the state as a regional tourism leader. While Minnesota
(now Chippewa) and Superior National Forests were initially established for
timber and watershed protection, the US Forest Service faced growing competition for the recreational dollar, as well as its lands, from the National Park
Service. The Forest Service's Term Permit Act allowed cottage construction in
national forests and extended leases for up to thirty years, starting at ten dollars annually, with renewal options. Promoting remoteness while also noting
that cars spurred tourist access to the Superior, the agency proposed, "The lake
region north of Grand Marais will no doubt be a mecca for people in search
of ideal summer-home sites upon the completion of the automobile road now
being built into that region."[18] The presence of two national forests along with
agency interest in recreation helped develop the tourist trade.

During the 1920s tourism transformed northern Minnesota's physical and
economic landscape. Resorts grew from a few hundred to nearly 1,400 as roads
opened previously inaccessible areas. Like other advocates, the MAA initially
hoped tourists would settle and become producers for a new industrial economy, commenting on the "proper type of tourists, attracted first for pleasure
and later, for permanency." But the MAA quickly realized tourism was tied to
transience rather than permanence, so the goal became attracting tourists as
consumers who contributed to the local economy.[19] It drew institutional members from area towns, encouraged reforestation and new roads, promoted va-

cations, and received inquiries. In 1926 the MAA conducted a naming contest for the organization, which had been known as the Northeast Minnesota Civic and Commerce Association. Pittsburgh's Odin MacCrickart, who coined the Arrowhead term because the region's boundaries appeared as such, emerged as the winner among thirty thousand entries. The Arrowhead name stuck, broadcasting a sense of adventure.[20] Promoters also reminded North Woods vacationers of the pine-scented air, cool summer breezes, lakes teeming with fish, and woods filled with wildlife. But the escape was not complete. Vacationers were immersed within a web of actors that regulated the experience and influenced how and where they vacationed. In 1931 Minnesota established a Tourist Bureau in the Department of Conservation to oversee promotional campaigns and assist regional organizations. The Highway Department addressed infrastructure and scenic roads while the Planning Board hoped to create jobs by attracting urban tourists.[21]

New Deal programs built on these earlier efforts to develop public recreational lands. Federal funds for roads, conservation, and the WPA American Guides promoted travel as a means to unity, recovery, and nation building. At a fundamental level these programs offered work during the Depression. But they also provided people with a new understanding of the nation's landscape and people.[22] Minnesota's state and Arrowhead guides reinforced the image of far northeastern Minnesota as a wilderness, with Superior National Forest offering fifteen canoe trips. Readers of Minnesota's guides discovered a place filled with outdoor adventure and learned about the region's industrial history.[23] In 1940 Governor Harold Stassen encouraged visitors to enjoy Minnesota's outdoor opportunities amid natural beauty, where residents offered cordial service and lodging while the state ensured a safe and pleasant experience.[24]

PRESERVING NATURE FOR TOURISTS

As urbanization proceeded apace, the idea of preserving natural resources emerged more forcefully among wilderness advocates, government officials, and North Woods residents. Born in Chicago and raised in northern Wisconsin, Sigurd Olson called northern Minnesota home starting in the 1920s. As a naturalist, writer, and educator, Olson viewed wilderness as both physical place and ideological concept. Olson, who operated a canoe outfitting and guide service, saw wilderness as an escape from the urban order. But that order clearly prescribed the ways people experienced and thought about wilderness. Guides still worked for vacationers, providing information on fishing spots and maneuvering through the landscape in exchange for payment. Justifying designating portions of Minnesota's Superior National Forest as roadless, Olson championed preservation.[25] He wanted people to experience it as voyageurs had. "Nowhere else can such beautiful lakes be found," he wrote in 1929. "Nowhere else can you find them close together enough to make what is known as a canoe country, and nowhere else is there so much beauty concentrated in one spot as here. It is the last area of its kind in the country. Are we going to sacrifice it to the ogre of commercialism?"[26] Olson joined other regional voices that saw spiritual and financial benefits in wilderness tourism.

The US Forest Service initially considered recreation a lesser use than timber production but with automobile touring increasing demand, it responded by providing recreational opportunities. In 1918 landscape architect Frank Waugh's report declared recreational forest use equal to that of timber, grazing, and watershed protection. Noting its "substantial commercial value" of $7.5 million annually, he encouraged the agency to further develop recreation.[27] The Forest Service's first landscape architect, Arthur Carhart, crafted plans for Superior National Forest that called for maintaining the area's pristine nature while providing easier access with backcountry lodges and marked canoe routes.[28]

Ernest Oberholtzer joined Olson, Waugh, and Carhart in battles over Minnesota's boundary waters. In 1927 Oberholtzer established the Quetico-Superior Council, whose conservation program brought together American and Canadian interests. Attacking industrialist Edward Backus's plan to build dams that would flood several thousand miles of international shoreline, Oberholtzer expressed outrage that private greed could destroy such a unique wilderness resource. He called for federal intervention to stop Backus, opining that "no conceivable benefits from such a plan can justify the wholesale impairment of public values in so large and so unique an area."[29] Backing came from the local Arrowhead Sportsmen's Association (ASA), which sought to protect its members' ability to hunt and fish; tourist entrepreneurs like Wilderness Outfitters owner Joe Pluth; as well as the Ely Commercial Club, a group of businesspeople convinced Backus's plan would harm Ely's economy.[30] To preserve scenic, recreational, and inspirational values, Oberholtzer advocated public control of the region's wilderness.[31]

Not simply a case of outsiders-as-preservationists versus residents-as-resource-users, the struggle over roads and Backus's dams saw local commercial clubs join sportsmen and wilderness advocates in fighting industrial expansion and landscape despoliation. Perhaps the most prominent of these sportsmen's groups was the Izaak Walton League, established in 1922 in Chicago. In April 1923 League president Will Dilg appeared at a Duluth public hearing, testifying that the Forest Service should abandon its road plans and telling those assembled, "Only God could make that forest and only man can destroy it. It must be preserved as wilderness for future generations of young Americans, and none of us have the moral right to destroy it as such." An *Izaak Walton League Monthly* article reported that Dilg helped convince the Forest

Service "that the biggest word in the Superior Forest dictionary is 'National'—that it is bigger than the forest itself."[32] The League's local Arrowhead chapter added its voice, fighting Forest Service road projects and pro-road interests hoping for "A Road to Every Lake." Vast road plans were ultimately defeated in 1926 when the Forest Service set aside three roadless areas. As a local organization, the chapter supported policies preserving fish and game for sport and suggested tourism offered possibilities for residents that other industries did not. Managed accordingly, it believed the region's "woods, waters, and wild life are not only its rarest distinction but promise to become its richest asset."[33]

As citizens advocated conservation policies aiding recreational development, Forest Service officials linked tourism and reforestation. Arthur Carhart's recreational survey of Superior National Forest conceived of the region's wilderness in terms of scenery and access.[34] He did not exclude timber cutting, but noted that in some locations, particularly shorelines, "the aesthetic qualities shall, where of high merit, take precedence over the commercialization of such timber stands."[35] Looking to balance forest uses, Carhart offered a compromise premised on the notion that scenic values were, in certain places, the most valuable and must be preserved.[36] His vision was not of a wilderness devoid of all other uses, but a recreational outlet in a forest guided by sound conservation principles. Carhart believed agency emphasis on timber production overshadowed his recommendations. Reflecting later on the importance of regional actors in spurring policy changes and vindicated by plans for Superior National Forest that drew on his endorsement of its recreation potential, he remarked, "I think we all know that the Forest Service was not ready to take a progressive, intelligent attitude toward the Superior and recreation until the people in Minnesota and the Middle West rather kicked them in the ribs."[37]

Local fears about losing control and access took root in diverse situations: against Backus's dam projects, over Forest Service policies in the boundary waters, and among those who opposed proposals to establish a national park. The Arrowhead Sportsmen's Association objected to a potential national park in the Superior National Forest, fearing it would make the area off limits to hunting and fishing. While the ASA rallied against the park proposal, Forest Service officials advanced forest recreation. Carhart believed "Ely has the opportunity of becoming a tourist town."[38] He suggested crowds could be handled without destroying recreational, wildlife, and forest values cherished by locals. Many Ely residents heard Carhart's call and catered to the growing tourist trade. These operators contributed to conservation debates since business depended on their clients accessing public lands.[39] The development of recreational facilities on public lands also brought government further into the tourist arena. State policies for fostering abundant fish and game dovetailed with promoters who viewed the outdoors as a commodity and knew selling retreats depended on having fish to catch and forests filled with wildlife.[40] Despite their diverging views, government officials, wilderness advocates, local citizens, and sportsmen's associations brought new approaches to land-use issues that helped establish Minnesota's North Woods.

Automobile access, lodging improvements, and the desire to escape cities during summer all contributed to a new vacation landscape. Helen Beebe of Minneapolis, who worked for Dayton's department store, wrote resort operator Roy Dunn about wanting to escape to the outdoors, explaining, "I am cooped up in a room the year round and when I am on vacation I want a change." Other guests had different motives, including fraternities using Dunn's for social functions and fishermen looking to catch walleye. Still others viewed the resort as a place for family vacations. Leslie Setzer, an employee at Saint Paul's West Publishing Company, inquired about accommodations and a ride from the Detroit Lakes train station for her and a friend. Guests came as individuals, families, and groups seeking refreshment and entertainment among Minnesota's northern lakes and woods.[41]

Vacationers also arrived to canoe in Superior National Forest. The Ely Commercial Club recommended outfitters and guides while Lake Vermilion resorts provided a location map. But who were commercial clubs and resort owners trying to attract? One group was clear, with one publication focusing on "the tired business man who must periodically get away from the rush and noise of the city and build up a run down and worn out body."[42] The tired businessman was a potential client but as guest lists, letters, bulletins, and registers suggest, so were sportsmen, single women, families, and students. Resorts explored many avenues to reach these vacationers. Roy Dunn contracted with Chicago's Curt Teich and Company, to create postcards for his resort. Dunn also worked with his legislative colleagues in hopes of garnering official business.[43]

THE NEW ECONOMIC LANDSCAPE OF NORTH WOODS TOURISM

While the boundary waters and the Gunflint Trail offered wilderness experiences, other operators joined the Ruttgers in providing for vacationers in Minnesota's Brainerd Lakes area, where nearly three hundred resorts marketed it as "Va-Ka-Shun Land" in 1929.[44] Resorts and lodges were one element of the lodging landscape. By the 1930s housekeeping cabins offered simple places to stop for a night and an economical alternative for longer vacations where guests cooked and cleaned for themselves. Tourist camps, state parks, and national forests provided camping at minimal expense.

While conversing with guests at their northern Wisconsin lodge, Dora Blankenburg and her son Russell discovered that some were heading to Minnesota's Arrowhead to enjoy Lake Superior's north shore and canoe the boundary waters. Others explored mining history, while boosters promoted curative properties among pine-scented air. Whether one sought health, adventure, or history, northern Minnesota offered opportunities for vacationers and potential resort operators. Russell visited Minnesota and bought land to establish Gunflint Lodge in 1925. After operating both places, the Blankenburgs sold Gunflint Lodge to the Spunners, friends who advised them on their Wisconsin resort purchase. Mae Spunner brought her daughter Justine to negotiate terms. Gunflint had three guest cabins without electricity or indoor plumbing, a small lodge, dining room, and a store. Justine made improvements, enlarging the dining room and adding a lounge. These additions faced the road rather than the lake, highlighting the lodge's orientation toward welcoming auto tourists on the Gunflint Trail. During the Spunners' first season in 1930, Mae planned meals and supervised staff while Justine handled administration and repaired equipment. Indian neighbors shopped at the store and worked as guides, housekeepers, and servers.[45] In 1933 the Spunners moved to the Gunflint permanently. That summer, Bill Kerfoot arrived on the Trail, camping on Gunflint Lake's west end. As Bill was eager for a job, Justine hired him in exchange for room and board. The following year Bill and Justine married and began operating the lodge together. Their newsletter, *Gunflint Gossips*, reported on lodge activities and created a sense of family among the guests. Originally male hunters and anglers came in spring, early summer, and fall. Gradually, more women and families visited.[46]

Gunflint Lodge joined the growing array of accommodations in Minnesota's Arrowhead. Like many who participated in constructing the tourist landscape, Bert Pfeifer entered the cabin business when he found himself living in an area drawing tourists. Born in North Dakota in 1915, Pfeifer came to work at his great-uncle's store near Minnesota's Itasca State Park. After marrying in 1939, Bert purchased housekeeping cabins, recalling, "No reservations. Most of them just come off the road and were just touring and looking for a place to stay." Cabin owners valued highway rather than lake access to attract visitors. In contrast to resorts, Bert's guests usually stayed only a night or two and included farmers wanting to fish and enjoy the state park. Bert added new cabins with running water and electricity. To make ends meet during slack times, he worked other jobs while his wife operated the cabins.[47]

While Pfeifer entered the tourist business after witnessing visitors in the area, others came north with clear intentions, including Jack Stedman's parents, who operated Pine Woods Overnite Log Cabins near Brainerd. Stedman lived in Rochester through ninth grade. The family traveled to Brainerd after a friend mentioned a cabin business was for sale. Pine Woods had electricity when they bought it in 1938, and they added indoor bathrooms in the 1940s, making it truly "modern." Pine Woods advertised through the local chamber of commerce and posted road signs. One of few Brainerd-area cabin businesses before the war, it offered privacy and reasonable rates. Most business was short-term, but the Stedmans encouraged longer stays by offering the last night free if people stayed a week.[48]

Places like Pine Woods received repeat business but at a lower rate than resorts. One school supply saleswoman made one of the cabins her home for several years. Other visitors included people building on nearby lakes who needed a place to stay during construction. Cabins often filled up on weekends but also attracted salespeople during the week. With gasoline rationing limiting travel during the war, Stedman's parents worked in Brainerd, but did not close the cabins. Since money was tight, the family handled all maintenance.[49] Unlike resorts where guests often communicated with the owner about their upcoming stay, people regularly arrived at cabins unannounced. As automobile travel increased, housekeeping cabins offered visitors flexibility and emerged as an important element of the tourist landscape.

Others entered the tourist business after vacationing in the region. Ed and Kay Gilman moved to Fifty Lakes to raise their family and run Gilman's Resort. Ed retained fond memories of fishing in northern Minnesota as a youth. After leaving the army and working for nine months, he and Kay came north on vacation and purchased an existing resort. Guests arrived from across the Midwest, including families friendly with the previous owners. Like other operators, the Gilmans depended on guests spreading the word, but they also belonged to the local resort association and advertised in the Twin Cities newspapers.[50] Carol Crawford Ryan's family was among those guests, visiting Gilman's starting in 1932 when another family owned it. Carol's father, an Iowa City postal worker, heard about the resort from a local gas station owner. The resort attracted a middle-class clientele, including civil servants and farmers. Like many young North Woods visitors, Carol's experience led her to purchase a summer place as an adult.[51]

The development of northern Minnesota's tourist landscape would have been incomplete without the Ruttger family. All four of Joseph and Josephine Ruttger's sons operated resorts—Alec at Bay Lake Lodge, Max at Pine Beach, Ed at Ruttger's Sherwood Forest, and Bill at Ruttger's Shady Point. Joe's oldest son, Alec, took over Bay Lake Lodge after returning from World War I, promoting the resort at outdoor shows armed with Oscar the Muskie—a mounted northern pike sporting a steel wool beard.[52] When Alec ran the resort in the 1920s, families arrived from the Twin Cities, but according to Alec's son, Jack, nearly 80 percent of business came from outside Minnesota. Interestingly, Jack compared operating a resort to farming: "You don't grow crops, but you still try and exist and raise your family and you work off the land." While Bay Lake Lodge attracted vacationers, it was not the only place on the lake. F. I. Boone, who stayed at Bay Lake Lodge in the early 1900s, purchased land and sold parcels to friends in Junction City, Kansas. For many, fishing and escaping the summer heat were key in choosing northern Minnesota for vacation.[53]

In 1927 the Ruttgers established Ruttger Brothers Incorporated, which included the Bay Lake Lodge, Bay Lake Store, and a Deerwood store. Max's brothers, Bill and Ed, who had worked in the mines and farmed, oversaw the Deerwood store. His brothers bought him out after a few years, giving Max the $2,500 needed to establish his resort. With help from local workers, Max built Pine Beach Lodge, twenty-five cottages, and employee housing by autumn 1931. During winter and spring, he added electricity and plumbing. A new lobby and recreation room were added in 1935, followed by six new cottages the next year. Max believed people were more likely to return if they befriended fellow guests and designed the architecture to foster community. Cottages lacked sitting rooms so guests socialized near the main lodge's fireplace, playing cards, bingo, and other games.[54]

The Ruttgers cooperated by placing large advertisements in metropolitan papers, and by Alec sending Bay Lake's overflow guests to Max at Pine Beach. Pine Beach guests generally vacationed for at least one week, and some families stayed the summer. All-inclusive rates were $35 per week the first summer and when bookings slowed, they dropped to $29.75. Saturday night music and dancing attracted locals and resort guests, helping endear the resort to the community because it provided both work and leisure. Food also attracted vacationers, and when Max hired a Minneapolis restaurant chef in the late 1930s, he received acclaim from his Twin Cities guests. In the mid-1930s Bay Lake Lodge developed a children's program, and Pine Beach followed. At Pine Beach guests used rowboats and wooden canoes before the resort added sailboats and a ski boat. Pine Beach opened in the midst of the Depression, but business continued to flourish.[55]

New tourist facilities also fostered opportunities. Bishop's department store in Park Rapids served locals but benefited from tourists visiting neighboring Itasca State Park by selling clothes to those unprepared for the weather. While some area resorts expressed displeasure with park competition, proximity ultimately boosted business at local establishments like Bishop's. Others provided entertainment, with Carl Warmington performing music at Breezy Point Resort and Tom Madden operating slot machines at area establishments. Tom's nephew Jack, who later opened his own resort, stayed at Ruttger's Pine Beach during its first summer to oversee his uncle's business. The Maddens also ran a golf course, selling tickets for resort owners to distribute to guests.[56]

While resorts and lodges offered entrepreneurial opportunities and places for tourists to stay, some people reshaped communities and the landscape by building or purchasing summer cottages. During the 1910s the Forest Service platted Star Island in the Chippewa National Forest and opened it to cottage development. The south shore offered a concrete sidewalk along the beach and a resort-like atmosphere, attracting businesspeople from the northern plains. The west shore was more rustic with no beach and swimmers using docks, while the east shore attracted professors and doctors. Summer residents brought with them the social markers of their place in urban society, creating a hierarchical structure in this vacation community. But residents also cooperated, establishing the Star Island Protective League in 1916 to encourage the government to establish ranger, telephone, and dock service for better fire protection. Despite constructing cottages on the island, which ultimately limited wider public use, residents claimed national forests were public playgrounds and worked with the Forest Service to control timber harvests.[57]

Star Island summer visitors also hired local residents. Des Moines attorney A. J. Starr contracted with a local concern, Lydick Mercantile, to build a summer cottage on Star Island. Far from luxurious, it was a haven in the woods for this professional man and his family. Between 1909 and 1918 Cass Lake residents helped build numerous Star Island cottages, and the island had one summer resort by 1926.[58] By 1941 Cass Lake advertised itself as the center of the forest. Visitors could stay in developed forest campgrounds or choose among 360 area resorts. The Forest Service continued issuing permits for summer homes, resorts, and camps, helping increase local work opportunities by generating tourist expenditures.[59]

While new accommodations changed the landscape and vacation experience, tourism also altered regional employment even as it aligned with a long regional tradition of seasonal work. Dunn's employed local women to clean, cook, and wait tables, while men served as guides and handymen. In 1937 sisters Ida and Ada Mellum of nearby Pelican Rapids wanted dining room work. For many, resort work provided money while in school or away from teaching. Inquiries also arrived from those wishing to summer in the area. School principal Albert Farnham inquired about employment for him and his wife and planned to bring their children. Others, like Margo Cairns of Minneapolis, were told to inquire at larger resorts like Ruttger's. By 1945, with war forcing resort closures, potential employees also turned to the State Tourist Bureau to see if northern resorts were hiring.[60]

Some expressed a desire to work at northern resorts and experience natural surroundings, but others found employment by chance. Ruby Treloar and her family farmed five miles from Ruttger's Bay Lake Lodge. As a girl, she recalled that the "Ruttger's name was always something real high-class and over and above all us farm people." This did not stop local residents from working summers at the resort, given the pay and opportunity to meet new people. Initially Ruby had little intention of working there. She wed in 1939, spent a year in Hinckley, Minnesota, and another in New York City before returning home in

1941. When Alec Ruttger went looking for local help, he persuaded Ruby to work for his family. She first worked for his brother Max at Pine Beach, but the following year returned to Bay Lake Lodge to lead the dining room staff and remained there for over fifty years.[61]

Across northern Minnesota operators marketed rustic outposts offering comfort and adventure amid natural beauty and rustic surroundings. The state's 1949 *Tour Guide to Minnesota* reported on tourist industry growth since 1900, capturing an enormous investment of time, money, and labor.[62] Lodging owners provided vacationers a place to stay and an experience that depended on a new workforce, landscape, and promotional language that linked city and country. Work remained connected to the land but instead of felling trees or mining ore and shipping goods to cities, it often meant caring for a new crop of tourists who consumed the region's forests and lakes. For many visitors, northern Minnesota came to be viewed as a place of leisure rather than labor, as a retreat from urban life rather than an economic storehouse of timber and iron ore.[63]

As one type of economic storehouse faded, another emerged. A second 1920s *Minneapolis Daily Star* cartoon portrayed tourists as quintessential consumers, paying for goods and experiences. Highlighting tourism's economic impact with the expression, "I'll take $100,000,000 worth of that!," the cartoon also aimed to convince residents that in-state vacations offered enjoyment at reasonable prices.[64] Allusions to escape appeared in promotional literature and highlighted how conservation efforts contributed to a new tourist landscape. In its 1919 *A Vacation Land of Lakes and Woods*, the US Forest Service emphasized opportunities on public lands, labeling the Superior National Forest an "ideal recreation ground for the camper, the fisherman, and the canoeist." Canoeists would see a country described in evocative language to entice visitors, which supposedly had changed little since fur company brigades plied these waters. The agency offered summer home sites for lease and worked to improve access.[65] Beginning in 1922, the *St. Paul Pioneer Press*'s annual vacation section, "Call of the Open," broadcast northern Minnesota to Twin Cities' residents, while federal agencies and regional organizations produced literature extolling the region's vacation virtues.[66]

Chambers of commerce and regional organizations also acknowledged tourism's economic benefits, informing their promotional efforts. Ely pitched itself as the Superior National Forest gateway, providing services and supplies to vacationers. In the early 1930s the Ely Commercial Club promoted rustic resorts and lodges with capacity for over one thousand people, suggesting it was a place where "breezes that fan its streets carry the perfume of the balsam, and its sky line is the jagged whip-saw of the pines in sharp relief against a clear blue." Such descriptions highlighted how natural environs could overwhelm the senses. Materials promoting Ely said little about surrounding mines, despite the fact that they employed 1,500 people with a monthly payroll of $200,000. No tourist would have been unaware of their presence, although local promoters emphasized natural beauty and concealed Ely's connection to the smoke-filled industrial city in order to sell the place to vacationers seeking escape.[67]

MAA promotion emphasized outdoor adventure, including a trip along Lake Superior's North Shore to enjoy fishing hamlets, the towns of Two Harbors and Grand Marais, and the expanding commercial roadside, including "summer places skirted by refreshment stands and filling stations." Others might venture to the national forests, iron ranges, or Itasca State Park.[68] The MAA enticed vacationers with the slogan, "Come to the Arrowhead, where the climate is sublime, come in the winter and in the summertime." It emphasized historical lore, with stories of Indians, trappers, fur traders, and lumberjacks. Canoeists could "hunt with a camera" to capture scenery. If roughing it in a canoe did not suit one's taste, the Gunflint Trail and its lodges catered to outdoor enthusiasts year-round.[69]

While many who promoted the North Woods lived and worked in the area, urban businesses also saw opportunities. Saint Paul's Schuneman and Evans department store published *Tips on Minnesota Motor Trips* to sell its camping gear, remarking, "It's all great sport—this getting away from civilization and losing yourself in the North Woods—but you can't get along without a few conveniences." Before leaving on a trip, the retailer urged purchasing equipment.[70] While the store sold tangible goods, the Ten Thousand Lakes of Minnesota Association viewed Minnesota's outdoors as a product to sell to vacationers. It suggested that the boundary waters remained "the Venice of the North" with waterways offering travel routes and assuring visitors that "it is a real wilderness—make no mistake about that." Canoeists would find serenity traveling through supposedly untouched waters.[71]

Newspaper writers also found their services in demand. In a 1938 brochure journalist and Minnesota Tourist Bureau director Ed Shave wrote an article in the form of a letter to a tourist interested in a Minnesota vacation. The "Long Bow" region, according to Shave, was a place for the explorer, voyageur, fisherman, and hunter. Tourists discovered the Cuyuna Iron Range in Deerwood and Crosby and encountered citizens festooned with long beards for the Paul Bunyan summer carnival in Brainerd, with its 362 resorts and five hundred lakes. Heading north to Nisswa, 75 resorts made vacations enjoyable and affordable. Shave told the story of one family who rented a summer cottage for less than the cost of remaining in the city. Farther north at Leech Lake, visitors entered the Chippewa National Forest with its Indian burial grounds and fishing. Looping back to Itasca State Park, vacationers could watch a pageant performed by CCC workers and Chippewa Indians. Shave closed by telling the potential tourist, "Here's hoping you've enjoyed the trip—but when you make it in reality you'll enjoy it ten thousand times more."[72]

Maps provided another vehicle to guide tourists through the landscape. MAA maps listed lodging and game laws, telling tourists to "Follow the Arrows to the Arrowhead" where they would discover a "climactic delight" and a "land of romance." Others offered maps with their guides, including the Ask Benn Wagner Service of Crow Wing County, which listed 267 resorts in the area. Minnesota's Cook County claimed the region as "nature's gift to the Arrowhead Country" and located all resorts along the Gunflint Trail.[73] The *St. Paul Pioneer Press* printed a state highway map in its 1934 vacation section. By 1938 the Gunflint Trail Association and Grand Marais Chamber of Commerce did the same, listing fifty-nine resorts on Lake Superior's North Shore Drive and the Gunflint Trail.[74] A diverse set of tourist advocates, ranging from the state with its official highway maps to local chambers of commerce, incorporated maps into the lexicon and helped tourists visualize and navigate the North Woods.

As consumers of leisure, North Woods tourists made decisions about how, when, and where to vacation. But their choices were also constrained by a consumer culture in which advertising framed such decisions. Tourist literature shared common themes like hay fever relief, avoiding urban congestion, adventure, and enjoying scenic beauty, all of which were increasingly possible in northern Minnesota during the interwar years. It presented vacations as an escape from routine, capitalizing on the degradation of work that led many urban residents to seek satisfaction through outdoor leisure. As such, vacations proved crucial in shaping consumer culture and environmentalism in Minnesota during the interwar years. Tourists from the Twin Cities and beyond followed the arrows to the Arrowhead to experience a region of rugged scenery and wilderness, adventure and tranquility, access as well as escape. In doing so, they joined the flow of people—local residents, sporting and wilderness advocates, tourist promoters, and state and federal officials—and commodities that helped create Minnesota's North Woods both as an idea and as a physical place.

PART II

The Twin Cities and the Built Environment

FOUNTAINS OF LIFE AND DEATH

A History of the Minneapolis and Saint Paul Water Supply Systems

John O. Anfinson

AT A BANQUET FOR AETNA LIFE INSURANCE COMPANY's medical director in 1902, physicians from Minneapolis and Saint Paul debated which city's water was better and therefore whose citizens should be cheaper to insure. Saint Paul, its doctors argued. As proof they cited the recent difference in typhoid fever deaths between the two cities. Minneapolis physicians countered by saying that anyone in their city who could afford insurance used spring water, not city water.[1]

After they began developing their public water supply systems in the late 1860s, both cities suffered decades of devastating typhoid fever epidemics. In 1890 their histories diverged. Saint Paul's death toll from typhoid began falling, while large outbreaks continued in Minneapolis. Although people at the time did not understand it, in retrospect we can explain the differing death rates in the two cities as a direct product of each city's different choices about water supply, as well as their respective campaigns to drive their residents onto city water and sewer systems. Saint Paul opted for a lake system, and Minneapolis chose the Mississippi River. For people in Saint Paul, abandoning their old water supplies and sanitary systems improved their health. For many Minneapolis residents, the shift increased their exposure to the typhoid bacteria, a

131

cause of illness and death. In the late nineteenth century, when city leaders made these decisions, officials in Saint Paul did not understand sanitary health or disease theory any better than those in Minneapolis. Neither city chose its water source or pushed its citizens toward modern water and sewer systems because they were more enlightened. In hindsight, their citizens became part of a natural experiment testing the relative merits of two very different types of municipal water supply systems, the outcome of which we can measure by the number of typhoid deaths each city suffered.

The Twin Cities' water and sewer systems had consequences beyond affecting the health of their residents. By funneling their wastes to the Mississippi River, both cities degraded the river's water quality. And because of their choices of water supply, Minneapolis focused more on water quality while Saint Paul focused more on quantity. Minneapolis moved its water intake facilities upriver four times in search of cleaner water, but it did not have to engineer the Mississippi to get the volume of water it needed. It could simply install more pumps. Saint Paul, however, continually looked for more water and had to transform the watersheds it relied on to get it. By 1910 Saint Paul had developed some twenty lakes into a connected system of reservoirs by building dams, canals, and conduits, while also dredging lake bottoms, removing aquatic vegetation, adding chemicals, and crossing watersheds. In some cases, the city even exerted power to prevent development around the distant lakes that it relied on for clean drinking water.

The Twin Cities therefore provide a unique comparative opportunity. Their choice of different sources for their water supply had life and death consequences for their citizens.

THE CONTEXT OF WATER SUPPLY

Geography influenced but did not determine the choices that Minneapolis and Saint Paul made about municipal water supplies. While the cities lie side by side, the Mississippi River carves very different landscapes through them. Downtown Minneapolis centers on Saint Anthony Falls. Above the falls, the river's banks—not bluffs—make the river accessible. Below the falls, the Mississippi enters a narrow, 8½-mile gorge, with high bluffs, through which the river drops about 110 feet. Nowhere else along its length does the Mississippi fall so steeply over such a short distance. Before locks and dams created a series of reservoirs, rapids or shallow water made steamboat navigation between the two cities dangerous to impossible. The drop at the falls, however, allowed Minneapolis to become the nation's leading flour-milling city from 1880 to 1930.

Where the gorge ends, at the mouth of the Minnesota River, the Mississippi widens and becomes a large floodplain river. Downtown Saint Paul lies another six miles downstream. While much of that city rests more than one hundred feet above the Mississippi, the land sloped gently down to the river on either side of the area that became downtown Saint Paul, which offered ideal steamboat landing sites. By 1857, steamboats had traveled to and from Saint Paul more than one thousand times, but only fifty times from Minneapolis.[2]

When Minneapolis decided to establish its water supply system, the Mississippi River offered all of the easily accessible water it could want, and Saint Anthony Falls provided the power to pump the water to the city's businesses, homes, and fire department. If Saint Paul wanted Mississippi water, it would have to pump that water out of the valley with powerful and costly steam engines. Instead, Saint Paul looked to the small lakes lying uphill and to its north. By tapping the lakes, Saint Paul let gravity do the work.

Before either city began to pipe water to its citizens, most people acquired water from their immediate surroundings. They dug wells, built rain barrels and cisterns to catch rain, or fetched water from the nearest stream or spring in buckets. Some drove a pipe into the ground and attached a hand pump to it for their use, ranging from a single family to entire neighborhoods. Vendors also delivered clean water to those who could afford it.

Getting clean drinking water presented one set of problems, while getting rid of wastewater presented another. Residents of both cities put their solid and liquid wastes where they lived. Cesspools and outhouses (privies) were the common receptacles for human waste. Cesspools were dug into the ground and lined with uncemented brick or stone. Houses or businesses with cesspools might have a water closet (bathroom) with a pipe discharging into the cesspool. Outhouses were small sheds over holes in the ground (privy vaults) that had no lining or plumbing. Both cesspools and privy vaults allowed liquid waste to seep into the ground, leaving the solid wastes behind. Once cesspools filled with solids, the owner or hired "nightsoil" contractor cleaned them out. Outhouse owners would dig a new hole and use the dirt to fill in and cover the old pit. Homeowners tried to keep their wells far from their cesspool or outhouse, since they knew that the sewage leaking out could contaminate their water, even if they did not fully understand how dangerous contamination could be.

Prior to the advent of urban water supply systems, city residents used an average of three to five gallons of water per day. Piped-in water boosted personal consumption to fifty to one hundred gallons per day, which overwhelmed existing systems for handling wastewater. In response, the Twin Cities began developing their sewer systems. In both cities sewers located in old stream beds carried their wastes to the Mississippi River. Saint Paul did not think much about polluting the Mississippi since its water came from upland lakes, but Minneapolis could not escape the consequences of contaminating its water supply.[3]

Medical science advanced significantly between the time Saint Paul and

Minneapolis started building their water and sewer systems in the 1860s and the second decade of the twentieth century. By then scientists knew that typhoid fever could spread through water and direct contact. They did not know exactly how and suspected a range of sources, some that would be proven right and some wrong. Both cities pushed their citizens to abandon private water and waste disposal systems and connect to city services, because they believed that municipal water sources were healthier than private alternatives and that their growing sewer systems were the best method for handling liquid wastes. The same limited understanding of disease theory governed both cities in these efforts.

Prior to 1880 most people ascribed to the "miasma," or filth, theory of disease. They believed that diseases originated in rotting garbage and other wastes and then became airborne. In his 1878–1879 annual report the Minneapolis City engineer pushed for developing the city's water and sewer system based on the miasma theory. In cesspools, he argued, "the filth accumulates and putrifies [sic], poisonous gases escape back into the houses, well water (which many are obliged to drink) becomes impregnated with their contents and they become the cause of fevers and disease."[4] That same year the Minneapolis health officer attributed the low number of infectious diseases to "the dryness of the season and the number of clear, sunshiny days, together with our natural healthful climate." He advocated for stronger nuisance laws to prevent diseases caused by miasmas.[5]

Two years later the State Board of Health published three facts about typhoid fever. First, it lectured, "typhoid fever is peculiarly associated with filth in the air and water, so much so that it is called the 'filth fever.'" Second, it said that typhoid could be contracted "by means of the discharges from its victims which gain entrance to the bodies of other people, chiefly through water, but also probably by soiled clothing or bedding, and possibly by means of the poison floating in the air." Third, "the peculiar poison of the disease probably escapes from the bodies of the sick in the discharges from the bowels alone."[6] The board insisted that any community following its rules could avoid typhoid epidemics, and it was a "disgrace" for any community to suffer one.[7]

In 1880 German pathologist Karl Joseph Eberth identified the typhoid bacterium (*Salmonella typhi*). His discovery and its subsequent confirmation undermined the miasma theory, but it took decades for doctors and public health officials in the United States to understand and accept what would become known as the "germ theory of disease." Today we know that humans are the only hosts for typhoid bacteria, which can spread via drinking water infected by sewage from sick individuals, or via contaminated food or drink handled by a person with typhoid. Symptoms appear one to three weeks after infection and include a high fever, headache, abdominal pain, and constipation or diarrhea.

Figure 7.1. In the late nineteenth and early twentieth centuries, before people understood the exact mechanisms by which typhoid fever spread, big cities in the United States wrestled to bring deadly outbreaks of the disease under control. The very different approaches that Saint Paul and Minneapolis officials took to developing their municipal water supply systems during this period unwittingly made their citizens part of a natural experiment testing the relative merits of two very different approaches to providing cities with clean, safe water. *Source:* "Typhoid Fever: A Disease That Can Be Prevented," *Virginia Health Bulletin* 1, no. 3 (September 1908): 120.

In the Twin Cities some patients died within a week of becoming ill; others lingered for over a month. The Minnesota Department of Health estimated the death rate between 1891 and 1910 at 10 percent. The long incubation period made determining the source of individual infections difficult. Complicating the effort, people who had mild cases or who had recovered from severe cases could still carry the bacteria and spread it. Despite the growing knowledge about germ theory after 1880, decades passed before the medical community, city administrators, and the public accepted it.[8]

In 1893 a Minneapolis committee on health and hospitals showed some advancement in its understanding of typhoid: "It is not considered abroad in the air, but communicable through waters polluted by the discharge from the bowels of typhoid fever patients."[9] By the turn of the new century, Minnesota's medical community appeared to understand the spread of typhoid far better. In a paper presented at the 1902 Minnesota Sanitary Conference, Dr. J. M. Robinson, the commissioner of health in Duluth, Minnesota, noted that until recently many still thought it came from dirt or filth, or that it spontaneously appeared. Now, he observed, medical professionals knew it came from a specific bacillus. Unless upstream water was contaminated by someone with typhoid, he concluded, a person drinking water downstream would not get typhoid. He noted that milk, flies, and vegetables fertilized with human manure could transmit the disease, but he argued that if a city had more than a few cases, the water supply was most likely the source.[10]

Despite advances in understanding, confusion about the causes of typhoid remained. A year later Dr. William E. Leonard, a Minneapolis medical inspector, still subscribed to the filth and miasma notion of how diseases spread. He proposed that anyone without a sewer connection should have to haul the contents of their privies or cesspools out and away by airtight systems to prevent sewer gases from spreading typhoid. If this was done, he claimed, "our epidemics would be easily confined to imported cases." He reported that the *Sanitarian* had published an article about the possibility of trains seeding typhoid fever along their rail lines. "This very likely cause of infection," he surmised, "would be especially operative in a great railroad center like Minneapolis and would tend to keep us supplied with cases even when we had none in our own borders."[11]

As late as 1910 the Minnesota Board of Health insisted that drinking water was not the primary source of typhoid infections. The board claimed "it causes but a minor portion of the infections, the greater number of cases being caused by direct infection from person to person (contact cases) or the infection of food by flies."[12] Confronted by such conflicting arguments, Minneapolis and Saint Paul pursued their own ideas and solutions.

MOVING TO AN URBAN WATER SUPPLY, 1867–1889

Exactly how each city's water supply affected the health of its citizens before 1890 is hard to discern, though the broad contours are clear. Minneapolis and Saint Paul both struggled to keep up with their rapidly growing populations and both suffered typhoid epidemics. During this period far more people relied on private water systems than used city water. After 1890, however, the effects of the two water systems became increasingly evident. The overall death rate declined for both cities as people shifted to urban water supplies, but Minneapolis experienced repeated epidemics while Saint Paul did not.

From 1867 to 1888 Minneapolis drew its water supply solely from the Mississippi River at Saint Anthony Falls. The city opened its first pumping station along a canal feeding mills on the river's west bank in 1867, primarily to meet the fire department's needs. Fearing the possibility of a monopoly if the city contracted with a private company to build the pumping station, citizens chose the city as their supplier. When completed, the new pumping station could supply 2.5 million gallons of raw river water per day.[13] Constant problems with the new pumps and the city's rapid growth soon led to demands for additional pumps and moving the waterworks to a new location. Responding to these demands in 1871, the city bought a stone building just upstream from the west bank canal above the falls and began putting in new equipment. They called this facility Station No. 1. While firefighting remained important, Station 1 sent increasing amounts of water to businesses and residents.[14]

As more people shifted to city water, the river's deteriorating water quality became an issue. In 1876 the *Minneapolis Tribune* criticized the location of Station 1 as a mistake, because it lay below Bassett Creek. The creek joined the Mississippi River one mile upstream and had become, the writer maintained, an open sewer. The *Tribune* predicted that sooner or later the waterworks would have to move upstream, above the natural and man-made sewer outfalls.[15] Still, the *Tribune* and others defended the city's water supply. For example, the newspaper published comments from Charles Gilfillan, the principal owner of the Saint Paul waterworks, praising Minneapolis water. Gilfillan proclaimed that a practical, scientific analysis showed that Minneapolis water "is the purest water, with the exception of that found in the granite districts of Maine, in the United States." Using the license granted by Gilfillan, the *Tribune* asserted that Minneapolis's water was purer than Saint Paul's and some of the best in the world.[16]

Three months later Professor S. F. Peckham, from the University of Minnesota, submitted a report to the city council providing a very different picture. Peckham showed that water coming from the intake at Saint Anthony Falls was severely polluted. He urged the city to move the intake above the city limits and warned the city council "that immediate steps [must] be taken in this matter before the city is scourged with an epidemic." The city's health officer endorsed Peckham's report.[17]

As experts argued over the quality of the city's water supply, adequate quantity also became an issue. Between 1870 and 1880 the city's population more than tripled from 13,066 to 46,887. The city also grew geographically. Until 1872 Minneapolis existed only on the west side of the Mississippi; the older community of Saint Anthony lay across the river. The two cities merged with the agreement that Minneapolis would supply water to Saint Anthony, which required additional expansion of the system.[18] The merger allowed Minneapolis to become the nation's flour milling capital in 1880.

To be a major industrial and population center, the city's boosters concluded that Minneapolis needed a clean and abundant water supply. On December 9, 1879, the *Tribune* reported that the supply issue had become paramount, and people had divided into three factions: those who pushed for a water supply upstream of Saint Anthony Falls to improve water quality; those who wanted to expand quantity enough to ensure that it could handle a major fire; and those who advocated for much greater capacity. The waterworks superintendent supported the latter, insisting supply had to match the city's growth.[19]

Problems grew acute during the 1880s. The population grew from 46,887 to 164,738, and city engineers and health officers simply could not keep pace. When typhoid epidemics struck Minneapolis in the early 1880s, the city blamed private wells, not city water. From 1881 to 1883 at least 453 people died from typhoid fever. The State Board of Health received so many complaints about the disease that it asked the city for an explanation.[20] Holding to the miasma theory, Dr. J. H. Salisbury, the city physician and health officer, replied that the ground had become "saturated with filth." "The entire city," he proclaimed, "is one vast hot bed for the propagation of disease germs." He asked the State Board of Health to get the city to ban open cesspools and outhouses, provide all its citizens with city water, and shut down all wells. The board was sympathetic, suggesting that the inability of the water and sewer systems to keep up with population growth had caused the typhoid epidemics.[21]

Minneapolis believed that its water supply was clean and healthy, especially compared to private water sources. So despite the epidemics and a sizable minority in favor of locating the city's water intake farther upstream, the city expanded Station 1 in 1883 and installed a new pump. As a concession to clean water supply advocates, Minneapolis moved the water intake from near the western shore out into the river, where a faster current, some believed, meant cleaner water.[22]

Even with the expanded station and a new pump, Minneapolis needed more water, and evidence kept mounting that water at Saint Anthony Falls was badly polluted. A committee appointed by the Minnesota Academy of Natural Sciences examined the city's water supply in 1883 and issued a scathing report. The committee took water samples from the original intake, the new intake, and a site above Bassett Creek. They found the water from the new intake somewhat better than the old one, but concluded that the water above Bassett Creek was the best. Since they had collected their water in February, they referred readers to water taken from a hydrant in June 1882, which, they reported, was more representative. This water, they said, "exhibited considerable quantities of animal and vegetable impurities . . . including *an unusually large proportion of parasitic worms.*" What came out of the fire hydrant also came out of household and workplace faucets.[23]

Population and industrial growth during the 1880s forced Minneapolis to make plans for a second pumping station. The city now had a chance to acquire less polluted water. It could shut down the old station and open a new one well above Saint Anthony Falls, or at least put the new station upstream, diluting the worst water with something better. The city, however, ignored or dismissed complaints from the public, the findings and recommendations of the Academy of Natural Sciences, and the warnings of its health officers. In 1885 the city completed Pump Station No. 2 on Hennepin Island, which straddled Saint Anthony Falls, so that it could continue to use the falls to pump its water. The new station could deliver another 10 million gallons per day of river water—with all its impurities—through a growing network of water mains.[24]

The 1880s ended as they began, with three consecutive typhoid epidemics. Deaths tallied 149 in 1887, 156 in 1888, and 114 in 1889. While the attack on wells intensified, some still understood the role wells played in the fast-growing city. The city's commissioner of health, Dr. S. S. Kilvington, called wells an old problem and claimed the highest death rates occurred in districts supplied by well water. Yet Kilvington recognized that his department could not be too zealous. For many, wells remained their only source of water.[25]

Recurring epidemics and rising demand for water forced the city to add Station No. 3 in 1889. Located on the river's west bank, four miles above the falls, the new station relied on two steam-powered pumps. The pumps ran on sawdust and edgings from the nearby timber milling industry, sometimes mixed with soft coal, and the station could deliver 30 million gallons of water per day. In his *Annual Report* for 1889, City Engineer Andrew Rinker acknowledged the disagreement over the river's water quality, but he would not take sides. Whether justified or not, he admitted, the argument for better water had forced the city to move upriver.[26]

Showing how the city had responded to the demands of its exploding population, Rinker reported that the length of city water mains had grown from 20 miles in 1882 to 141 miles in 1888. The waterworks provided water to vastly more people than it had only six years earlier. While Rinker and others thought the opening of Station 3 would allow the city to shut down Stations 1 and 2, problems with the new station and the high demand for water kept all three operating, mixing the better water from Station 3 with the more contaminated water from Stations 1 and 2.[27]

Saint Paul began thinking about an urban water supply system even before Minnesota became a state. On December 25, 1851, James M. Goodhue, editor of the *Saint Paul Pioneer* suggested the city needed an urban water supply system, since it was hard to get water in the winter, and many did not have their own wells or cisterns.[28] Five years later, the territorial legislature chartered a private water company. Rather than looking at the Mississippi River, its investors examined Lakes Como and Phalen, north of downtown Saint Paul. The new Saint Paul Water Company chose the lakes because gravity could deliver the water to the city. Charles Gilfillan and other entrepreneurs bought the water company in August 1857, and the editor of the *St. Paul Daily Pioneer and Democrat* newspaper expected the waterworks to open within a year. The company's timing was poor, however. In 1857 the United States entered a depression, and before the economy could recover, the Civil War began. Both events stalled progress on a waterworks for Saint Paul.[29]

Following the war, the Saint Paul Water Company continued its effort to find investors and struggled to begin work on its water supply system. The company finally began laying water mains to Lake Phalen in the fall of 1868. As the work progressed, the company encountered a problem that would plague Saint Paul's effort to expand its water supply for the next three decades. In this instance owners of milling rights on Phalen Creek objected that drawing water from the lake would damage their operations. To ensure sufficient water for the mill owners, the state legislature authorized the company to connect surrounding lakes to Lake Phalen. The company began work on canals, dams, aqueducts, and gates later in 1869 and largely completed the work by year's end.[30]

By 1880 Saint Paul's population had grown to 41,473. The private water company had made steady progress, but it could not keep up with the city's growth. As of 1881 the company served about 1,800 consumers—just 4 percent of the population—and many residents pushed the city to take over the waterworks. Gilfillan offered to sell, but only under conditions established by the legislature, which he would soon propose as one of the state's senators. Gilfillan's bill passed on February 10, 1881, and in November Saint Paul's residents voted to purchase the Saint Paul Water Company. The sale became final on August 10, 1882.[31]

Saint Paul quickly began expanding the lake system up through the Phalen Creek watershed, which drained into the Mississippi River just below downtown Saint Paul. In 1882 the city started pumping water from Vadnais, Sucker, and Pleasant Lakes. But like Minneapolis, Saint Paul's population growth continued to outstrip its water system, and the city began worrying about the volume of its supply. In its 1885 *Annual Report*, the Board of Water Commissioners noted that it would have to undertake "considerable work north of Pleasant lake this season, to increase our supply in case of a dry season."[32] And in 1886 Simeon P. Folsom, in his *Attorney and Engineer's Report*, envisioned a grand expansion. Under his plan, the city would increase the available quantity of water by four-fifths.[33]

During the 1880s Saint Paul's population grew by over ninety thousand, putting still greater stress on its water supply. In 1887 the Board of Water Commissioners reported that of all the water connections made in Saint Paul over the previous eighteen years, one-half had been made in the last three alone.[34] Two years later Saint Paul received only fourteen inches of rain, less than half the normal rainfall for any year over the past thirty years. The city kept careful track of the yearly precipitation, knowing rain and snow were essential to replenishing its lakes.[35] This led the board to expound on the nature of urban water supply systems and define the water quantity problem Saint Paul faced.

> The water supply for cities in this country is to become the first study of engineers, and the price of great sums of money. Water is a commodity that unfortunately decreases with the demand for it. That is, the *city* grows, the streams and lakes about it, its natural and cheap sources of supply, with the improvement and settlement of the country, diminish. Added to this difficulty in the planning of water works is the great and disproportionate increase in the use of water over the increase in population. This is the situation now. Water works are never finished. The investigations into possible future needed supplies should never cease. It is not enough to feel safe on calculations of the average rain fall, but the extreme of an occasional drought has to be provided against.

Without mentioning Minneapolis, the Board of Water Commissioners observed that "none but cities depending upon great lakes or rivers of the first magnitude can avoid occasional shortage in their water supplies." If needed, the board said it could draw from the Mississippi or Saint Croix Rivers.[36]

Like Minneapolis, Saint Paul suffered its worst epidemics in the 1880s and also ended the decade with three consecutive epidemics. Typhoid killed 141 people in 1887, 135 in 1888, and 101 in 1889. In his 1886–1887 and 1888 reports, Saint Paul's commissioner of health, Dr. Henry F. Hoyt, cited wells, a lack of sewer connections, and poor plumbing for the epidemics.[37] In 1888 only 7,500 (12.5%) of the sixty thousand houses in Saint Paul used city water.[38] Hoyt praised the previous health officer for beginning a crusade against well water that Hoyt was continuing.[39] The epidemics did not shake Saint Paul's faith in its water supply. Instead, like Minneapolis, it made city officials more determined to deliver city water to all residents.[40]

DIVERGENCE, 1890–1910

After the epidemics of the late 1880s, the number of deaths from typhoid fever in both cities steadily declined relative to their population size, but whereas Minneapolis continued to suffer massive outbreaks, Saint Paul did not. Typhoid deaths in Saint Paul dropped to 74 in 1890 and 65 in 1891. The number would not rise above 60 again. During the 1890s Saint Paul averaged 47.9 deaths per year, and during the first decade of the twentieth century the average fell to 29.5 (see Figure 7.2). As typhoid deaths dropped and diphtheria, tuberculosis, "cholera infantum," and other diseases caused more deaths, the city health commissioner and other city officers paid less attention to typhoid fever. Saint Paul's greatest worries over the next two decades would not be re-

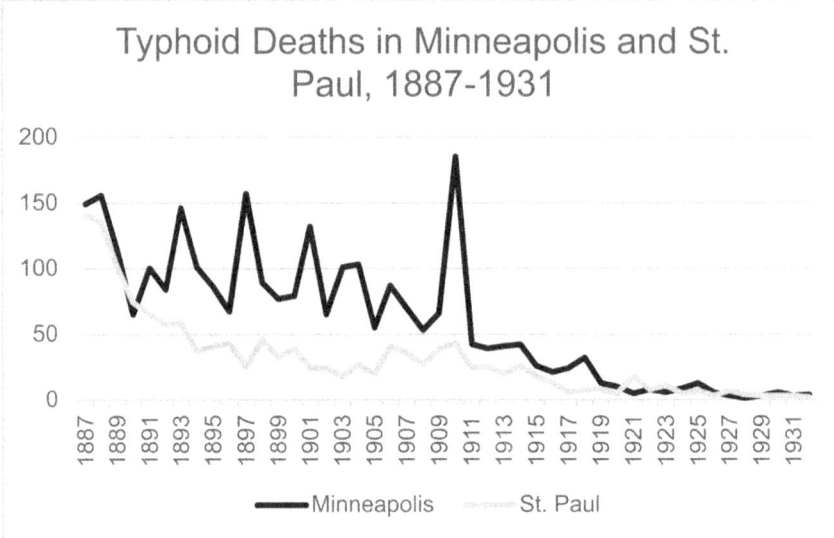

Figure 7.2. The data in this graph come from a number of sources that overlap in dates. For 1887–1891, see *Third Biennial Report, The Vital Statistics of the State of Minnesota*. For 1890–1891, arranged and edited by the Secretary of the State Board of Health and Vital Statistics (Minneapolis: Harrison & Smith, 1893), 132. For 1888–1909, see *Third Biennial Report (New Series) of the State Board of Health and Vital Statistics of Minnesota, Sanitary Engineering Division*, 1909–1910 (Minneapolis: Syndicate Printing, 1911), 48. For 1900–1932, see Minnesota Department of Health, *Division of Vital Statistics* (bound unpublished mimeograph).

lated to its water purity. Instead, its greatest tests would be expanding the water supply in the face of opposition from those with conflicting rights to water and land and transforming the natural lakes it relied on into a well-managed reservoir system.[41]

In 1890 Saint Paul's quest for water led it to Baldwin Lake, on the Rice Creek chain of lakes. Taking water from the lake not only angered a sawmill owner on Rice Creek, but the Board of Water Commissioners crossed the powerful milling companies at Saint Anthony Falls. Since Rice Creek flowed into the Mississippi River above Minneapolis, the millers sued, claiming Saint Paul was taking water from their mills. Saint Paul had undertaken what may have been the first inter-watershed transfer in the state's young history. The Hennepin County District Court judge ruled in favor of Saint Paul, and the millers appealed to the Minnesota State and US Supreme Courts. Both courts supported Saint Paul.[42]

The search for more water continued into the twentieth century. Saint Paul added more lakes to its system, and by 1911 the city relied on a watershed consisting of 137 square miles and over twenty lakes.[43] By the early 1920s Saint Paul had reached the limit of its lake system. With no more room to expand, the city began drawing water from the Mississippi River above Minneapolis.

Saint Paul loved comparing its water supply to its cross-town rival, emphasizing the clean water it drew from its lake system. Yet Saint Paul's system contained more vulnerabilities than it admitted. In 1911 Caroline Bartlett Crane exposed some of these in her *Report on a Campaign to Awaken Public Interest in Sanitary and Sociological Problems in the State of Minnesota.* During her stay in Saint Paul, she toured Gervais Lake, one of Saint Paul's many water supply lakes. The city owned land around many of the lakes, but she found two hotels on private land on Lake Gervais. She thought the drainage from the sewage systems of both hotels could enter the lake. Given these examples, she wondered how the city monitored other lakes. Two men, she reported, worked year-round policing the lakes, and the city added three more from April until fall. She doubted, however, whether they could do so adequately. While Saint Paul had issues with pollution entering its water supply, the low number of typhoid fever deaths supports the city's claim that its water was safer than that of Minneapolis.[44]

Unlike Saint Paul, Minneapolis continued suffering significant outbreaks of typhoid fever. Worried about the reputation of its water supply, the city blamed other sources for the frequent epidemics. Wells became the primary scapegoat for city officials. In 1891 one hundred people died from typhoid fever, and the city's new health officer argued that the city should force its citizens to connect to city water and sewer as soon as either was available, no matter what it cost them. He lectured that the "main factors in this unfortunate mortality, literally 'a slaughter of the innocents' in some sections of the city, is the cesspool system of the average householder, and, more commonly, the use of well-water polluted by drainage from an adjacent privy vault or cesspool." Unless the city forced people off their wells and onto city water, the typhoid epidemics, he warned, would continue.[45] Minneapolis also blamed nonresidents for many of the cases and deaths attributed to the city. City officials claimed that workers from the lumber camps in northern Minnesota and Wisconsin and from the grain fields of the Dakotas brought in many cases. The city did not consider the number of cases it exported.[46]

When another typhoid epidemic erupted in 1893, the city established a committee to examine wells. The committee studied the wells of those who had become ill and found them all badly polluted, shutting some down. But the committee also reported something that cast doubt on both the city's campaign to demonize wells and the quality of city water. Of the 1,073 suspected cases of typhoid fever, the committee found well and city water equally at fault. Well water accounted for 529 cases (49.3%) and city water for 523 cases

(48.5%).[47] Consequently, the committee made five recommendations. The first two insisted that people close their cesspools and connect to the city's sewer system. The third urged the city to run water mains into areas of the city with polluted wells. The last two addressed the quality of city water itself. The committee recommended that the city consider building a reservoir and filtering system, and that it close Stations 1 and 2 as soon as possible.[48]

Minneapolis was not ready to accept the mounting evidence against its water supply, however. From 1894 to 1896 the number of typhoid cases dropped, but the proportion linked to city water remained high.[49] The city engineer wanted to test a filtration system, but the committee on waterworks "did not think it necessary to hurry the matter." Louisville, Kentucky, had begun testing one, and the committee thought it could wait for the results.[50] Minneapolis could have looked to the acclaimed Lawrence Experiment Station, in Lawrence, Massachusetts. In 1893 that station had discovered that a well-built slow sand filter could reduce typhoid bacteria contamination. As a result, many cities had adopted the slow sand filtration system.[51] Filters were expensive, however, so Minneapolis delayed the decision.

In 1897 the city experienced the highest number of deaths from typhoid since it had begun keeping records: 157 people died. Dr. P. M. Hall, medical inspector, admitted that "the cause seems to have been the city water," but, he added, the number of cases caused by well water had also increased. He also claimed that eighty-seven cases had been imported.[52] The epidemic occurred despite Minneapolis's decision that year to build two reservoir basins in Columbia Heights, above the city limits and east of the river. Water from Station 3 began filling the reservoirs at the end of December. Each one could hold 47 million gallons, and gravity carried the water into Minneapolis. The reservoirs were only settling basins, where some of the sediment could drop out, but some hoped they would reduce or stop the epidemics.[53]

In 1900 the city's population stood at 202,718, and it calculated the number of people using city water at 95,000 (46.9%). Two years later the city estimated its population at 225,000 and claimed the number served had risen to 135,000 (60%). As more people used the city's water, it became harder to blame wells and other sources for its unrelenting typhoid fever outbreaks.[54]

The city reached a turning point in 1903–1904. The Minneapolis Department of Health acquired the knowledge and equipment it needed to conduct new bacteriological tests in 1903, and immediately began sampling water from the three Minneapolis stations. The tests showed that the water taken from the lower two stations "is dangerous and absolutely unfit for drinking purposes." Water from Station 3 and the reservoirs, however, was safe to drink. The department did not think this would be true for much longer, however. They were right.[55] Due largely to pollution at the two lower stations, the city decided to build a new plant and purchased a site in what is now Fridley, on the river's east bank, just above the city limits. After many delays, Station No. 4 opened in early 1904. The city then shut down Stations 1 and 2 at the falls, eliminating the worst sources of contaminated water.[56] This completed Minneapolis's pre–pure water pumping system.

Unfortunately, Station 4 did not open soon enough for the estimated 79.2 percent of the city's population using city water. Epidemics struck the city in both 1903 and 1904, killing 101 and 103 people, respectively. One of the water mains crossing the river from the west bank to the east had broken in November 1903, and the city had relied on Station 2 to supply the Second and Ninth Wards. By January 1904 typhoid fever had become rampant in these wards. The city reported, "The people were aroused by mass meetings and the council was called upon to appoint an expert commission . . . to recommend some method of purifying the city water supply." The committee advocated for sand filtration.[57] Between 1904 and 1905, however, the voters of Minneapolis rejected a one-million-dollar bond proposal to fund a sand filtration system. Two councilmen, Frank H. Castner and James H. Duryea, led the opposition. They insisted that a filtration system was too expensive, and that Minneapolis should explore a system of deep wells.[58] Again the city delayed, hoping that new water from Station 4 and the decision to abandon the two lower stations would solve the problem.

The number of deaths from typhoid fever dropped by more than half in 1905, but the city still found "colon bacilli" (fecal coliform bacteria) in water samples taken from the two upper stations. Minneapolis suspected cities upstream were the source.[59] Confirming its suspicion, the number of deaths from typhoid fever rose to eighty-seven in 1906, but the city still resisted expensive filtering and treatment. Dr. P. M. Hall, the city's health commissioner, prophesized that "some great epidemic will open the eyes of people to the necessity of an adequate filtration plant, and at the cost of thousands of lives Minneapolis will someday have a pure water supply."[60] Skeptics, however, saw other signs over the next several years. From 1907 to 1909 the number of deaths per year fell below eighty, with only fifty-three typhoid deaths in 1908.[61]

Though Minneapolis suffered the worst reputation in Minnesota for typhoid fever, it was not very different from other large American cities that sent their wastes into the rivers and lakes they relied on for their drinking water. Cholera, yellow fever, dysentery, and especially typhoid fever ravaged their populations. In Massachusetts, Lowell and Lawrence both drew their water from the Merrimac River, and in 1890–1891 suffered severe typhoid epidemics. One hundred and thirty-two died in Lowell and seventy-four in Lawrence. The Allegheny and Monongahela Rivers supplied Pittsburgh, and from 1894 to 1906 the city averaged more than five thousand cases each year. In the fall of

1901 typhoid fever epidemics broke out in Boston, Baltimore, Cincinnati, New Orleans, Philadelphia, Pittsburgh, Saint Louis, and Washington, DC. Chicago suffered the worst epidemic in its history.[62] All struggled to improve their water supplies.

Early in 1909 the Minneapolis city council authorized a Citizens' Commission to study ways to improve the city's water. The commission evaluated four options. Two they quickly rejected. They determined it impractical to draw water from a well system and also rejected building a pipeline to Lake Superior. They decided to consider siphoning water from Mille Lacs Lake, in central Minnesota, and to continue to draw water from the Mississippi River. Several factors weighed against Mille Lacs Lake, but opponents of Mississippi River water advocated for further study.[63]

Minneapolis also hired Rudolf Herring, a nationally known waterworks engineer, to review the city's options. After examining water in the river and the two reservoirs, he reported that "the colon bacillus is constantly found in every c. c. of water examined, indicating considerable fecal pollution." Herring emphasized that over the previous two years, fifty water suppliers in the United States had successfully begun using hypochlorite of lime (chlorine) to purify their water. With appropriate sedimentation, filtration, and purification, he argued that the Mississippi River was the city's best source of water. Herring issued his report in March 17, 1910, two months after another massive typhoid fever epidemic had begun.[64]

It had become impossible to defend the untreated Minneapolis water supply. "Like the outbreaks of 1893, 1897, and 1904," the Department of Health admitted, "this one was directly traceable to the city water." Since Stations 1 and 2 had closed, the department had to acknowledge that river water coming from above Stations 3 and 4 carried the typhoid bacteria.[65] Minneapolis realized it could no longer delay and, on February 25, 1910, decided to build a water treatment plant and immediately began cholorinating the city's water. Still, 185 people perished in the 1910 epidemic. On January 10, 1913, the Columbia Heights purification plant opened, which combined rapid mechanical filtration with coagulation and chlorination. Like other cities around the country using these methods, the infection and death rates from typhoid plummeted.[66] That same year Saint Paul began treating water at Vadnais Lake with copper sulfate to eliminate algae and improve taste and odor. Confident in the purity of its water supply, Saint Paul did not use chlorine until some point between 1918 and 1920; it did not begin constructing a water treatment plant until 1920.[67]

When city officials in Minneapolis and Saint Paul launched their attack on wells, outhouses, and cesspools, they targeted an old and obviously inadequate system of water supply and wastewater disposal. As the Twin Cities grew and density increased, sewage from outhouses and cesspools leaked into wells. Believing that all polluted wells contained typhoid bacilli, both cities pushed their citizens to the new city water systems as a way to separate the link between pollution and water supply. Because they made different choices for their water supply systems, this worked for Saint Paul, but Minneapolis simply exchanged individual polluted wells for a polluted municipal system. As a result, massive typhoid epidemics plagued the citizens of Minneapolis, while the rate of infection and death in Saint Paul declined.

Both Minneapolis and Saint Paul experienced typhoid epidemics before they began supplying city water, and in both cities residents contracted typhoid from infected wells and other sources. However, while typhoid bacteria in a well often infected only a few families, the same bacteria in the public water supply could and did reach tens of thousands of city water users. Without adequate knowledge of disease transmission, Minneapolis officials inadvertently replaced older water and waste disposal methods with a modern system of pumps and sewers that themselves became a vector for spreading disease.

Drinking water protection has come a long way over the last century. Federal and state governments now play leading roles. The Clean Water Act of 1972 and the Safe Drinking Water Act of 1974 are two examples. Minneapolis and Saint Paul have also greatly improved their water supplies. In his book, *The Blue Death: The Intriguing Past and Present Danger of the Water You Drink*, Robert D. Morris calls out the new microfiltration plant in Columbia Heights: "There, on September 1, 2005, water began to flow into one of the most advanced water treatment systems in the world and out to the people of Minneapolis."[68] And today, Saint Paul has over 1,100 miles of water mains that supply the city and some suburbs. The city's reservoirs can store enough water to meet its daily needs for approximately a month.[69] Today citizens of both cities are able to take the quantity and quality of the water flowing from their taps for granted.

URBAN ENVIRONMENTAL HISTORY AND LORING PARK

How Cultural Views of Nature Influence Recreational Design

Karen Wellner

IN 1883 A TRACT OF SOGGY LAND, situated outside the western boundary of Minneapolis, was sold by its owner, Joseph S. Johnson, to the city of Minneapolis for $147,125.[1] The thirty acres contained two small bodies of water, which before that point had been visited mainly by cows and horses sent out to pasture. Shortly after its sale, the land became the scene of men, muscle, and machinery. Work crews excavated and reshaped the ponds, sculpted the landscape, planted trees, and established walkways. Within one year they had transformed the former farmland and forest into Minneapolis's first urban park. In that same year, the Minneapolis park system went from paper sketches to reality, and the newly designed thirty-acre parcel, first known as Central Park and later renamed Loring Park, played a substantial role in this transition.

The transformation of the landscape, however, was not unique to Minneapolis. In the late 1800s cities across the United States were busy altering patches of nature to make public space more appealing to urban residents. The leader in urban park design at the time was Frederick Law Olmsted, a park planner whose best-known achievements include Manhattan's Central Park and Boston's Emerald Necklace parkway. Olmsted and other park advocates praised the potential of open space within cities to cure residents from the stresses of hard work and high urban density. They also believed that parks inspired moral values and helped maintain social order. Olmsted argued that pastoral landscapes, characterized by large rolling meadows, combined with picturesque landscapes in which the topography was a bit shaggy and rough, was the best way to provide park users with an uplifting and soothing consciousness. To create parks that could achieve these goals, some of the uneven edges of nature needed smoothing, with alternating areas of sunlight and shade and easy approaches to ponds and lakes. The goal was to make open space look as natural as possible, while remaining as safe as possible.

Just as particular social, intellectual, and moral forces have shaped and changed cities, so too have the same forces altered the "natural" design of urban parks. Those forces have changed over time, evolving through five major park eras defined by sociologist and park designer Galen Cranz: Pleasure Ground (1850–1900), Reform Park (1900–1930), Recreation Facility (1930–1965), Open Space (1965–1990s), and Sustainable Park (1990s–present). In each of these eras the influence of shifting social and cultural attitudes toward nature and recreation clearly emerges.[2] Yet Cranz's analysis examines only three large urban parks in New York City, Chicago, and San Francisco. Does Loring Park, a smaller but equally urban landscape, exhibit similar patterns of design that fit with Cranz's national model?

THE PLEASURE GROUND PARK, 1850–1900

Cranz identifies 1850 as the beginning of professionally planned American urban parks. Frederick Law Olmsted and Calvert Vaux were busy overseeing the construction of Central Park in New York City. This first large pleasure ground park opened in 1857, although it took nearly twenty-five years to complete. Olmsted's vision of weaving nature into the city soon became the standard for park planning. The limited number of American landscape architects during this era emulated Olmsted's bold ideas of designing parks to fulfill the physical and psychological well-being of city dwellers.[3]

The pleasure ground idea, characterized by meadows, lakes, and walkways for strolling, was well entrenched in park planning by the time Minneapolis voters approved the establishment of an independent park commission called the Minneapolis Park Board. Citizens charged the board with purchasing or condemning land for park development, levying taxes, and overseeing all monies for park use. This freedom allowed the board, then and now, to raise its own money, primarily through a property tax enacted in 1883. In most American cities park finances are part of the overall city budget, which occasionally leads to money originally tagged for park use to be transferred to other city departments. In Minneapolis the park system continues to operate semiautonomously from the rest of the city's government.

Figure 8.1. Minneapolis's first urban park—Loring Park—required extensive work in the 1880s to transform a soggy piece of land on the city's western boundary into the picturesque crown jewel of what soon became a renowned urban park system. In the century and a half since, city leaders remade Loring Park's landscape and identity time and time again, reflecting shifting social and cultural attitudes to nature and recreation. *Author's personal collection.*

In 1883 the Minneapolis Park Board commissioned Horace Cleveland, a Chicago landscape architect and friend of Olmsted, to design a park system for the small but rapidly growing city. Like other early landscape architects, Cleveland embraced the idea that everyone innately appreciated nature. No one, Cleveland wrote, would value parks more than young children, and the poorer the child, the greater his or her need for nature. Wealthy residents already had summer homes where nature was abundant, but Cleveland argued that public parks were like public schools: important civic institutions that the rich were obliged to provide for all children, including the poor.[4]

Cleveland's original plans for Loring Park included the excavation and connection of two wetlands, Jewett Lake and Johnson's Pond (later Loring Pond). Several circuitous walking paths allowed people to maximize their views of the small park (Figure 8.1), even if they were instructed to walk only on designated paths. Compared to the rigid urban grid soon to surround Loring Park, Cleveland's winding walkways made users feel as if they were truly in a natural setting. Cleveland argued that since Loring was the closest of all parks to the heart of Minneapolis, it was the city's most civilized park, and its upkeep would be most important.[5]

The designation of Loring Park as the city park system's "grand jewel" is apparent in annual reports published by the Park Board commissioners. In 1906 newly hired park superintendent Theodore Wirth wrote, "Loring Park is in some respects the show park of the Minneapolis park system and should be kept at all times, up to the highest possible degree of perfection."[6] In order to achieve perfection, the Park Board allotted more money for flowerbeds in Loring Park than any other Minneapolis park. For example, in the Park Board's *Annual Report* in 1910, the floriculture department tallied 40,017 bedding plants at a price of $2,255. The number of plants, and their price, was almost double to that of Minnehaha Park, the next most manicured park in Minneapolis.[7]

In practice, city parks developed in areas where they had little competition from agriculture, housing, or industry. This was true of Loring Park, constructed out of part of a five-hundred-acre extensive marsh and bog ecosystem. Excavation began in November 1883, with native plants planted on it for a natural effect. Workers also graded around the ponds to form a gentle slope to the water's edge.

Loring Park presented several challenges to Cleveland, including the fact that the site had so few trees. Cleveland made do by transplanting fast-growing trees and shrubs from nearby woods and laborers planted many large trees along the perimeter of the park to give the impression that Loring went far beyond its property lines. A postcard picture shows a combination of trees, paths, and lawns with large residences surrounding the park (see Figure 8.1). Nearby residents perhaps held the view that Loring Park was simply an extension of their own front yards.

Park activities in pleasure ground parks included leisurely exercise and psychological restoration, but this did not mean that parks were entirely passive. Loring Park was a place where people were often seen running, sledding, tobogganing, and ice-skating. A skating rink was improved in 1890, complete with a warming house and a horse-drawn team that pulled a large "planer" to scrape the ice and keep the surface level for skating. Ice-skating was well patronized and the skate rentals often brought in a small profit for the park system.[8] The idea of more active use in Loring Park slowly gained acceptance, but this was not so for public speeches and sermons; city leaders did not consider parks to be a place for politics and religion, and park superintendents did not allow speeches or meetings in parks.

The Park Board also discouraged ornamentation in the form of signs, banners, and statues. To park commissioners, Minneapolis's parks were to represent "nature" rather than human handiwork, although certainly a great deal of human handiwork had gone into the sculpting of the parks themselves. Public musical performances, on the other hand, became accepted park practice because music provided culture and pleasure. In 1883 the Minneapolis Park Board declared that summer-month concerts would occur as often as every other weekday evening in every one of the large and moderately sized parks, including Loring. High-class opera and classical music were offered to park users, while vaudeville and variety shows, considered low class, were not allowed.

Park landscaping soon became an area of concern. To some landscape architects, colorful combinations of nonnative flowers presented excessive visual stimulation and interfered with a region's natural beauty. During the Pleasure Ground era and spilling into the early years of the Reform Park era (1900–1930), park experts weighed in on their preferences, most often in articles published in the *Bulletin of the American Institute of Park Superintendents*. Theodore Wirth, Minneapolis parks superintendent from 1906 to 1935, favored flower beds at Loring if they were well located, well arranged, and harmonious in color. Such decoration, Wirth declared, would be more beneficial than "any amount of native shrubbery and perennials that you may crowd into the same or in a ten times larger space."[9]

Wirth believed that formal flower beds helped keep park users on walkways and that the working class appreciated flowers more than wealthier classes, who often already had flower beds at their homes. Such attitudes represented the commingling of natural landscapes and gardens, with park users seeing both as the same.[10] Cleveland disagreed and complained to Olmsted as early as 1893 that Americans did not understand the purpose of a park, thinking that parks were amusement areas such as Coney Island, where spectacular exhibitions rather than natural surroundings were the chief attraction.[11] Perhaps more alarming to Cleveland was that too few people were using the parks, especially during the summer months. Park commissioners compared Minneapolitans to European park users, wondering if perhaps Europeans had a more "highly developed sense of the beauty of nature and the charms and uses of outdoor life."[12] The commissioners felt that America was still in its early years of appreciating outdoor art and that possibly the lack of park users was due to a lack of attractions. To bring more people to Minneapolis parks, park commissioners offered suggestions such as shows of plants and flowers or ball games.

In 1900 several members of the Park Board suggested that carriages and bicycles be allowed in Loring Park. Park Board president, Charles Loring, nixed the idea entirely, believing that the park was a sacred institution and not a thoroughfare. Nature lovers who desired peace and quiet may have won that round, but commerce—in the form of automobile showrooms springing up where stately residential row houses once stood—was already beginning to invade the garden as the Pleasure Ground era ended.

THE REFORM PARK, 1900–1930

Loring Park did not remain a pleasure ground park for very long. On the heels of the sanitary reform movement, women in particular encouraged park planners to bring the mental and physical benefits of wholesome recreation to urbanites, especially children. Reformers called for a more utilitarian use of nature where parks would be places to gather and stay, and not just stroll through. At the turn of the twentieth century planners nationwide began to scrutinize who was using parks. Following suit, Minneapolis park superintendent Wirth examined the number of park users and their activities, and declared that Loring Park was not living up to its potential. To Wirth, it was not enough that parks looked beautiful; they also had to be practical. The role of the park was about to change, and with it, how nature would be used. This shift marks the beginning of what Cranz describes as the "Reform Park Era," when city planning aspired to more scientific approaches to urban development and governance.[13] Parks did not escape the scrutiny of planners who, like Wirth, were influenced by the progressive conservation movement's quest for a more efficient use of nature.

Whereas the pleasure ground park was viewed as a solution to the lack of personal free space, the reform park helped serve as a defense against the chaos of too much free time. Many adults now worked shorter weeks with higher wages, and there was growing concern about the healthy upbringing of children. All of this saw more people with more free time, and especially youngsters, now playing in the streets.[14] The immediate fears were that wandering children would be injured by cars or streetcars and that they risked exposure to criminal elements, especially gambling and drinking. Removing children from the streets and relocating them to parks began with the idea of public playgrounds. The idea that children's playgrounds should be added to parks was sound, wrote the president of the 1910 Minneapolis Park Board, although the location needed to be carefully selected, due to the children's noise.[15]

Nature, which had previously been viewed rather passively, was now a "container" where children played in groups, exercised in groups, and competed against one another in gymnastics and swimming. Even though physical aspects of the park changed, as slides and swings replaced trees and grass, the objective remained the same: to shape a cohesive urban moral order.[16] Children, proclaimed psychologist G. Stanley Hall, learned good social and democratic skills by playing together. They should congregate in natural settings, he said, because earlier civil societies had developed in rural areas.[17] The Park Board could not send hundreds of Minneapolis children to rural areas to play, but it could make Loring Park more amenable for this purpose.

During the Reform Park era, two professional park organizations emerged to help guide park advocates: the American Institute of Park Executives, founded in 1898, and the Playground Association of America (later the National Recreation Association), founded in 1906. The Playground Association of America envisioned parks as giant playgrounds armed with recently invented special play equipment and staffed by play directors who represented a new layer in social engineering efforts. The organizations' rational call for play persuaded many park commissions, including the Minneapolis Park Board, to strike down its restrictive use policy, allowing children to use previously off-limit areas for skiing, sledding, and tobogganing.

In Minneapolis, park superintendent Wirth established more flexible park use rules and gave instructions to the park police "not only to permit the use of hillsides in any park for sledding and skiing, but to assist and properly supervise the employment of park property for such activity."[18] While many citizens welcomed the new instructions and the removal of park signs that read "Keep Off," older park users, still operating in the sphere of proper pleasure park use, came into conflict with children who were straying from the walkways.[19]

Women also became more noticeably present in parks during the Reform Park era. During the earlier Pleasure Ground era, women most often visited parks accompanied by men. As more women worked outside of the home, however, an increase in independence and leisure-seeking attitudes occurred. Working-class women began frequenting parks to enjoy nature and to partake in activities that were open to women at the time, including bicycling and tennis at Loring Park. As women's clubs became involved with park design, they advocated for playgrounds and persuaded park commissioners to purchase play equipment for the parks. While Loring Park never had the large-scale gymnastics bars that areas such as Powderhorn Park had, by 1911 Loring Park was home to two sandboxes, one merry-go-round, one seesaw, and one swing set.[20]

Children who had once walked and played *in* nature at Loring were now using nature as a backdrop for physical play on constructed exercise equipment. Soon Loring Park would accommodate more people and more "things," such as shuffleboard courts, horseshoe pits, and a new shelter house, paid for by Charles Loring and used as a warming center for winter skaters. This building highlights the civic role that Loring Park was intended to play in bringing children and adults together in an area where they might otherwise never meet.[21]

Nearby residents, however, soon complained about the children who used the new playground equipment: they were too noisy. Park superintendent Wirth stood firm and wrote in the park minutes that "statistics fully bear out the statement that playgrounds properly conducted under supervision have a tremendous influence in directing along right lines the activities of young people and eventually helping to make them better citizens."[22] By 1916 Wirth was scrambling to obtain adequate funding for all the playground equipment requests: "The demand for additional playgrounds and facilities of all kinds, together with supervision, is becoming so pressing and the playground movement is progressing to such an extent that a special tax levy for playground operation will have to be secured through legislative enactment."[23]

At the end of the Reform Park era, the 1930 US census revealed that the population density around Loring Park, especially at its eastern and southern boundaries, was higher than most other areas of Minneapolis. The census also revealed that 10–19 percent of the Loring neighborhood consisted of children less than fifteen years old, which was not as high as other Minneapolis park areas.[24] Along with Loring Park's small size, this fact may account for why Loring Park had less playground equipment than Powderhorn Park and Lake Nokomis.

Nonetheless, accompanied by sandboxes and merry-go-rounds, the quality of experience in Loring Park during the early 1900s became markedly different from that which prevailed in the earlier Pleasure Ground era park. Rather than quiet and serene, the park was now noisy with children and teenagers playing in organized activities. As the 1930s began, the park would become even noisier and less well organized.

THE RECREATION FACILITY PARK, 1930–1965

By the early 1930s the close tie forged between urban parks and social reform began to sever. Recreation remained important as a municipal function, but the types of recreation now had little or no need for organized leadership.[25] Nationally the American Institute of Park Executives abandoned the idea of parks for reform and substituted it with a new motto: "recreation for all." This change marks the beginning of what Cranz calls the Recreation Facility era, a period characterized by ball fields, basketball courts, parking lots, and increased budget-making decisions. While Loring Park did experience a reduction in organized play, it did not become a "ball field" park due to its small size and the fact that the nearby Parade athletic fields already offered areas to play tennis, cricket, basketball, baseball, and football.

The Recreation Facility era was also characterized by a growing tension between park utility and park appearance. This tension no doubt arose from a lack of park funds as the weight of the Great Depression took hold. Playground supervisors were dismissed, and park users were now a mix of children and unemployed vagrants. Between 1930 and 1936, at the height of the Great Depression, mortgage foreclosures around Loring Park were among the highest in the city.[26] Once considered a high-class residential area, wealth now drained from the Loring Park neighborhood. Large single-residence homes were converted into apartments, and the automobile business district on the park's northern edge suffered from a lack of car buyers. As the neighborhood changed, so did its park.

Some of the changes to Loring Park were genuine improvements. During the Depression, cities procured Works Progress Administration (WPA) relief for maintenance and construction, but unlike national and state parks, where new trails and trees appeared, WPA workers in Minneapolis were limited in what they could accomplish. Two things that WPA workers could do were to build swimming and wading pools and construct or repair buildings, both of which characterized Loring Park during the Recreation Facility era.[27]

The Minneapolis Park Board secured WPA funds in the mid-1930s and allocated part of them for improving Loring Park. In 1936 the WPA undertook repairs to the shuffleboard and outdoor basketball courts. In 1937 WPA workers redesigned a roque court (an American version of croquet played on a hard sand court), finished an addition to the Loring Park shelter, and completed the construction of a wading pool. WPA workers, perhaps to keep busy, also laid concrete beneath the swing sets at Loring.[28] These improvements were not made to change or repair the natural setting of Loring, but to improve man-made features that were coming to dominate the Recreation Park era.

The WPA also launched an examination of traffic snarling near Loring Park, especially where Hennepin Avenue met Lyndale Avenue to form a giant bottleneck. Traffic backup was tremendous and numerous accidents occurred as drivers collided with other drivers and occasionally with electric streetcars that also ran through the intersection. WPA workers constructed a model of the Loring Park area and put the model on display in the 1936 Minnesota State Fair. The display foreshadowed future planning work that would address the roadways around the Loring Park neighborhood.

During World War II Loring Park hosted first-aid instruction, pageants, and other patriotic activities and, like many other parks, began to experience the effects of teen delinquency. As fathers went off to war and mothers took jobs in war industry factories, a lack of parental supervision stoked fears about teen delinquents.[29] Teen centers or "canteens" were built and used for organized and chaperoned parties, dances, and social gatherings. During this same time, the elderly competed with teens for the focus of park programming. In addition to teen centers, another dedicated building type emerged: the senior center. At Loring Park, the WPA-expanded shelter accommodated both age groups.

By the late 1950s cities realized that parks by themselves could not eliminate crime, delinquency, and disease. Vandalism and lack of maintenance caused many urban parks to fall into disrepair, giving new meaning to the term "urban wilderness." This, however, did not stop young adults from taking part in concerts, talks, and sit-ins at Loring. As the tumultuous 1960s approached, park users seemed increasingly unconstrained and free to do whatever they wanted.

THE OPEN SPACE PARK, 1965–EARLY 1990s

According to Cranz, the "Open Space Era" developed in the mid-1960s and slowly returned to the idea that parks could help curb urban problems, particularly urban decay and rioting.[30] Unlike previous park eras, Cranz's open space model reflects more of a change in park use rather than changes in physical design. For this reason it is a bit harder to recognize elements of the Open Space era at Loring Park itself. The term *open space* became popular in the 1960s when planners realized that recreation could take place in areas other than parks and schoolyards. Just as streets had served as play areas for children in the late 1800s, streets reentered the park movement in this park era. In the early 1970s the city called for pedestrian access to the downtown area from nearby neighborhoods. Part of the new urban park aesthetic took shape with a pedestrian greenway linking Loring Park to Nicollet Street downtown. Here extensive use of concrete, fountains, and green plantings funneled pedestrians into Loring Park, fostering the park's transition from a sit-and-stay park to a passing-through park.

In the mid-1960s large numbers of people moved from downtown Minneapolis to the suburbs to live on their own one-half acre of nature. The days when nearby Loring Park residents were elites living in stately mansions around the park were long gone. A new generation found Loring to be unsafe, littered, and once the sun started to set, deserted. Loring Park no longer received accolades; instead, the press reported about its potential for violence and disruption.[31] With such negative publicity, people no longer went to the park seeking nature. More often it seemed that park visitors went to the park looking for trouble. In response to the potential for rioting and violence, park landscaping during the Open Space era included low shrubbery (to minimize the nuisance of vandalism), and the classic sculpture of the era: sturdy metal play equipment. Planners now changed the park in response to what they saw as changes in human behavior.

During the unrest of the 1960s, Loring Park's natural setting served as a backdrop for those protesting the Johnson and Nixon administrations. Similarly, today Loring Park has been the site of civil unrest protests. This should not be surprising given that Loring is one of the few public spaces in downtown Minneapolis large enough to hold mass demonstrations. Park policies concerning how Loring Park could be used during the Open Space era now made it possible for diverse groups to interact and express their differences, maybe more so than Horace Cleveland had ever intended.

While large groups were using the park to demonstrate for change, power players outside of the park were making their own dramatic changes to downtown. In the early 1960s urban renewal became the method of choice to revitalize downtown areas, and it was during this time that I-94 freeway construction would forever change the surrounds of Loring Park. The 1958 Housing Renewal Act guided revitalization, resulting in an overall program of redevelopment for the city. The program identified the entire central section of Minneapolis, including Loring Park, as needing either minor rehabilitation, major rehabilitation, or complete redevelopment.

Planners identified the Loring Park neighborhood as one of the most attractive urban areas for residential redevelopment.[32] Private investors wanted to fund small, well-built tower apartments for a more upscale clientele as part of Minneapolis's urban renewal programs. Whereas the Loring area was originally touted as a desirable place to build a house because of its proximity to the park, developers now emphasized the area's proximity to easy freeway access as a selling point.

Even though the Federal-Aid Highway Act of 1968 included a special provision that no federal money could be used to build roads through publicly owned parks, the law did not prohibit building a freeway right *next* to a park. With the completion of I-94 in 1971, what once was an easy pedestrian crossing of Hennepin and Lyndale Avenues was now much more difficult, dangerous, and unpleasant. Nonstop freeway traffic cut off skating rinks in Loring Park from the ball fields in the nearby Parade Grounds. An underpass (Lowry Hill Tunnel) carried traffic below Hennepin and Lyndale Avenues, but transportation planners made little effort to minimize the freeway's visual or audible impact on the park. Nature no longer looked or sounded like it had in previous park eras. Strolling in the park now required visitors to contend with a barrage of freeway noise and a sea of gray concrete (Figure 8.2).

As lengthy construction projects progressed, residents around Loring Park became more assertive in envisioning how nature in Loring Park could fit their needs, which differed substantially from those of residents near parks like Lake Harriet or Lake of the Isles. Beginning in the late 1960s, the voice of the Loring Park neighborhood grew louder, demanding improvements in safety and facilities.[33] In the early 1970s the park commissioners decided to raze the antiquated Loring shelter, declaring that the structure needed too-costly repairs. Nearby residents decried the decision, banded together to protest, and eventually saved the historic shelter from demolition. This action by neighborhood residents eventually led to the formation of Citizens for a Loring Park Community in 1972, which remains active to this day.

In the late 1960s the Minneapolis Park Board changed its name to the Minneapolis Park and Recreation Board (MPRB). Along with the name change, a charter amendment was passed to reduce the number of park commissioners from sixteen members to nine. Prior to this, thirteen commissioner seats were filled by citywide elections and three seats were held by the mayor and two city council members. Beginning in 1969 the MPRB became more accountable to every section of the city, as commissioners were now voted into office by their district constituents. The MPRB continues to consist of nine members, six elected by park district voters, and three members voted in at-large. This is

Figure 8.2. View of Loring Park (left) and I-94 (foreground). Even though the Federal-Aid Highway Act of 1968 included a special provision that no federal money could be used to build roads through publicly owned parks, the law did not prohibit building a freeway right next to one, and transportation planners made little effort to minimize the freeway's visual or audible impact. Following I-94's completion in 1971, strolling in Loring Park required visitors to contend with a barrage of freeway noise and a sea of gray concrete. *Photograph by author.*

unlike most American urban park agencies, where members are appointed by a mayor.

While the previous Recreation Park era saw provision of recreation *facilities*, park planners during the Open Space era emphasized the need to provide recreation *experiences*—for everyone, including minorities and other marginalized groups. Figuring out how to get everyday park users back to the park, however, was difficult. In response to crime and safety concerns, park designers began to articulate principles that were at odds with Cleveland's original rationale for good design. For example, *A Visual Approach to Park Design* (1981) urged landscape architects to design parks where criminal elements could easily be seen and reported.[34] Planners removed walls and shrubbery that blocked the views of onlookers into play spaces, often replacing them with utilitarian chain-link fences. Park designers became familiar with terms such as *defensible space, territorial design*, and *sociopsychological agendas*, while park superintendents planned sporting, theatrical, and musical activities that would draw residents, workers, and poor urban youth to parks to revitalize the city. Part of the revitalization involved keeping downtown Minneapolis alive after 5:00

p.m. and stabilizing deteriorating neighborhoods, including those around Loring Park.

THE SUSTAINABLE PARK, 1990S–PRESENT

In the early 1990s many urban parks were more than one hundred years old. Faced with an aging infrastructure, operating costs for parks skyrocketed. Concerns about landscape management and the emerging idea of sustainable city design led to a new urban park model. These present-day parks, including Loring, typify Cranz's fifth categorization of park design—the ecological or sustainable park. Although there is no one agreed-on definition of a sustainable city, most city planners agree that environmental criteria (e.g., waste management and energy saving plans) and quality of life issues (e.g., amount of green space and public transit) are important factors in urban sustainability. Once again park design is viewed as an important way to achieve sound ecological and human health, but unlike the Pleasure Ground era, when parks were intended to help citizens tolerate city life, the aim of sustainable parks is to enhance the overall sustainability of cities. To do this planners must "reintegrate" the landscape so that its ecological functions approach that of its original "natural" functions.

Along with social and psychological services, nature in Cranz's fifth model is seemingly rebranded and "put to work" to provide environmental services such as air and water purification, wind breaks, noise filtering, and summer cooling.[35] More recently, the MPRB's 2007–2020 Comprehensive Plan embraced an ecosystems approach to make urban parks more appealing and useful.[36] Although Horace Cleveland originally designed Loring Park with nature in mind, he focused on aesthetics rather than improving the park's ecological functions. Many current landscape ecologists, restoration ecologists, and landscape architects have adapted their urban park design principles to emphasize ecosystems and ecosystem services by using native plants, restoring natural habitats, installing on-site stormwater retention basins, and integrating appropriate green technologies.[37] Moreover, sustainable approaches provide economic advantages. Significant quantities of fertilizers, herbicides, pesticides, and labor are needed to keep flowers blooming and lawns green. Highly controlled nature is expensive to maintain.

Loring Park is now transitioning to a sustainable park, aided by city and state programs. For example, natural springs originally fed Loring Pond until the installation of city sewer lines in 1915 blocked the springs. The resulting stagnant water could not naturally cleanse itself. In 2012 the Minneapolis Park and Recreation Board decided to remove nonnative plants around the pond and replace them with native iris and other species to form a one-thousand-square-foot "rain garden" or stormwater collection site. This action, combined with sealing the lake bottom to prevent seepage and an oxygen aerator installed in the pond to prevent oxygen depletion, has led to some mitigation of the low-oxygen conditions. Returning nature to its former state, however, requires constant oversight. Reintroducing native plants requires the removal of other successful plant species and *keeping* them removed for native vegetation to gain a foothold. At Loring and other Minneapolis parks, narrow-leaved cattails have again moved back into riparian areas. With its main competitors also removed, the cattails have proven particularly resilient.

The ideas behind sustainable park management are not just based on ecological principles, but also draw on the principle of cultural sustainability, a concept developed by landscape architect Joan Nassauer, who believes that landscapes and activities that attract people are more likely to survive than landscapes that no one visits or cares about.[38] With this in mind, many decisions about Loring Park involve community-based stewardship. The Loring Park neighborhood has been instrumental in planting more trees, designing a central garden, building bike paths, fencing off parkland for a dog park (a park within a park), and renovating the Loring shelter. Such neighborhood input would have been unheard of during the Pleasure Ground era, where only outside experts were called on by the Park Board for recommendations.

While Loring Park is transitioning to a sustainable park, an important remnant of how citizen groups used the park during the Open Space era, remains in the form of mass protests. Recently the park has again returned to a political space much like that of the 1960s. Loring Park has lately served as a site for Occupy Wall Street protests, the Black Lives Matter movement, and tent encampments as COVID-19 forced many low-income and unemployed residents to form a tent community, before they were evicted by the city.

Loring Park, although small and sometimes ignored as Minneapolis's original grand jewel park, uniquely reflects Galen Cranz's descriptions of park history and design in the United States. Originally designed as a pleasure ground park in the late 1800s, Loring simulated nature with ponds, trees, and curving walkways. The nationwide movement to make parks more practical during the Progressive Era saw playground equipment and a park shelter built at Loring. The shelter symbolized a built form in nature where children would be safe, organized, and formally trained. Growing out of the Great Depression, parks soon became less responsible for organized play and character building but remained areas where citizens could still exercise and wander on their own. During the latter half of the twentieth century, parks became part of the Open Space movement where recreation took place in parks and in less "natural" areas around them. This is evidenced at Loring Park with the building of the pedestrian-only Loring Greenway to connect parts of downtown Minneapolis with the park and the Irene Hixon Whitney Bridge, opened in 1988, to connect Loring Park with the Minneapolis Sculpture Garden. Finally, Loring Park is transforming into Cranz's definition of a sustainable park, distinguished by native vegetation plantings, water quality restoration projects, an off-leash dog area, and a renovated shelter for education and community meetings.

The five park eras that have come to characterize Loring Park resulted from relatively homogeneous reactions to urban problems across the United States. Whether Loring was designed to assist with the goals of public health, to offer an economic incentive to the city, to provide a social coherence for urbanites, or to be more ecologically useful, this small park has shaped and reordered the importance of nature in a variety of ways.

In sum, the history of Loring Park shows us that not all residents thought the same about nature; some wanted Loring Park quiet and "natural" while others wanted to engage in play. Others wanted to use the park as a site for protest, and still others wanted to remove natural elements to make Loring appear safer. From the Pleasure Ground era with its park bureaucrats, all of whom were industrious businessmen who served as a sounding board for prominent landscape architects, to the current Sustainable Park era, where Loring Park has seen its landscape divvied up for a multitude of old and new uses, nature at Loring continues as part of a shifting cultural landscape that is engineered, managed, and repeatedly re-created.

"AWHEEL FROM CHICAGO TO THE TWIN CITIES"

Legacies of Turn-of-the-Century Bicycle Paths in Minneapolis and Saint Paul

James Longhurst

IN 1902, BEFORE THE PAVING OF MOST URBAN STREETS or rural roads in America and before the dawn of the automobile, Saint Paul residents had already built a network of hard-surfaced bicycle trails. The city was "the proud possessor of about 115 miles of excellent cycle paths," boasted the *Saint Paul Globe*, and cyclists could "ride to almost any place in the corporate limits without having to wheel through a sea of mud." Riders could follow "an interurban path between St. Paul and Minneapolis . . . on the Marshall Avenue route," or take the scenic route between the cities along West Seventh Street. The paths also connected Saint Paul to recreational sites and hinterlands outside the city: "You can go to almost any corner and lake in Ramsey County without having to be afraid of the muddy roads that used to greet the cyclist a few years ago," the journalist observed. For their part Minneapolis cyclists were proud of their "seventy-five miles of bicycle paths in the city and perhaps as much more outside the limits and connecting with city paths." Using this separate, bicycle-specific network rather than unpaved roads, cyclists could ride into Hennepin County, to Lake Minnetonka and around Lake Harriet, along both banks of the Mississippi, and to Saint Paul by a variety of routes.[1]

In a short burst of excitement lasting less than a decade around the turn of the century, the people of the Twin Cities imagined a landscape connected by bicycles and paths built expressly for bicycles. They actually built some of this network before the automobile's surging popularity changed the nation's direction and instead demanded cities and road networks designed for the car. In this same era Twin Cities cyclists participated in fantastical plans for a network of interstate bicycle paths heading south to Saint Louis and east to New York, beginning with a link to the nearest neighboring metropolis: "There is probably no more favorable part of the country in which to build a long-distance bicycle path than between Minneapolis and St. Paul and Chicago," wrote one hopeful observer.[2]

More than a century later, the Twin Cities are once again nationally known for their paths and on-street bicycle facilities. The editors of *Bicycling* magazine pronounced it the premier riding spot in the nation in 2010: "It's a cold, hard fact," wrote journalist Steve Friedman. "The unforgiving and frigid city of Minneapolis is the country's top spot to be an urban cyclist." The modern network of paths is one factor: "Between them, Minneapolis and St. Paul have 84 miles of dedicated bike paths and 44 miles of designated bike lanes on streets." The Twin Cities have regularly ranked in *Bicycling*'s top-ten in the decade since, including number 4 in 2018. International recognition has followed, and the Copenhagenize urban design firm recognized Minneapolis as a top-twenty global bike city in 2015. It was only the third US city to make the biannual list; there has been none since. In 2021 *National Geographic* declared that "if you love to bike, all roads lead to Minneapolis," since it possesses "a chain of 250 miles of on- and off-street bikeways that help make Minnesota's largest city one of the top spots for urban cycling." In the twenty-first century the Twin Cities are exceptional, at least in comparison to the auto-dependent sprawl of most American cities.[3]

It would be tempting to believe that these two points in time are directly linked—that the "115 miles of excellent cycle paths" of 1902 and the "250 miles of bikeways" in 2021 are one and the same, a continuously existing piece of transportation infrastructure. But the reality is far more complex. The turn-of-the-century network of paths largely disappeared in America, subsumed under the widened pavements of the Good Roads movement and then forgotten in the decline of the cycling fad of the 1890s. The new twentieth-century bicycle infrastructure had to be built almost from scratch.

This disjuncture raises significant questions for historians: Why were these early twentieth-century bicycle paths built? Why were they abandoned? And why are their legacies slightly more persistent in the Twin Cities than in the rest of the nation?

Partial answers to these questions come from placing the Twin Cities in context with a largely forgotten, nationwide movement from the golden age of cycling. Beginning in the middle of the 1890s bicycle boom, groups of cyclists in cities across the United States attempted to build bicycle-specific paths—or "sidepaths"—alongside the unpaved roads of the era. The legal and political model for this movement began in western New York in the late 1890s but moved to the Midwest and elsewhere by the turn of the century. The larger movement fell apart by about 1910 due to weaknesses in the funding mechanism, a reality exacerbated by fading interest in cycling and eclipsed by the success of the Good Roads movement. While cycle paths in the Twin Cities went through the same boom-and-bust cycle, the exceptional success of Minnesota path building left infrastructural legacies for the region.

Historians in general are uninterested in cycling as a topic, but even historians of the bicycle have overlooked sidepaths. As such it is not at all surprising that current residents of the Twin Cities have also missed the existence and impact of this nationwide movement. The city's 2010 bicycle master plan acknowledges a long history of cycling, but never uses the word sidepath, and misidentifies sidepath tags as "bicycle registration" while incorrectly claiming that licenses were required for all cyclists. Similarly, the cycle path tags in the Minnesota Historical Society are incorrectly labeled as "bicycle licenses." These distinctions are important: "license" and "tag" imply completely different systems of funding that were hotly contested in their day. So while Minnesotans today seem to know that paths existed more than a century ago, they are missing the national context of sidepath history and its unique legal and fiscal underpinnings. This context provides some clues as to why the paths were built and subsequently abandoned.[4]

But why is the Twin Cities experience different today? Unlike the rest of the nation, the Minneapolis–Saint Paul region did not entirely dismantle its early experiments in path building. Instead, cycling paths provided an early template for parkways and roads that interconnect the cities and their parks. Before the Twin Cities and their parks were linked by the "Grand Round" of paved roads and parkways for automobiles, they were connected by a complex network of cycle paths. While other cities paved over their separated paths and made them part of a combined traffic road network, some paths in Twin Cities parks and along the riverbanks were maintained throughout the twentieth century, providing a foundation on which the more recent bike boom has been built. This legacy left the region more interconnected by parkways and paths than most others and set it on a different path through the twentieth century.

CYCLE PATHS IN THE TWIN CITIES, VIA NEW YORK

The craze for building bicycle-specific paths in the Twin Cities was a part of the golden age of American cycling in the late nineteenth century. While the boom had multiple causes, the introduction of the "safety" bike was the centerpiece of the bicycle fad of the 1890s. Named for its most obvious advantage over the precarious high-wheels that preceded it, the increasingly affordable safety bicycle also opened cycling to both men and women.[5] While much riding was for leisure and recreation, in a world before cars, the bicycle briefly seemed to be the future of individual transportation.

Road conditions significantly limited bicyclists in the 1890s. Rural American roads in the nineteenth century were almost all unimproved; their unpaved surfaces lacked adequate drainage, were built without surveying or engineering, and were nearly impassable after rain or snow. The parlous state of nineteenth-century roads was a symptom of an underlying institutional cause: an essentially local, decentralized system of financing for road materials and labor that avoided general taxation for road construction.[6]

Responding to these limitations, bicycle riders in Minnesota and throughout the nation clubbed together to build cycle paths on their own. Outside of New York State, the Twin Cities was home to the most successful of these projects, built mostly for recreational purposes and largely dependent on voluntary donations, club membership, or "subscriptions." The earliest of these featured the Minneapolis city engineer overseeing preliminary work "late in the season" of 1895, producing nearly six miles of paths, mostly along Lake Street and Minnehaha Avenue. These paths were financed through private funds supporting city workers; while $4,000 was pledged from the city, the engineer reported that "it was finally found impossible to spare the amount for next year, and the riders will have to depend on further subscriptions." Construction was continued in 1896 with "a first-class gravel path" connecting the two cities along Marshall Avenue, as well as much additional construction along Humboldt, Fifth, Sixth, Tenth, and Eighteenth Avenues; and Lake, Fifth, and Fiftieth Streets. City residents also began a path connecting their homes to Lake Minnetonka, nearly twenty miles to the west of the city.[7]

While there were some unspecified contributions from individual wards, private citizens funded most of the work: "Several efforts have been made by the authorities to set aside a special fund," reported the city engineer, "but the public demand for lower taxation has invariably defeated the proposition." As such, the 1896 city ordinance allowed "such private individuals . . . to construct such paths or roads without expense to the City," and while those private individuals were allowed to use city machinery, the completed paths "shall immediately become the property of the City and such roads shall be under the full control of the City the same as any other portion of the public street," according to ordinance.[8]

Beginning in western New York in 1896, cyclists began experimenting with countywide attempts to raise funds through taxation or mandatory subscription to build entire networks of bicycle-specific paths. Members of the Niagara

County Sidepath League originally built short paths voluntarily funded by club dues for all cyclists to use freely, but eventually found this approach "a slow and unsatisfactory method." In 1896 they drafted a state law permitting Niagara County supervisors to tax all bicycle-owning county residents and build paths with the proceeds.[9] While taxation worked in Niagara County, it proved incredibly controversial when introduced in neighboring Monroe County, and so the statewide version of the law allowed any county to create a quasi-state commission and charge a user fee for those who chose to ride on the path, rather than taxing all cyclists or all property owners. Under New York's General Sidepath Act of 1899, commissions were empowered to charge at least 50 cents for "a license, badge, emblem, or device suitable to be affixed to a bicycle" to fund path construction and upkeep. Path riders without tags could be fined, or even arrested and imprisoned, by local police.[10]

This 1899 law authorized New York's sidepath commissioners "to construct and maintain sidepaths along any public road, or street" with the approval of elected officials. These paths were to be not "less than three feet or more than six feet wide . . . [and] constructed within the outside lines and along and upon either side of such public roads and streets." These new paths fit in the liminal space between unpaved roads and before the addition of concrete curbs or sidewalks built through city assessments. While the sidepaths were to be built within established legal rights of way, they were segregated from both the adjoining road and from existing sidewalks. Later court decisions made it clear that wagons, carts, and horses could not use them. They were to be a separate network, set apart from foot and vehicle traffic, reserved solely for bicycles.[11]

New York's user-fee sidepath commissions were widely adopted across the nation, with varying levels of success. Twenty New York counties, mostly upstate, built sidepaths under the state law. Monroe and Niagara Counties were exceptionally successful, with Monroe completing 150 miles of paths by 1900. Maryland, Ohio, Rhode Island, Connecticut, and Florida all passed laws on the sidepath model in 1900 and 1901. Massachusetts, New Hampshire, Vermont, and Wisconsin all debated New York–style sidepath legislation but did not immediately pass their own laws.[12]

While Minnesota cyclists knew about the New York sidepath experiment as early as 1897, their own interminable debate over paying for paths in the Twin Cities was just getting started. Were sidepaths privately funded resources for public use, publicly owned utilities to be funded by all, or some quasi-state combination limited to dues payers? An unusually opinionated Minneapolis city engineer theorized in 1897 that paths should be financed by general taxation: "While . . . the custom is to build cycle paths by subscription in different cities, this appears to me a hardship since the bicycle is a vehicle subject to personal tax equal to that of buggies, and is a maker of good roads and as there are at least 25,000 of them in daily use during the season in Minneapolis, it occurs to me that this should be borne by the people as a whole." This proposal was an outlier in the debate, particularly since Saint Paul was already experiencing success with the interim solution of blended public-private funding. Nearly $4,000 was raised in 1897 and 1898, evenly split between private donations raised by the Saint Paul Cycle Path Association and contributions from the city of Saint Paul and Ramsey County. The national League of American Wheelmen (LAW) reported approvingly that the city was off to an impressive start, with fourteen miles of paths built in 1897 out of this public-private partnership.[13]

But New York's success with sidepath legislation inspired more ambitious plans. On March 11, 1900, the *Saint Paul Globe* predicted that Minnesota legislators would soon take up "a law taxing the bicycles in every county," in order to build paths "under the direction of a side path commission similar to the New York commission that has been in existence several years." The perceived problem was that the existing system had a number of what political scientists would today call "free riders," inspiring the journalist to worry that "the cost of side paths is not equally shared. . . . A small number do all the work. Last year 2,300 St. Paul wheelmen contributed . . . while 7,500 received the same benefits without lending any assistance whatever." Therefore new state legislation and city ordinances sought a more "equitable distribution of the cost." Another article on March 25 echoed national bicycle advocates' hopes that "as many states as possible will be persuaded to follow the example of New York . . . making the construction of such paths obligatory upon the county boards of supervisors of the public roads, and imposing a side-path license tax upon the rider."[14]

Thus in 1900 the two cycle path associations of Saint Paul and Minneapolis—volunteer, independent civic organizations—attempted to arrange their own replacement by a quasi-state sidepath system. This led to controversy as the associations got ahead of supporting legislation: advocates announced plans in early 1900 to create a reciprocal system that would mandate the purchase of cycle path tags to ride on paths in both cities, issuing tags without the backing of law. Visiting his counterparts in Saint Paul in March, the secretary of the Minneapolis Cycle Path Association announced that henceforth, all "cyclers using the paths must pay a license of 50 cents." That plan quickly fell apart when Minneapolis mayor James Gray vetoed the city council ordinance intended to codify the fee. Expressing approval, the *Saint Paul Globe* described the ordinance as "radically obnoxious" and objected to the aspects of the plan that smacked of "class legislation": "The public thoroughfares are public property, and their use should be free and unrestricted," they opined; "to exclude [cyclists] from the use of the cycle paths because they have not paid a fee is clearly unjust discrimination and against all true public policy."[15]

By the summer of 1900 the Minneapolis City Council overcame the mayor-

Figure 9.1. In the late 1890s and early 1900s, before the automobile's surging popularity changed the nation's relationships with city streets, bicycle sidepath tags like this one from 1899 provoked fierce debate in the Twin Cities. At stake were important questions about the legal and fiscal underpinnings of bicycle sidepaths, which provided an early template for the parkways and roads that would later interconnect the cities and their parks. Courtesy of the Minnesota Historical Society.

al veto and passed a city ordinance requiring tags for cycle path riders, though without the explicit reciprocity. There was a "tacit understanding that St. Paul wheelmen . . . will not be molested, but the Minneapolis police do not all understand it," complained the *Globe* in May. The council threw in an additional $7,000 for construction in June, reportedly "a direct result of the new ordinances which requires every wheelman who uses the path to pay a fee." At the same time, Saint Paul continued to sell tags as voluntary accoutrements for cyclists but approached the county commissioners for $5,000 for general construction and to improve the leisure path to White Bear Lake.[16]

For all involved it was unclear how much of the cost of cycling amenities should be borne by society and how much by the individual. "The cycle tag in St. Paul has represented little more than an individual expression of appreciation," said the *Globe*. "The problem is how to make it more than that without invoking compulsory public agencies." Thus tags were for some time a voluntarily purchased symbol of charity issued by a quasi-public association—sort of like an NPR tote bag today—and attempts to shame riders into buying tags appeared in local newspapers. Back in 1899 the secretary of the Minneapolis Cycle Path Association had made "an urgent appeal for everyone who rides a bicycle to reflect on the good the paths are to him, and immediately satisfy his conscience" by joining. The *Globe* did "not think that any self-respecting lady or gentleman ought to be without his cycle tag in this town," and repeatedly attempted to identify and shame non-purchasers through the next year.[17]

As the bicycle boom continued, the hortatory solution reached its limits. Saint Paul's city engineer reported that the association was tiring of constantly stumping for donations in 1901, and that it "wishes to have the entire work in the future taken care of by the city, as the balance of roadways and sidewalks are." Cyclists felt that "if the city can provide sufficient funds from the general tax levy to keep the path construction up to the needs of the public that will be the most satisfactory condition." Minneapolis's Reverend Isaac Houlgate was reduced to a bit of linguistic legerdemain, calling a license a tax in order to advocate for general taxation: "Theoretically . . . the tax is not imposed because you ride over the road or path; but rather because you *demand* a good road or path to ride on; so it is only fair that you should pay your part." While cycle riders might have wished for paths built by existing taxes, broad support for such a policy did not exist. But there was still the possibility that the public might be brought around to support the user-fee license system.[18]

THE TWIN CITIES AND AN INTERSTATE BICYCLE NETWORK

In their efforts to rally support for the sidepath tag, some boosters celebrated plans for an interstate bicycle network to connect the Twin Cities with the rest of the nation. This visionary transportation network would be "a complete system of side paths from Maine to Texas, and from the Atlantic to the Pacific." Cyclists supported the LAW's 1900 plan, aiming to build that continent-spanning network of paths to allow cyclists to "go from New York to any point in Maine, Florida or California on smooth roads made especially for them."[19]

The first part of this imagined coast-to-coast network was to be a central spine of surfaced paths stretching from Minneapolis to New York City, and connecting Chicago, Gary, Buffalo, Rochester, and Albany along the way. Supporters referred to these as "the 'trunk lines,' i.e., the paths intended to connect the centers of population, even at great distances." But additional linkages were to split off from these central lines, including cutoffs to Omaha. In February 1900 one Saint Paul journalist predicted the ability to travel "awheel from Chicago to the Twin Cities" in soaring rhetoric, depicting the connection of existing local paths as "a consummation of perhaps the greatest project ever undertaken by wheelmen in the Northwest, and anywhere in the United States, for that matter."[20]

The plan to connect the Twin Cities paths with others ran into logistical complications, however. The first problem was that while the strategy depended on all counties and municipalities along the route to create sidepath commissions as successful as those of central New York and Minnesota, only a few managed to do so. The second problem was Wisconsin, which didn't yet have a sidepath law, and whose southern Driftless region was surprisingly hilly. The original plan featured a poorly thought-out Mississippi River crossing, requiring cyclists to descend steep bluffs immediately before climbing again on the other side.

The plan started to skid out of control after the defeat of Wisconsin's statewide sidepath bill in 1900. Without that law, it was not safe to count on individual Wisconsin counties to build their own parts of the interstate path. But the plan did not die entirely: Minnesota cyclists pitched a revised connection to Chicago in 1901. By angling southward rather than directly east from the Twin Cities, the interstate route could avoid the difficulties of climbing the Mississippi bluffs while simultaneously skirting pathless Wisconsin. Still, Minnesota cyclists persisted in the dream of a bicycle path connection to the east, arguing that "it is safe to assume that if the route were a satisfactory one and the path good, thousands of wheelmen would pass over it every year." In the height of their dreams, it seemed as if they were laying the foundations for the paved interstate highways to come: "It is believed that a large part of the path could be laid out with a view to future widening for the common use of automobiles and bicycles."[21] This most ambitious dream of the path builders depended on organizing the local projects into dependable creators of cycle paths, which would require a Minnesota state law enabling sidepath commissions.

SIDEPATH COMMISSIONS COME TO MINNESOTA

The 1900 explosion of Minnesota path building—and the difficulties of a shared system of funding—resulted in a 1901 push for statewide sidepath legislation based on New York's, judged by the *Globe* as being "so nearly perfect as to form a model for all other states." In January cyclists were lobbying "with several legislators known to be in sympathy." By February state representative and Saint Paul resident Rukard Hurd submitted legislation "to provide for the appointment of side-path commissioners, to define their powers, to provide for the construction, maintenance and preservation, and to regulate the use of bicycle side-paths and for licensing bicycles."[22]

In Hurd's original draft, all counties were granted the ability to create sidepath commissions, but "in order to secure favorable consideration from the house," that power was later limited to the three counties with populations greater than fifty thousand—Ramsey, Hennepin, and Saint Louis. The bill also excluded any municipality with a preexisting sidepath ordinance, thus applying to Hennepin County but not to Minneapolis itself. Like New York, Minnesota created a legal foundation to fine those who failed to buy an annual bicycle tag but continued to ride on sidepaths. Broadly supported in the house, the bill unanimously passed the Senate in March.[23]

Saint Paul cyclists began petitioning for the sidepath commissioners to be named in April 1901. The *Globe* felt that the transition from a strictly charitable funding stream was an improvement that addressed the free-rider problem: "Considerable money has been raised [before the state law], but it was a noticeable fact that the tags were only bought by the enthusiastic wheelmen, and that a great many wheelmen did not think [it] necessary to pay a dollar for a tag," a writer pointed out. By the next year the Saint Paul Cycle Path Association had declared itself superfluous, and transferred its funds as well as its officers to the new county sidepath commission.[24]

By contrast, Minneapolis seemed pleased with its previously established ad-hoc alliance: "City, county, and park board are to unite in an effort to make it a season of extraordinary expansion," predicted journalists optimistically in 1901. There was loose coordination under a "Cycle Path Committee" of the City Council, with "Alderman C. B. Holmes, himself an ardent wheelman [as] chairman," reported the *Journal*. The positive view of this collaborative system eroded support for a county sidepath commission. The *Journal* noted that the existing system was already under cyclists' influence: "The committee will consult frequently with the officers of the Wheelmen's association and be controlled largely by their counsel." With an anticipated $7,000 from the county commissioners, a "mite" from the Minneapolis Park Board, and perhaps an additional $15,000 from tag sales, cyclists had ambitious plans: "Instead of two routes into the country, there will be not less than four, besides a notable development in the business paths."[25]

The following year was the last great season for sidepaths in the Twin Cities. Saint Paul's success was celebrated locally and nationally, with an incredible 115 miles of paths built for the use of a reported twelve thousand cyclists, supported with club dues and government subsidies. Saint Paul riders could use the West Seventh Street path, "the route usually taken by cyclists riding to Fort Snelling," as a recreational connection to Minneapolis. Years of city directory maps of Saint Paul, culminating in 1902, show a complex network of bicycle paths throughout the city and into the countryside. Cycle path maps of Minneapolis show a remarkable radial network of more than 50 miles of paths extending from downtown, and at least another 75 into the surrounding county, connecting city dwellers with parks, lakes, and neighboring Saint Paul. At the turn of the century, bicycle paths connected the people of the Twin Cities to each other and to the rest of Minnesota.[26]

Figure 9.2. At the turn of the century, bicycle sidepaths connected the people of the Twin Cities to each other and to the rest of Minnesota. This 1902 map showed a remarkable radial network in Minneapolis of more than fifty miles of paths extending from downtown, and at least another seventy-five into the surrounding county, connecting city dwellers with parks, lakes, and neighboring Saint Paul. Courtesy of the Minnesota Historical Society.

DISCONTENT AND DECLINE

Even as path building peaked, weaknesses were becoming apparent, ranging from internal divisions to waning public support. While officially cooperat-
ing with each other, the Twin Cities were unofficially in a cycle path cold war:
"Minneapolis, as usual, is slow in repairing its end of the [University Avenue]
path," complained the Saint Paul newspaper in 1901. "If Minneapolis was as
prompt as St. Paul in repairing the [Minnetonka] cycle paths, the Twin City
riders would appreciate it. Every season St. Paul is ahead of Minneapolis in this
respect." Disagreements also divided Minneapolis along class lines. The *Journal*
reproduced a letter from an East Side resident who complained that "no one
seems to give a thought to the needs of hundreds of workmen" of that part
of the city, and requested that a route on Main Street be given consideration
"some time when the path engineers are all through building their pleasure
drives." Complaints about the East Side were common, resulting in circular
accusations that East Siders did not deserve paths since they did not buy cycle
path tags, and also that East Siders would not buy tags because they did not get
their fair share of path construction.[27]

The much sought-after sidepath commission seemed to be in danger in
Hennepin County, as the *Journal* began running a series of articles predicting doom if the distributed path-building system was consolidated in a county
commission. Late in 1901 the county board threatened that "if the new commission does finally secure control, there will be a long and strong string attached":
the county did not want to give up its newly built path on the Excelsior Road.
A week later the paper ran a story worrying about legal liability, predicting that
"the change in the management of cycle path affairs will prove a mistake." The
concerns resulted in cancellation of the plan to hand over control to a sidepath
commission in Hennepin County; the previous distributed arrangement would
continue.[28]

Overshadowing these petty disagreements, by 1903 even *Globe* journalists were realizing that the bicycle boom was fading, and that the Good Roads
movement was replacing the push for sidepaths: "There is no getting around
the fact that wheeling has dropped off considerably in the last two years and
there are some that make the claim that it will be almost entirely abandoned
within the next five. . . . It is hardly likely . . . that any more mileage will be added
to the present lists of paths. Improved streets and roads are becoming the order
of the day and the paths are not really needed." Minneapolis likewise reported
declining interest: by May 1904 city wrecking crews were pulling out paths
on Humboldt and Cedar Lake Roads and Penn Avenue. "The property owners
want trees in front of their land, and, as no one seems to care to have the path
maintained, it will disappear," reported the *Journal*. "In many places the paths

are made unnecessary by new paving, while others are used by little." The original private funding of the paths was used as a rationale for their removal: "The city spent over $50,000 on the cycle paths," the *Journal* reassured its readers, "but as most of the money came directly from the wheelmen thru the sale of bicycle tags, the city is not out."[29]

The wheels were starting to come off the whole enterprise. Plans in 1904 to boost the cost of Saint Paul's tags to $1 to pay for maintenance and expansion faced strong resistance and fractured overall support for paths, resulting in a flood of negative coverage. Opposition to the plan "has been so pronounced that the five members of the cycle path commission have almost concluded to resign as a body and leave the paths at the mercy of the general public," wrote the *Globe*. Some upstart hobbyists were even volunteering to take over the paths for their so-called automobiles.[30]

Even though opponents succeeded in returning the price to 50 cents, by late June the unthinkable happened: Saint Paul cyclists stopped buying sidepath tags. "The sidepath commission is discouraged and is ready to let the paths go by default, and the members say it will be nobody's fault but the wheelmen themselves," reported the *Globe*. "There is no more money available and the commission is becoming disgusted." By the beginning of 1905, even as Ramsey County path construction was consolidated under a revised ordinance, support was waning. "The tags are going very slow," the *Globe* reported. "It is claimed that the method of using the money . . . last year" triggered disaffection.[31]

The collaborative partnership and the paths themselves deteriorated from that point on. It had never been a seamless relationship: "There has been no branch of the work of this department that has caused us so much bother and called forth so much criticism as the cycle-path work," reported longtime city engineer Andrew Rinker in 1906. "The wheel people are the most unreasonable in their demands and the most unjust in their criticisms. It would be a great relief . . . if the council should take the cycle-path work out of our hands." While the city spent thousands of dollars annually repairing paths, they were damaged every winter. Additionally, the volunteer nature of the sidepath system left it an orphan project, according to Rinker: "There is nothing we can do at present, as no funds are available . . . until a considerable number of bicycle tags are sold." By 1906 the assistant engineer thought repair unlikely "as there are no funds available for the purpose. The paths should either be put in shape for use or abandoned and removed." The public chose the latter.[32]

THE PERSISTENT LEGACIES OF TWIN CITIES PATHS

While poor nineteenth-century road conditions and the bicycle boom explain the creation of sidepaths, and the failures of limited funding explain their decline, the diversified Twin Cities commitment to path building at the turn of the century helps to explain the unique success and lasting impact of sidepaths in the region. The failure to unify control in Minneapolis contributed to some physical persistence: the Minneapolis Park Board's bicycle paths did not fall to city wrecking crews, and distributed responsibility between the city clerk, city engineers, and county institutions kept some path work limping along with available funds until around 1910. County paths built along railroad rights-of-way and recreational locations were left unmolested: "These paths were not built in a day and ought not to be torn out unless their usefulness has entirely passed," wrote the *Journal* in 1905, and the city continued collecting 50 cents for licenses until at least 1909.[33] On the other hand, in Ramsey County as in upstate New York, the concentration of responsibility in the hands of the volunteer sidepath commissions meant that when the commissions fell apart there was no government agency with any responsibility to continue their work, and the paths were simply paved over for expanded roadways in the next several decades of Good Roads engineering.

Minneapolis's failure to create a Hennepin County sidepath commission actually seemed to foster a more successfully decentralized network for the maintenance of paths. This sense of shared responsibility had been a part of city path building for some time. Reflecting on the ad hoc arrangement in a *Journal* article, Minneapolis city engineer G. W. Sublette had predicted in 1902 that "the council could very gracefully appropriate from time to time . . . such small amounts as might be necessary." According to Sublette, the city "should at least be as generous as the park board, as long as the wheelmen have shown a commendable disposition to help themselves."[34]

While conflicts among the Minneapolis city engineer, the city council's sidepath committee, the county board, the volunteer sidepath association, and the Board of Park Commissioners continued to make news, the fact that all these entities had some responsibility for paths seemed to make it possible for some paths—particularly in parks and outside of city limits—to persist locally even as the sidepath movement faded nationally. The essential difference was that while other locales had only a quasi-state sidepath commission that vanished when volunteer interest waned, the Twin Cities continued to distribute responsibility for paths among many overlapping institutions. For example, the City Council of Minneapolis had its own Committee on Bicycle Paths with political appointees who heard complaints and directed the city engineer's actions. But the city clerk also collected tag fees on behalf of a nongovernmental Cycle Path Association, which then funded paths constructed by city employees. For its part Saint Paul kept its quasi-state Ramsey County Sidepath Commission; unlike other locales, it funneled city funds for path maintenance through this commission, keeping it afloat even as volunteer subscriptions declined.[35]

This distributed responsibility included the Minneapolis city engineer taking over much path building, supported by both private donations and public funds. This was unlike Saint Paul and most of upstate New York, where county sidepath commissions oversaw construction while they existed. Thirteen years after New York's sidepath commissions disappeared, the Minneapolis city engineer continued to repair existing paths. Combining construction and repair into a single budget, the engineer estimated over $67,000 had been expended on bicycle paths by the city as of 1909.[36]

Even though little money was spent on path building by 1910, it did not mean that bicyclists themselves had disappeared; they were merely riding on newly paved roads rather than on separate paths. An engineer's survey at the corner of Nicollet Avenue and Second Street counted more than a thousand cyclists passing on individual days in June, July, and October of that year. Unlike many of his counterparts, the Minneapolis engineer still appeared to include accommodating bicycle travel among his responsibilities.[37]

Likewise, the Twin Cities' expansive commitment to public parks involved a large and powerful Board of Park Commissioners that ended up preserving many turn-of-the-century bicycle paths. Historian David C. Smith has noted that Minneapolis possesses "an independent public entity that retains dedicated tax-supported funding," making it a rare institution with the single responsibility of "owning and managing parks." Partly because of this, "Minneapolis is unique in the number of bodies of water within its borders and the fact that the public owns them." As early as 1896 the parks board built an eight-foot-wide gravel path exclusively for bicycles around Lake Harriet, "really a part of the boulevard system of the park board," as one observer noted.[38]

Commitment to lake parks and the inclusion of paths connecting them made maintenance a project of this independent entity. The park commissioners in 1900 described the parkways as "the unique feature of our system formed by carrying roads and bicycle paths around our intra-urban lakelets." Even paths that the city or private associations built would occasionally become a parks board responsibility: for example, the board took over construction of the bicycle path on the parkway to Minnehaha Falls in 1902. Once a part of a parkway or lake park, the paths were likely to be maintained even as cycle paths along roads were paved over. They might be briefly transformed along the way: the bicycle paths built around Lake Harriet after 1896 were made into a bridle path after 1914, and thus persisted into the late twentieth century, when they again became bicycle paths. The same thing happened to the Minnehaha Creek path. Built alongside the unpaved parkways, they were preserved as horse trails in the 1910s "after interest in bicycling declined," and were then rebuilt at the end of the twentieth century alongside the now-paved parkways. Other agencies occasionally contributed; the Works Progress Administration refurbished paths between the parkway and the west bank of the Mississippi in the Great

Depression, extending path life span enough for them to be refurbished for bicycle use in the late twentieth century.[39]

One additional factor may have contributed to the infrastructural persistence. Perhaps more so than in other cities, Minneapolis and Saint Paul committed themselves to an ideal of public ownership that made their bicycle path infrastructure partly the responsibility of government rather than solely a voluntary expression of charity. Compared to western New York, where uneasiness with public responsibility for path funding led to the creation of a quasi-state sidepath solution, some Twin Cities residents seemed to be more receptive to public ownership. Minneapolis engineer G. W. Sublette and Reverend Isaac Houlgate had both advocated for paying for paths out of general taxes—a remarkable argument for the time—and actions taken by the parks board, the city engineer, and the city clerk seemed to indicate a broad commitment to public responsibility. These expressions came within the context of campaigns against government corruption and perceived dishonesty stemming from the private ownership of public utilities. In many midwestern cities, Progressive Era reformers built political alliances to take public control of utilities, shift control of government agencies to engineers and experts, and increase the power of municipal governments.[40]

COUNTERFACTUALS

While the sidepath movement was a brief craze, forgotten by the nation with the coming of the automobile in the first decades of the new century, it still shaped the future of the Twin Cities. In 1974, when the city of Saint Paul completed a feasibility study for bike lanes on city streets, the chosen routes all followed sidepaths that had been built seventy years before, and that had subsequently been paved over. The author of the 1974 study was somewhat aware of the old sidepaths, noting that the 1969 funding for a recreational path around Lake Phelan "was the first such funding of any sort of bikeways in the City of St. Paul since the turn of the century."[41] But while it seemed that bicyclists in the 1970s were starting from scratch, the late twentieth-century bicycle paths of the Twin Cities actually had impressive head starts: existing paths in parks scattered across the urban landscape, around lakes, and along the parkways that connected parks and followed the river. Even these small scraps gave the Twin Cities an advantage in creating alternative transportation networks in auto-obsessed America. This prompts a counterfactual question: What would have happened if more of the nation's sidepath network had survived through the twentieth century and into the present? It seems likely that many more cities could have shared Minneapolis's present-day status as a top bicycling city, and more Americans might have a choice of how they move through their daily lives, possibly not limited solely to personal automobiles.

THE SUBURB OF MINNEAPOLIS

Defining the City's "Urban" Form

Robert S. Thompson

IN THE EARLY TWENTIETH CENTURY city planners contemplated strategies to get people out of Minneapolis's crowded core. The city's ordained visionaries deplored the New York style of skyscrapers and tenement housing and developed plans for a city of suburbs and extensive green spaces. Although perhaps not following the planners' particular vision, the city's developers and residents cooperated to colonize the Minneapolis landscape with single-family homes that spread from downtown for miles in all directions. In the first half of the century Minneapolis's population boomed, spreading out from downtown across a gridded collection of subdivisions and streetcar suburbs. The city bulged at its seams even as its residents moved farther and farther from downtown.

As in other cities, streetcar lines influenced Minneapolis's development from the late nineteenth century into the early twentieth. The lines stretching out in a hub-and-spoke model from the central city made suburbia possible. Streetcars provided solutions to the problems that early urban planners deplored as "congestion" and "overcrowding." Working just as they were designed, trolley lines resulted in diffused populations centered on streetcar intersections. In the 1920s, as automobile ownership accelerated, they slowly began to replace effective streetcar operations.[1]

Despite the prominent role of streetcars in the suburbanization of Minneapolis, urban-minded Minneapolitans in the early twenty-first century have been tireless in their efforts to reestablish streetcar lines and revitalize neglected streetcar intersections as commercial nodes to foster greater urbanism. Ironically, Minneapolitans are attempting to create higher-density neighborhoods by using an urban planning form designed to decentralize the population from urban centers. At the heart of their vision for Minneapolis's future is a rejection of what modern Minneapolitans define as suburbia, by idealizing areas that were developed to be quintessentially suburban and have identified automobile-centric development as the root of Minneapolis's perceived urban ills.[2]

Beginning in earnest in the late 1990s, the city of Minneapolis embraced new strategies to redefine and market itself as the progressive urban center, not just within its metropolitan area but within the greater Midwest. Minneapolis, of course, was not unique in pursuing urban rebranding. The core of this vision rested on a belief that an ever-extending suburbia is unsustainable, and by the early twenty-first century a Sustainable Urbanism movement had emerged, oriented around the shared concepts of New Urbanism, smart growth, and green building. At its heart, of course, Sustainable Urbanism is part of a legacy of urban planning, one which exalts the urban core and its potential for dynamic growth while rejecting automobile-centered suburban forms. And Minneapolis, like other cities, is consumed with growth—population growth, tax-base growth, economic growth, growth for growth's sake.[3] Viewing overdependence on automobiles as the antithesis to urbanism and smart growth, however, Minneapolitans who embrace Sustainable Urbanism are now attempting to eliminate some of what they consider suburban-type sprawl. Although sprawl may be an ill-defined concept meaning vastly different things, among certain Minneapolis officials and neighborhood groups it has come to mean cul-de-sacs, superblocks, and surface parking lots. These development types have now become the focus of the city's efforts to combat sprawl.

PLANNING FOR AUTOMOBILES

Reshaping parts of the city as idealized public-transit-oriented spaces is at the core of plans for sustainable development in Minneapolis. Although the city grew on streetcar lines, many of the city's sustainable urbanists look back wistfully at a city that never was; from its origins Minneapolis used transit as a means to expand ever outward into suburbia. The first generations of Minneapolitans embraced technologies and regulations that allowed people to move out from the center city. As streetcars expanded the city's perimeters at the close of the nineteenth century, Minneapolitans embraced policies that define suburbs today, especially residential-industrial zoning, which is designed to separate living and working spaces. Concurrently, automobiles began appearing in larger numbers within the city. Despite nascent traffic concerns, as Minneapolis planners confidently explained, "City planning is the anti-toxin of congestion" and embraced motorcars and their expected centrifugal effects on population density.[4] "There are new elements in municipal life which tend to the diffusion of population," Minneapolis's 1917 Civic Commission declared. "It is our obvious duty to take utmost advantage of them. The automobile should result in spreading the housing of the people."[5] The planners were prescient: automobiles quickly took over the metropolitan landscape. They enhanced the inhabitants' proclivity for moving away from the city's historic core while reshaping the city's traditional low-density, suburban form.

By the 1920s automobiles and detached single-family homes defined Minneapolis and its nearby suburbs. As a result, the distances between the center city and its suburbs, retailers and their consumers, and employers and employees expanded. Regardless of where automobile drivers traveled, they built a collective expectation that they should have a free and convenient place to leave their unattended cars on reaching their destination. Municipalities thus faced the problem of what to do with vast numbers of mostly stationary automobiles. In this regard, outlying commercial nodes and suburbs found themselves in an advantageous position. With copious undeveloped land to dedicate to parking, these areas possessed distinct advantages in automobile-centric development. Coupled with a post–World War II surge in car ownership, suburban growth spread across the state and the nation, leaving cities and their central business districts in obvious decline.

Minneapolis responded to the ascendance of suburbia by establishing auto-centric development in the city. Real suburbs, however, had the two things that the city did not: a vast majority of residents entirely dependent on cars and plenty of space to store the cars. By the 1960s and 1970s, hoping to reverse a postwar population decline by mimicking developments at the suburban fringes within the city, Minneapolis embraced the automobile-oriented planning that prevailed in suburbs and dedicated itself to superblocks, cul-de-sacs, and surface lots within the city. The developers of Lyn Park in near–North Minneapolis, for example, broke the street grid and built a cul-de-sac community. Downtown developed its suburban-inspired Nicollet Mall. Nicollet-Lake embraced plans for a superblock parking lot. Within this auto-centric context, drivers demanded convenience, and developers and the city conspired to give it to them.[6]

SUBURBAN CITY

Historian Robert Fishman, in *Bourgeois Utopias*, defined suburbia as a paradox where exaltation of individualism and private property are built on a foun-

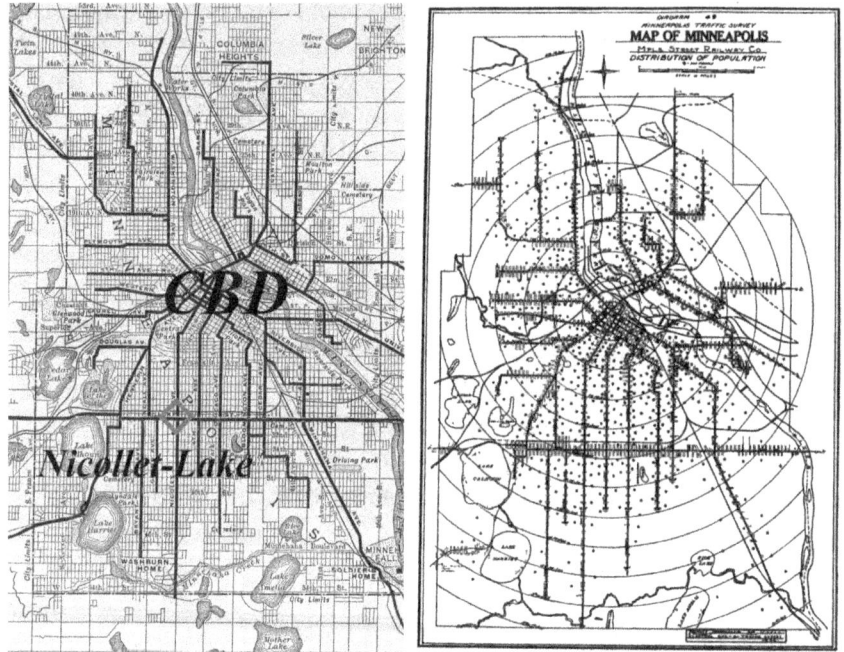

Figure 10.1. The map on the left depicts Minneapolis's Downtown–Central Business District (CBD) and the expanding street grid and trolley lines, circa 1913. The city expanded southward in 1927 with the annexation of part of Richfield. The map on the right was created for the Minneapolis Civic Commission to demonstrate growth along the trolley lines, circa 1917. *Sources:* Frederic Nicholas, ed., *McGraw Electric Railway Manual: The Red Book of American Street Railway Investment* (New York: McGraw Publishing Company, 1913), 122 (left); Edward H. Bennett and Andrew Wright Crawford, *Plan of Minneapolis* (Minneapolis: Civic Commission, 1917), 12 (right).

dation of land speculation and exclusion of urbanism.[7] To follow Fishman to a logical conclusion, an urban area might well be described as possessing traits quite the opposite of suburbia, a place where urban masses and public spaces are sublime and grounded on community ownership and the exclusion of suburbanism. A more realistic version of urban reality, however, can be found in a nondescript shopping plaza in South Minneapolis, approximately two miles south of downtown, which was once home to a Kmart and SuperValu grocery store. The building obstructs Nicollet Avenue's right of way for two blocks to accommodate the store and its supersized parking lot, and has become one of the major battlegrounds in defining Minneapolis's urban identity.[8]

The story of this intersection (often referred to as Nicollet-Lake) and parking lot is much bigger than the store's oversized and paved setback. At its essence, it is a story of a city's aspirations for "growth" and the twenty-first-century Minneapolitans and planning professionals who want to "de-suburbanize" Minneapolis so that they can market the city as a regional model of urban vitality. A closed big-box Kmart with a nearly empty five-hundred-space surface lot that breaks the city grid does not fit this vision. Replacing a parking lot with an old-fashioned street grid whose origins lie in land speculation has, ironically, become a primary focus of Minneapolis's commitment to urban sustainability.[9]

Located at the confluence of two major South Minneapolis arteries, to some urban Minneapolitans the Kmart epitomizes many of the main features of suburban "blight." To these urban enthusiasts, the Kmart complex, including a large SuperValu store, is a microcosm of how a tradition of top-down urban planning and subsidized automobile use destroyed urban Minneapolis. It is a box store surrounded by a large surface parking lot, bordered by an interstate highway, on a superblock. The Kmart plaza succinctly represents suburbia in the city.[10]

So how did the Nicollet-Lake intersection become an entrance to a Kmart and its accompanying massive surface parking lot? The story begins with Minneapolis and its long romance with suburbanization—building ever outward. By the early 1900s the blocks surrounding the intersection bore the telltale characteristics of suburbia. A 1914 plat map detailed numerous "additions" or subdivisions named for developers keen on speculation. Single-family homes dominated the landscape in these gridded blocks as well as the Minneapolis Millers' baseball park, the Twin City Rapid Transit Company's streetcar yard, and a lumberyard. By 1950 over 520,000 people lived in Minneapolis, but the following decade represented the beginning of a slow decline in the city's population as Minneapolitans moved into new homes outside the city limits. Nicollet-Lake's important position as a commercial hub also began to wane.

MINNEAPOLIS'S TRADITIONAL LOW-DENSITY DEVELOPMENT

The tradition of perimeter growth began not after World War II, however, but with the city's 1856 founding. Minneapolis's river-aligned street grid appeared around Saint Anthony Falls in 1854, and in 1855 Congress granted squatters' claims to land on what was formerly part of the Fort Snelling military reserve. As early as the 1880s, to accommodate thousands of new single-family homes, Minneapolis's grid had expanded well outside the growing city boundaries and far beyond the reaches of the streetcar system, prompting two major annexations in the decade. That the grid fast outpaced city boundaries created indistinguishable areas where the city ended and suburbia began.[11]

As the twentieth century progressed, commercial and industrial development in the expanding downtown district further pushed residential areas to the urban fringes. Accommodated by an ever-expanding streetcar system, real estate developers frantically divided and subdivided their holdings at the outwardly expanding edge of the city. For ease of sale and rapid construction, speculators divided Minneapolis and its near neighbors into an almost perpetual collection of five-acre blocks, 330 feet east–west and 660 feet north–south. Foreshadowing the post–World War II suburbanization of professional sports following highway development, in 1896 professional baseball moved to what was then the edge of town along a major streetcar route. The Millers chose for their home an emerging neighborhood near the intersection of Lake and Nicollet, which was in the process of becoming one of the city's busier streetcar intersections. By the first decade of the twentieth century, streetcars not only connected Minneapolis to Saint Paul but also to the immediate suburbs of Saint Louis Park, West Minneapolis (later Hopkins), Richfield, Robbinsdale, Fridley, and Columbia Heights—each, at least in part, extensions of Minneapolis's grid. A western line even extended to Excelsior, connecting Minneapolitans to the leisure and recreational opportunities of Lake Minnetonka.[12]

The 1902 establishment of the Automobile Club of Minneapolis signaled a new era in urban transportation. Automobile ownership surged in the first decades of the century as urban planners tried disparate means to accommodate the new technology. Not shy with its lobbying efforts, the American Automobile Association encouraged urban parking reforms, warning that "decentralization of business" from the traditional downtown center would result from its "inadequate parking." To planners, where to put parked vehicles became a constant challenge. As more people traveled downtown in their automobiles, on-street parking began to cause substantial downtown congestion. Although marketed as tools of mobility, to city planners it became evident in the early twentieth century that automobiles were stationary far more often than they moved.[13]

Large surface parking lots are often depicted as entirely within the bailiwick of American suburbia. Many interpret the vast patches of asphalt that surround commercial "box stores" and malls as the first stages of suburban and exurban development that is known simply as "sprawl." Sprawl-type parking lot development, however, is not confined to the suburbs. In the 1930s the perfect storm of a stalled economy and growing automobile use led urban developers to raze under-occupied buildings to reduce tax assessments and to establish in their place "taxpayer" lots that could generate enough income to cover tax bills until the economy improved. As automobile use soared, streetcar ridership plummeted. Between 1920 and 1932 ridership on Twin Cities' trolleys dropped nearly 50 percent. Far from being a suburban invention, the surface lot first thrived in the Depression-era city, where sufficient center-city development drew large numbers of people looking for places to park their automobiles. By the 1930s the state of Minnesota had established some general regulations for on-street parking, but these regulations provided no guidance regarding off-street parking. More and more cars descended on the downtown commercial district, and by 1945 the Minneapolis City Planning Commission warned that insufficient downtown parking could lead to further decentralization and an economic downturn.[14] The Great Depression kept public transportation in business, but after World War II Minneapolis took advantage of federal policies that incentivized automobile use and car-centric development in the suburbs— including the development of large surface parking lots. Competition from outlying commercial centers, which offered the convenience of free and plentiful parking, left downtown struggling to attract shoppers to its paid parking lots.[15]

Throughout the first half of the twentieth century, garages and driveways sprouted up on Minneapolis private property to accommodate rising numbers of automobiles. In the first decades of the twentieth century, barns and carriage houses had served as passable substitutes for automobile-specific storage buildings. It did not take long, however, for garages to begin sprouting up in Americans' backyards, although these early versions were typically small and unassuming. Recognizing demand, by the 1910s precut-home retailers offered garage kits similar to house kits. Increasingly extravagant garages soon speckled American cities. By the 1930s Sears offered attached garages, proclaiming that "the successful designing of the garage as part of the main building is desired by every homeowner." Simply safeguarding automobiles had given way to maximizing the convenience of storing one's automobile in an accessible way.[16]

THE SHRINKING CITY

Haphazard planning and city growth prompted the 1917 *Plan of Minneapolis*. Based on the *Plan of Chicago*, Minneapolis's planning advocates imagined a city of grand civic monuments and broad parkways leading directly to a sublime city core. The *Plan* included a projection that by 1960 the city's population would surpass one million. Perhaps too ambitious in its civic mindedness, the city pursued few of the planning advocates' loftier goals. Minneapolis did, however, embrace the *Plan*'s major housing recommendation: build at the perimeter to reduce the congestion of the center city. In order to foster single-family homeownership, "areas of undeveloped land must be continuously brought into reasonable 'time-distances' of the center of cities; otherwise an unnecessary crowding of the population on land will result, with all the accompanying evils of congestion." The city thrived as it grew outward, and following the tumult of the Great Depression and World War II, postwar Minneapolis planning began with similar ambitions. With fixed city boundaries, however, growth had pushed far beyond city lines. Without space to grow, Minneapolis planning soon devolved into a mere hope that the city might survive rapid suburbanization. The city's first growth era had ended.[17]

To accommodate the city's surging population in the first decades of the

twentieth century, interwar urban planners dreamed of tearing down perceived slum districts in an attempt to populate the city with magnificent ed-
ifices and public spaces. Postwar Minneapolis certainly achieved the first part
of that dream. By the 1950s Minneapolis began a slum-clearance program in
the downtown Gateway District, but the rebuilding phase stalled. Within the
next decade federal dollars financed the wrecking balls that advanced toward
downtown from all directions in the form of interstate freeways. With a juxta-
position characteristic of America's postwar era, these limited-access highways
cut urban neighborhoods in half as they promoted the growth of federally subsidized housing construction at the urban perimeter. Federal highway
dollars, however, provided infrastructure only for moving cars, not stationary
ones, leaving the state and municipalities to fend for themselves to provide
parking.[18]

As more cars and parking lots appeared throughout the metropolitan area,
fewer and fewer people called Minneapolis home. As this housing boom coincided with a spike in automobile sales, attached garages became a popular
suburban selling point. Minneapolis for the most part offered homes with detached garages or converted carriage houses. People who moved to suburbia,
however, often did so for the simple reason that federal policies encouraged
developers to build new suburban housing. In 1953, for example, when Ger-
trude and Jerome Ulrich bought a new home, they found out they lived in the
first-ring suburb of Richfield south of Minneapolis only when Mayor Hubert
Humphrey did not appear on their voting ballots. Richfield developers, working in a suburb with no planners, simply continued replicating the Minneapolis
grid as they moved south, so only the newer housing stock differentiated suburbia from the city. With such fast-paced peripheral development, Minneapolis
faced a slow population bleed in the decade following World War II. Between
1960 and 1980 the population loss accelerated; Minneapolis lost approximately
112,000 residents, or nearly a quarter of its population. During the same period
the Minneapolis–Saint Paul metropolitan area grew tremendously, with most
of the growth occurring in the first and second rings of suburbia.[19]

Developments in the 1950s foreshadowed a major population shift to suburbia. With federal and state subsidies for road construction, automobiles became the transit mode of choice for those with the means to afford it. As in
most cities, public transit ridership plummeted as World War II gas rations
ended. That the Twin City Rapid Transit Company shuttered its streetcars in
1954 in favor of buses helped in decentralizing the metropolitan population,
and Minneapolis seemed intent on making sure the trend continued. In 1955
the Millers played their last game at Nicollet Park and moved to the near-south
suburb of Bloomington for the 1956 season. There the team played at the newly constructed Metropolitan Stadium, with its fifteen-thousand-space surface

parking lot, located at East Eightieth Street and Cedar Avenue South near the southeastern limit of the Minneapolis-based street grid. Funded primarily by Minneapolis to attract a Major League Baseball franchise, in 1960 the local team folded after the Washington Senators agreed to move to Bloomington and play as the Minnesota Twins. The Minnesota Vikings joined the Twins in Bloomington in 1961 as members of the National Football League.[20] Minneapolis willingly ceded the near future of Minnesota's professional sports to suburbia. The postwar suburban-style ballpark, surrounded by rings of parking, required far too much land for an urban infill project.

In another move that accelerated suburban migration, Minneapolis chose to raze thousands of homes for the interstate system's right-of-way and for other "blight" clearance programs. In 1957, for example, Minneapolis received federal urban renewal funding to redevelop its Gateway area, the traditional center of downtown Minneapolis. Between 1960 and 1962 the Minneapolis Housing and Redevelopment Authority razed all the buildings in the project area—fully a third of the entire downtown district. Various other projects cleared additional "blight" from downtown. To compete with expanding suburban parking lots, Minneapolis established numerous surface lots of its own. To further accommodate driving, the city's 1963 zoning ordinance included Minneapolis's first minimum parking requirements, which mandated parking minimums for all new development.[21]

Downtown Minneapolis's plans for parking seemed less like competition with suburban development than a reaction to it. Thirty-six blocks south of Nicollet-Lake, at the intersection of West Sixty-Sixth Street and Nicollet Avenue South in Richfield, for example, twenty-four retailers occupied the HUB Shopping Center, built in 1954. The one-stop shopping experience offered "unlimited free parking." Shopping experiences changed even more dramatically in 1956, when Southdale Mall opened as the country's first enclosed shopping center near the southwest terminus of the Minneapolis grid at West Sixty-Sixth Street and France Avenue South in the first-ring suburb of Edina. It boasted a climate-controlled interior and over five thousand extravagantly wide, angled parking spaces, which spanned fifteen surface lots. The asphalt sea's developers established a parking-lot map but no street-to-mall sidewalks. The "free" parking that suburban developers offered certainly came with costs, including a lack of dynamic land use, but consumers embraced the nominally heightened costs they paid for ease of access.[22]

Local downtown retailers and their customers flocked to the new mall. Despite a high-modernist vision predicated on automobile use, Southdale Mall architect Victor Gruen imagined Southdale as a corrective to urban sprawl. Endowed by local residents with zoning changes to enable his vision, according to Gruen the mall and its nearby complex of service and residential areas would become a community where people could comfortably live, work, and shop. To the architect this utopian community would bring people together and stop the seemingly endless outward growth of suburban strip malls. Rather than creating competition for downtown Minneapolis, Gruen believed that Southdale should serve as a model for downtown redevelopment. He was wrong; the 463-acre complex became a model for suburban mall development rather than downtown revitalization.[23]

KMART AND THE "SUBURB IN THE CITY"

As suburbs changed land-use norms, fostering the growth of an automobile-centric environment, Minneapolis attempted to mimic zoning standards that prevailed in Twin Cities suburbs. The combination of minimum parking requirements with strict separate-use zoning was in hindsight an example of city officials admitting defeat regarding population loss. Intended to replicate the attractive, low-density development that drew people to the suburbs, strict new land-use regulations that included parking minimums more often than not made transit-oriented urban development nearly impossible. With self-imposed limits on its own development, by the mid-1960s the massive Gateway project in downtown Minneapolis had slowed considerably. With title to the land and no development plans in sight, the Minneapolis Housing and Redevelopment Authority established additional surface parking lots throughout the numerous vacant blocks.[24]

As downtown faced a daunting task of redevelopment, Minneapolis embraced the prospect of the interstate highway system and its accompanying federal dollars. In the case of connecting South Minneapolis with southern and southwestern suburbs, the primary direction of the city's postwar population movement, planners envisioned a new north–south freeway running parallel to Nicollet Avenue two blocks east between Second and Stevens Avenues for nearly thirty blocks. With little input from the affected neighborhoods, construction on I-35W began in 1964, which required razing or moving every home and business in its path.[25] To the north the construction of I-94 accomplished comparable results. As the freeways plowed their way toward downtown, the city's residents continued their outward move. Minneapolis responded to the fleeing population with piecemeal developments such as "the suburb in the city" of Lyn Park, a place of cul-de-sacs and convenient freeway access where young professionals could have "their own slice of suburbia adjacent to downtown," but the outmigration seemed unstoppable.[26]

As the city used federal highway dollars to tear through the interior of Minneapolis neighborhoods, successive Minneapolis mayors solicited other federal dollars in an attempt to rehabilitate these same communities. In 1967 Minneapolis began its participation in the federal Demonstration Cities Program. Better known as the Model Cities Program, the city nominated portions of the near-south Lyndale, Phillips, and Powderhorn communities for federal assistance. As early as the 1910s city planners had complained that the blocks near the Minneapolis Institute of Arts leading south toward Lake and Nicollet suffered from "blight," a decidedly catchall term.[27] The 1960s streetscape at the corner of Nicollet-Lake housed numerous small businesses, including several popular music clubs and dance halls, although at city hall these venues occasionally caused considerable tension. The first African American woman to obtain a Minneapolis tavern license, Mattie Johnson, opened Mattie's Barbecue at Twenty-Ninth Street and Nicollet amid the local officials' apparent concerns about "mixed" dancing.[28] By 1967 city planners classified the neighborhoods bordering Nicollet-Lake as "unstable" for their "transient non-family population, lacking primary relationships and socially acceptable outlets for self-expression" and "rapid population-racial change." The Minneapolis Planning Commission targeted the Nicollet-Lake area as a primary focus for redevelopment.[29]

As part of the Model Cities Program, by the early 1970s the city began purchasing and razing buildings near Nicollet-Lake. The city spent over $8 million to obtain title to and clear buildings from a broad swath of low-rise commercial properties surrounding Nicollet-Lake. Intent on rebuilding the tax base, the city's leaders envisioned building a commercial center that would pull the community out of its perceived decline. With a new array of empty lots within blocks of a newly constructed freeway, city planners believed they were preparing the area for commercial success. In an attempt to rehabilitate the troubled intersection and its adjacent communities, by 1972 city planners and their private partners imagined the area as a suburban-inspired mall. The Nicollet-Lake Development District Plan included renderings of a large, covered courtyard and two parking facilities, presumably to house cars arriving from the next-door freeway. For decades the corner had served as a South Minneapolis neighborhood commercial center, but in 1975 redevelopment in the "project" area stalled. When Nicollet-Lake Associates and its plan for a suburban mall fell through, the city had few good options available.

By late 1975, with no major development agreements in sight, an Atlanta-based real estate business promised Minneapolis a solution. At that point, only a few neighborhood businesses remained open, most notably adult entertainment magnate Ferris Alexander's bookstore and theater, which stood on the northeast corner of the intersection. The city, however, remained determined to radically alter Nicollet-Lake.[30] In this context, in March 1976 Financial Properties Incorporated made an offer: sell the land to Financial Properties, which would then build a major retail store operated by an S. S. Kresge Company "anchor tenant." The plan, however, had a catch. According to Kresge, building a Kmart discount store required over two blocks of contiguous land. The company demanded that the city create a superblock to accommodate the development. Despite objections from some council members and neighborhood organizations, Kmart officials would not budge from their "winning formula," which included construction of a massive surface parking lot. Hoping to enhance the area's tax revenues by revitalizing a commercial center, the Minneapolis City Council voted to support the company's plan.[31]

Financial Properties proposed a radical reorganization of the streetscape: close Nicollet Avenue north of Lake Street through Twenty-Ninth Street to the railroad cut. Designed as a box-store plaza, the private firm had lined up Kmart and SuperValu as its tenants for the nearly eight-acre site. Accepting the plan in 1976, Minneapolis closed portions of Nicollet Avenue, Elroy Street, and Twenty-Ninth Street in the summer of 1977 and created a suburban-style superblock in the middle of the city. Behind a massive asphalt setback, the Kmart and SuperValu opened in 1978. One city council member's dream of building a complex "equivalent to the suburban 'dales'" malls died with it.[32]

The new plaza was like nothing around it. As geographer Joseph S. Wood wrote in 1988, "suburbanization means creating the appearance of non-urbanized space" that is "homogenous" and "hygienic." The Kmart plaza certainly fit the bill. Having the massive parking lot swept four times a week, Kmart's manager, Paul Feller, contrasted his store with the surrounding area by surmising, "Ours is probably the cleanest block in the neighborhood." Despite the protests of some local community members, decades of Minneapolis planners and city council members had dreamed of the city in the new Nicollet-Lake form. In the midst of a grid that spread miles in all directions, a freeway led to a center-city suburban oasis. Alderman Keith Ford welcomed the new development, citing the "34,000 people in my ward who won't have to drive to the suburbs" for a suburban-type shopping experience. Serving the role of grand entrance, the parking lot provided the supposedly necessary buffer between the development and the city that surrounded it. The sheer size of the parking lot told a story in itself. The shopping plaza was a destination for the auto-mobilized, not the neighborhood's pedestrian traffic.[33]

The backside of the massive horizontal building, complete with a railway trench blocking plaza access, told a similar story. When Whittier neighborhood residents requested a backdoor for pedestrians, Kmart officials declined. Feeling the city and Kmart had literally turned their backs on the local community, local artists convened to express their dissatisfaction. After an agreement with Kmart, community artists competed to design a mural for the 161-foot back wall. The winning entry was a mural that, through a series of imagined doors, exposed a battleship fixing its guns upon the neighborhood. The dramatic painting revealed community contempt for the suburban-style monument, and it left open for interpretation who might be the ship's admiral—the city or its corporate partner.[34]

MARKETING A GRID

By the 1990s Minneapolis began an attempt to break free from a century-old planning mindset that encouraged low-density development. Minneapolis began a high-wire act of fostering organic neighborhood development while still centrally planning the city's core structure. The new Minneapolis city planning, in its simplest form, became a marketing campaign based on rejecting suburbia—where people lived in preordained residential neighborhoods and shopped in predetermined commercial districts—but in actuality it was a rejection and denial of the city's own development past. To these new urbanists, what made cities better than suburbs was interconnectedness. No city thrives by marketing empty parking lots, so Minneapolis began selling vitality and convenience.

Sustainable urban living, while a laudable urban-environmental goal, soon became part of the city's growth-centered marketing strategy. Although the repetitive grid that now defines Minneapolis was a legacy of the city's early commitment to ever-expanding low-density development, Minneapolis began promoting the grid as a critical part of its "urban" environment. Despite this marketing, Minneapolis's past is built on a foundation of growing outward, not fostering community and neighborhood development. The city, in other words, was never the city that modern Minneapolitans imagine it once was. Interestingly, the city now began imagining that the rigidity of the urban grid would enhance economic development in the form of walkable and diverse neighborhoods.[35] In 2009 the City Council voiced its intention to "maintain the street grid, reconnecting it where possible, and discourage the creation of superblocks that isolate pedestrians and increase walking distances." When establishing the grid, however, the city's founder, Joseph Stevens, had no intention to create a walkable city. He designed a city for easy commercial accessibility to the Mississippi River. As the city expanded, the grid no longer followed any natural landmarks or land contours. In the 1870s the city squared the grid to a north–south orientation to more easily adapt it to land sales aligned with the section map.[36]

Minneapolis's grid is, of course, not unique in being built on a legacy of commodification of land and landscapes. Historian Kate Brown explained the grid as a vehicle of conquest and domination over both land and peoples. Political scientist James C. Scott viewed it as an abstraction of topography and ecology that allows for easy purchase and sale of the landscape, all within the framework of exalting the center city.[37] The environmentalism of sustainable urbanism, juxtaposed against Minneapolis's commitment to an imposed and arbitrary street grid, is better understood as a rejection of suburbanism rather than a whole-hearted embrace of sustainability. As historian Kenneth Jackson opined, suburbs generally "emphasize their distinctiveness from rather than their relationship with the metropolis." After years of emphasizing perimeter growth, and then attempting to replicate suburbia, the city now hopes to distinguish itself from suburbia.[38] Much of this has to do with rejection of stereotypical suburban box-store parking lots and residential "garagescapes." Where in the 1930s a home designer described the garage as "the motor entrance to the house, just as the front doorway is the pedestrian entrance," requiring that it be "easily entered from the street, and readily accessible from the living quarters of the house," by the early 2000s garage design enthusiast Kira Obolensky described front garage communities as places for cars to park rather than places where people lived.[39] By rejecting this suburban landscape, and despite its own repetitive grid, Minneapolis now markets its city streetscape as "urbanism" rather than recognizing that it is nearly as dominated by detached single-family homes as any of its near suburbs.

By looking at the transformations of Minneapolis's built environment, it becomes clear that defining a city as urban is more complicated than comparing grid versus cul-de-sac or choosing public transit over surface lots. The lines remain blurred, especially in the interstices between the city and its neighbors. In South Minneapolis, for example, the city ends and Richfield begins at Sixty-Second Street South. In this area of Nicollet near the Crosstown Freeway the tangle of raised roadways and strip-mall development makes it difficult to discern the line between the two cities. With such a contested relationship between Minneapolitans' urban vision and the city's nearby suburbs it seems odd that the divide is not in stark contrast.

A NEW PARADIGM OF URBANISM

In 2000 the Nicollet Avenue Task Force determined that "the Nicollet-Lake intersection typifies the incongruity that can result when a suburban development pattern is forced onto an urban setting" and called for reopening Nicollet Avenue. Defining *urban* as "high density and high amenity development . . . in unique two- to four-story buildings that clearly define the street and sidewalk edge," the task force imagined the intersection as the confluence of a Nicollet "commercial corridor" with its "community corridor." While stressing the importance of reduced parking, the task force concurrently imagined a restored Nicollet Avenue with "increased traffic levels."[40] The disconnect between the task force's strategy for growing traffic while reducing parking was an example of how parking, to some extent, remained dissociated from driving. The heart of the problem, of course, was that many policy makers and business owners continued to conflate increased traffic with profit and growth. Removing parking options would obviously limit traffic growth, but the Nicollet-Lake parking lot had become so underused that it became an easy target for elimination. Removing the plaza and its surface lot was an easy sell to urban-minded Minneapolitans.

According to Nicollet-Lake planners, fully realizing the avenue's urban potential meant street-facing surface lots had to go. The 2009 *Minneapolis Plan for Sustainable Growth* summarized the city's ambition: "The goal of this plan is to demonstrate that Minneapolis is, and will remain, the heart of the upper Midwest region in terms of residing and working, and a premiere destination for dynamic urban living."[41] Suburbia was the place for setbacks, surface lots, and anti-pedestrian development. Nicollet would be walkable from downtown south to Richfield, but only if the city could reconnect the grid at Nicollet-Lake, in Mayor R. T. Rybak's words, by "busting right through the back of that Kmart." To Minneapolis's recent urban promoters, the grid defines urbanism. As David Frank, the Minneapolis director of transit development, said in 2012, the Kmart parking lot is "an important real and symbolic hole in the street grid."[42] In 2015 the city's plans took a step forward when it acquired the western side of the lot for $5.25 million, allowing it to close the Super-Valu grocery store, and another step forward in 2017 when it purchased the land beneath the Kmart for $8 million. Then, following the 2018 bankruptcy of Transform, which owns Kmart, the city found itself on the doorstep of finalizing its dream of gaining control over the remaining portion of the superblock and reconnecting the grid.[43] Yet the site's future was once again thrown into doubt in 2020 when a wave of civil unrest swept the city after Minneapolis police officers murdered George Floyd, during which the US Postal Service's Lake Street Station was burned to the ground. The city leased the old Kmart building to serve as a temporary replacement facility, rendering its future again unclear.[44]

What Minneapolitans are attempting to do at the Nicollet-Lake intersection is redefine the city's history and its traditional form. They envision the city not as a place of low-density development, but one committed to pedestrians and mixed-use communities.[45] Wrapped up in the battle over what to do with Kmart are idealized views of urban living and a continuation of suburban stereotypes. Rejection of "bland," automobile-centered suburbia is certainly political, though it is expressed as devotion to sustainability, and the Nicollet-Lake Kmart and its parking lot are one of the city's prime redevelopment projects.[46] Although the city is committed to population growth, the city boundaries are not expected to change, making areas such as Nicollet-Lake likely sites for higher-density urban development. The city's newest strategy for Nicollet-Lake seems almost as radical as the 1960s–1970s plan, which built a mixed-use urban oasis in the middle of a low-density commercial corner. As the city mulls its options for reshaping the intersection, the people most intimately affected by the change will continue walking from the Nicollet-Lake bus stop across a flat, black monument to one of Minneapolis's past attempts at neighborhood revitalization.

THE CAMPUS AS WATERSHED

Urban Sustainability and the Pedagogy of Place

Joseph Underhill

ENVIRONMENTAL SUSTAINABILITY PRESENTS ONE of the most pressing challenges of our day, and a key dimension of this challenge is managing water supply. The impact of humans on the earth's ecosystems continues to grow, and our collective consumption of resources has yet to reach a level that could be considered in balance with the earth's ability to supply them. The exponential growth of human population, resource consumption, urbanization, and their resultant environmental impacts all call for fundamental changes in how we organize our societies. In terms of water use, problems manifest in several ways: availability of adequate supply, the dangers of too much water and flooding, and the threats of water pollution and contamination. To varying degrees, each of these interrelated issues will need to be addressed by cities, businesses, and institutions of all sorts (including colleges), if they are to take sustainability concerns into account and remain viable in the long term.[1]

In the past few decades we have finally begun to see new ways of incorporating sustainability initiatives into many institutional frameworks. As humans have profoundly shaped their natural surroundings, in turn human institutions have been shaped by their environments.[2] For the first one hundred years of its existence, Augsburg College—a small liberal arts college in Minneapolis—

displayed a lack of clear, conscious connection between its institutional struc-tures and the ecological realities within which it operated.[3] But, beginning in the late 1960s, it began a fundamental transformation in its relationship to both its sociopolitical and environmental contexts. This marked a paradigm shift in the college's relationship to the natural world, part of the larger soci-etal transformation that emerged with the development of the environmental movement.

In the 1960s Augsburg College began to think of itself not just as an in-stitution of higher education, but as an institution that existed in a particular ecological context and, as such, having to take the particulars of that place into account. Following the lead of early environmental thinkers such as Henry Da-vid Thoreau and John Muir, a number of scholars have argued that the move toward sustainability requires close attention to the specifics of place.[4] This at-tention to ecological context requires an epistemology grounded in lived reali-ties and the experiences of the people and ecosystems that constitute the world to which we are connected.[5] This approach is at the heart of the critical peda-gogy of place, or "place-based education." Drawing on a mix of the philosopher John Dewey's pragmatism and ideas such as bioregionalism and social ecology, a pedagogy of place asks us to pay attention to how we live—to the needs, ex-periences, and wisdom of all members of our community—rather than just to abstract and ostensibly universal theories and data.[6] Augsburg College has not rejected either its theological roots or the modernist rational-legal professional disciplines that are at the core of most colleges and universities today, but it has begun to recognize the limits and problems of those approaches and begun to embrace a set of practices that view the campus as intimately interconnected with its human and ecological surroundings.[7]

This essay explores the relationship between Augsburg's institutional struc-tures, pedagogies (especially its embrace of place-based education), and water management practices on campus in three eras. The college began in 1869 as a "premodern" and "pre-urban" Norwegian *presteskole* (priest school), lacking the machinery, power, and infrastructure of the modern industrial metropo-lis. The early campus consisted of a few simple buildings, including a barn in which the campus's farm animals were kept, and a hand-dug well from which the Norwegian Lutheran pastors-in-training drew water for their daily needs. There were no laboratories, no science classes, no electricity. The campus had a minimal impact on the local watershed. Though it was devoid of machinery or urban infrastructure, it rested on a moral foundation—the Lutheran imperative to do good works and serve one's community. This ethical heritage would later provide impetus for the adaptation of an ethic of stewardship and conserva-tion. The campus was located on Indigenous homeland, known to the Dakota as Mni Sota Macoke (the land where the water reflects the sky), a fact that the institution would largely ignore until much later in its history. In this sense, the campus also rested on an intrinsically immoral, settler colonial foundation as well.

Whereas the first fifty years of the college's existence were characterized by frugality and pietism, by the turn of the century the college had become integrated into the booming "Mill City" of Minneapolis and had evolved from a seminary into a liberal arts college.[8] In 1900 it acquired running water, electricity, and its first laboratory; the primary institutional imperative of its second fifty years was gaining the approval of the professional accrediting bodies for higher education; it acquired the structure of a liberal arts college, with an increasing diversity of electives and majors that reflected the complexities of modern American society and increased specialization in the academic disciplines. Reflecting generic academic standards that were not related to place, its curriculum sought to impart the universalized logic of modernity, with science labs that began to utilize the technologies of its urban, industrial context, including indoor plumbing and electricity. Attracting students with needs that went beyond the monastic asceticism of the early Lutheran pastors required that the campus modernize by providing plentiful water and a campus landscape dominated by the lawn monoculture that had become the standard for homes and campuses alike.

The third phase of the college's life saw a fundamental institutional shift that increasingly drew attention away from its earlier inward cultural values to a more outward-looking stance that engaged the many problems occurring in the communities and ecosystems forming the context within which the college's educational work took place. By the late 1960s and early 1970s a series of social, ecological, and political crises and challenges had risen to the fore, and faculty began developing courses that addressed these issues. Having largely addressed earlier demands of assimilation and cultural maintenance, as well as the external demands of accreditation, environmental and other social concerns began to receive greater attention from the college's faculty. With greater economic security, a set of post-materialist courses and topics entered the curriculum. Environmental awareness began to emerge within the curriculum by the turn of the century in the form of both an environmental studies program and a new emphasis on more sustainable practices. At the same time water became increasingly expensive and commodified. College life, mirroring the wider world, became ever-more expensive, presenting new challenges to the sustainability of the campus's water management regime. Within a rapidly evolving urban context, the college began to transform itself into an institution that took its relationship to the watershed seriously, yet remained enmeshed in the carbon-intensive, consumerist economy of the new millennium. The developments at Augsburg College reflected both the wider trends in higher education and the evolving techno-economic context in the Twin Cities. At the same time they were the result of its own unique history and character. The resultant water management regime is an amalgam of both these wider trends and its unique story.

PRESTESKOLE ON THE EDGE OF THE PRAIRIE, 1869–1900

Three years after Augsburg's founding in 1869, August Weenaas, the president and lone professor of the Augsburg Seminarium, left Marshall, Wisconsin, to relocate to the burgeoning city of Minneapolis, joining the large influx of Norwegian immigrants settling in the Upper Midwest's commercial centers.[9] The search for greater religious freedom and objection to the relatively hierarchical Norwegian state church spurred pietistic Haugean Lutherans to seek residence in the United States, where they saw greater opportunity for practicing their faith "unhampered by human traditions and erroneous conceptions and ideas."[10] The largest wave of immigration came in 1866–1873, following passage of the Homestead Act in 1862, and it was as part of this wave of migration that the Augsburg Seminarium was founded.

Weenaas's mission was to share with aspiring Norwegian immigrant pastors the "example of the master himself and his first devoted followers . . . in order that they may become imbued with the Truth and its saving power, and be filled with the desire to declare it before all men in order to satisfy the greatest wants of and heal the deepest wounds in the human soul."[11] Although this calling made a claim to the timeless and universal, what brought Weenaas to Minneapolis was its "potent combination of waterpower, lumber, railroads, wheat, immigration, a vast hinterland, and Yankee enterprise."[12] Weenaas—along with an assistant, twenty-four seminarians, and his wife and two daughters—came to Minneapolis because of the water and the economic activity it shaped. The remains of the last ice age and the channel dug by the glacial River Warren made Saint Paul the effective head of navigation on the Mississippi River, and endowed Minneapolis with the energy of 3 trillion gallons of water a year, dropping eighteen feet over Saint Anthony Falls—more than enough to power the mills that sawed the region's white pine and ground its Durum wheat.[13]

Upon arrival, and with the help of local business and church leaders, they built the campus's first building on the corner of Twenty-First Avenue and Seventh Street, just a few blocks from the banks of the Mississippi, in what was then described as part prairie and part swamp, with the only other structure on the land being a "haunted shack" on its southern edge.[14] This oak savannah ecosystem consisted of a rich diversity of plant and animal species that were adapted to the roughly twenty-nine inches of rain received each year. Below the prairie were the complex geological layers—the vadose zone through which water passed before seeping into the Platteville Limestone, Prairie du Chien Formation, and lower aquifers, which absorbed, channeled, stored, and eventually discharged the area's rainfall.[15] The rain collected by the thirty-eight acres of the Augsburg campus amounted to an average of 30 million gallons of free water received each year.[16] Students hauled water that they collected by hand from campus wells into their on-campus residences, where it was used for cooking and bathing. Their relationship with the water was thus local, free, embodied, and direct. They knew where the water came from since they gathered it themselves. Campus life was characterized by a high degree of self-sufficiency, with all the labor on campus done by the students themselves, with no hired staff, cooks, or groundskeepers. They heated the buildings with firewood gathered by the "wood boss," a student whose challenging job was to oversee the gathering and storage of enough wood to make it through a Minnesota winter.[17] By almost any measure, this resource management regime was sustainable, efficient, frugal, and in keeping with the ecological realities of place. It also clearly required a lot of physical labor, something that became increasingly anathema to the cultural expectations of a modernizing city.

In the context of this initial phase of urbanization, what we would now think of as water management problems manifested themselves in the form of concerns about hygiene and public health. Early urban sanitation presented clear challenges that affected the water supply. In 1881, for instance, the city reported having to remove 2,317 dead horses, mules, cows, cats, and dogs from the mostly unpaved city streets.[18] Like most American cities in the nineteenth century, the Twin Cities faced outbreaks of diphtheria, cholera, and typhoid.[19] Complicating matters, experts did not understand the nature and causes of these diseases, with some doctors denying that they had anything to do with the water supply, instead attributing the problems to the so-called miasmas or to the city's "rotten sidewalks." Gradually, however, awareness that diseases spread via contaminated well water took hold, with newspaper editorials from the time warning that "disease and death often lurk in the old oaken bucket or the iron pump handle."[20] These cases were linked to well water from locations near outhouses, generally in the tenement houses in lower-income parts of town—an early manifestation of environmental injustice.

Two years before Augsburg's founding, Minneapolis formed its Water Department. Despite the misgivings of many about the obviously polluted water of the Mississippi River, the city began drawing its municipal drinking water from just above Saint Anthony Falls in 1874. Originally using the water's own power to run the pumps, the city drew 146 million gallons a year from the river, with that amount increasing rapidly to 10 billion gallons by 1920. It took decades for the water mains to reach the edges of the city, and for the first twenty years of its existence the Augsburg community continued to draw on its own supply of well water.

As part of a frontier community, the college's most pressing need at the time was to provide spiritual guidance and service to the new immigrant community. Georg Sverdrup, the second president of Augsburg College, had a deep desire to create and maintain a college that would instill in its graduates the values of the Norwegian Lutheran Free Church, a particular culture and faith to which he had a strong attachment. His desire was to "focus attention away from the shifting cultural climate of our time to the simple truth that alone can give the heart peace and the spirit purpose."[21] His primary goal for the college was maintaining Norwegian culture (by teaching Old Norse and Norwegian literature and history), assimilating to their new home (most fundamentally by teaching English), and training pastors to serve and minister to new immigrant communities. The student newspaper was published in Norwegian, and the school's exclusively white, male students took courses such as the Old Testament course on "God's infinite love and man's perverted ways on account of sin."[22] They advocated piety, temperance, and abstinence, which ran counter to the materialist culture that began to take root in an increasingly wealthy city. There was an absence of environmental concern in the curriculum and in the writings of college leaders at the time.

Although the students' daily work was connected to the land, their course of study was not. The curriculum was set as a counter to local knowledge and Indigenous culture, which was seen as uncivilized. At the core of their study was ancient Greek and Christian theology, which connected students to the spiritual traditions of the Levant, and Norwegian language and culture, which sought to maintain their connections to their home country. The "deepest wounds in the human soul" that the graduates sought to heal did not encompass the damage being done to the environment or the forced removal of the Dakota.[23] This would slowly begin to change as the seminary began to shift toward a liberal arts college that included the study of the natural and social sciences—disciplines that laid the groundwork for an understanding of the limits on humanity's ability to extract resources from the natural world.[24]

LABORATORIES, STORM DRAINS, AND ACCREDITATION, 1900–1970

From 1870 to 1890 municipal water lines and sewer and stormwater drains spread out from the core of the city in a dendritic pattern, and the machinery of modernity began to arrive on campus. By 1892 the Minneapolis Water Department reported, with a fairly remarkable degree of precision, that the city had 158.732 miles of water mains and 83.218 miles of sewer lines in place. Instead of hauling water from the campus wells, the students would now be supplied with water at a metered price, with the work required to bring it into the buildings no longer supplied by Lutheran muscle power, but by the two "15,000,000 gallon capacity Worthington pumps," which each consumed eleven tons of "Lehigh Buck Mountain vein coal" per day. This coal, shipped by rail from central Pennsylvania's aptly named Carbon County, for the first time tied the water consumption on campus to the nascent fossil fuel economy.[25] It also linked the campus to the Upper Mississippi River Basin, an area of land of approximately twenty thousand square miles that included farmland and forests, as well as the effluent of cities such as Saint Cloud and Grand Rapids.[26]

By the turn of the century Augsburg College was no longer on the outskirts of town, but right in the midst of an industrial boomtown whose population had grown to just over two hundred thousand. In 1890, when the Augsburg campus connected to the municipal water supply system, the primary source of water for the college shifted from rain- and well water to the river water pumped to campus through the city supply system. The water from the Upper Mississippi River was siphoned out of the river in Fridley, Minnesota, run through a sand filter, and piped into buildings. Although many cities around the world are currently running short of water, such is not the case for Minneapolis. The flow in the Mississippi at Saint Anthony Falls varies annually from a low of around 1,100 cubic feet per second (cfs) to highs of around 50,000 cfs during floods and spring runoff.[27] Although not limitless, this amounts to almost 3 trillion gallons of water per year, more than sufficient for the city's water demands—currently around 25 billion gallons a year, or less than one percent of the annual flow.

The first building on campus with a direct connection to this supply of river water was the new Main Building, built in 1900 and touted for its modern construction, abundant natural light, a boiler room, and lavatories with "tile floors and . . . tub and shower baths and every other modern accessory."[28] All baths, the 1901–1902 catalog bragged, were furnished with hot and cold water. There were sewer connections and even electrical lighting. Reflecting the shift in physical activity from manual labor to recreational pursuits, the building had the campus's first gymnasium. Instead of chopping firewood or hauling water from the well, students could now play basketball. Even so, financial limitations on campus forced Augsburg students, through their self-run Medical Aid Society, to limit their water use to "one bath a week (tub or shower), free of charge."[29] Over the course of this second phase of the college's history, the life of students became more separated from the daily practical concerns and manual labor that had been part of campus life in the nineteenth century. Health services, previously the responsibility of the Medical Aid Society, shifted during this period to being provided by doctors. Hired staff prepared food, and a grounds crew rather than students performed most maintenance work.

By 1940 the original prairie, and the farm that had replaced it, were both completely gone. A 1941 photo of the campus shows Memorial Hall set in a barren landscape, with a lone tree and a patchy turf lawn. By the 1960s the campus had been separated from the neighborhood to the south by the great trench of I-94, and paved streets and automobiles came to dominate the campus. The cities continued to draw on the river for drinking water and use it for waste disposal. Water quality in the river reached its nadir in the 1950s and 1960s when the combination of untreated sewage, industrial waste, and the series of dams built along its length transformed the Upper Mississippi into a series of virtually lifeless, fetid cesspools.[30] People on campus could smell the stench of the river. As the city began to address problems of water pollution, spending millions on water and sewage treatment, the college likewise began thinking about its place in the city and as part of the ecosystem. Changes in the college's educational mission facilitated this shift.

The transition from seminary to liberal arts college began around World War I with the presidency of George Sverdrup (son of Georg Sverdrup). Demand for Lutheran pastors for the Norwegian immigrant community had begun to dwindle, and it was becoming clear that the financial viability of the college required an expanded institutional mandate. At the same time businesses demanded more highly trained workers, which generated pressure for institutions of higher education to professionalize. From the 1920s on the college had to take account not just of the needs and standards of the Lutheran congregations that supported it, but also of accrediting agencies and the voracious capitalist economy. In order to be accepted as a modern institution of higher learning, the college needed to acquire the machinery and attendant curriculum and infrastructure of the natural sciences. In the period following World War I, the college started a small museum (with artifacts gathered on mission trips) and acquired a new chemistry classroom, with "hot and cold water, [and] gas and electric current." By 1919 the college had four courses in chemistry and in 1922, for the first time, students could pick one of four tracks: religion, teaching, medicine/engineering, or business/law. Previously, the only option had been general education as preparation for ministry. In 1930 the first courses in what might be seen as a place-based or ecological pedagogy appeared, with a new course in Rural and Urban Sociology and one in field biology, which included study of "ecological localities."[31] The first course in the new field of ecology showed up in the 1939–1940 catalog.

The college continued to diversify its departments and offerings, and the percentage of faculty with PhDs or other terminal degrees steadily increased. After World War II the Business Department was added and would grow to have the largest number of majors on campus. The college continued its process of differentiation and a more diversified organizational structure with the creation of academic divisions in the humanities, social sciences, and natural sciences. The 1948 catalog mentions the location of the campus as across the

Mississippi River from the University of Minnesota, but generally there continued to be a lack of any explicit connection to place or local ecosystem. The role or significance of water specifically or environmental issues more generally were not discussed at all, and are not evident in campus planning or curricula. The priority of the college at this time was on gaining institutional recognition within the larger cultural context of higher education, but as the campus became increasingly intertwined with a larger, increasingly polluted urban environment, the campus began to rethink its fundamental approach to education.

NEW PEDAGOGIES AND THE COMMODIFICATION OF WATER, 1970 TO THE PRESENT

By the 1950s the triangle of land that constitutes the Augsburg campus had become a complex mosaic of campus streets, sidewalks, aprons, lawns, gardens, buildings, and city streets—all artificial forms designed to meet the need for mobility, habits, work, and recreation, without much thought to how they integrated into local ecosystems. Reflecting the broader trends in urban land use, a 2005 land-use survey showed that approximately 60 percent of the campus was covered in paved, impermeable surfaces.[32] There was a steady addition of artificial environments to the campus, such as the domed athletic field with its artificial turf, three indoor ice arenas, and a system of skyways and tunnels, which further separated the campus community from the natural world. The structures and land use on campus removed rainfall as quickly as possible through storm drains, while securing usable water from the virtually limitless municipal supply of treated river water. The three dimensions of water management—supply, flood control, and contamination—all presented challenges for the campus. As water consumption became more expensive, problems of flooding on campus also increased. The combination of an increasingly impervious landscape and intense summer downpours led to flooding in one of the campus's buildings on several occasions, causing thousands of dollars in damage. In the Upper Mississippi River, water contamination—most notably excess sediment and nutrient loading from nonpoint sources—presented the greatest challenges to water quality. The river saw dramatic improvements in its levels of lead, DDT, and nutrient loading as a result of legislation such as the 1972 Clean Water Act. Around the turn of the millennium studies began to detect trace levels of "contaminants of emerging concern" such as pharmaceuticals and endocrine-disrupting compounds in the river, raising new concerns and uncertainties about the health and safety of the water, both for human consumption and for aquatic life.[33]

Over this period the city and region made great strides in reducing the amount of pollution entering the river. They took seriously the Clean Water Act's mandate to make the nation's rivers and lakes "fishable and swimmable."

Perhaps the most basic (and dramatic) improvement in this regard was the virtual elimination of the direct discharge of untreated sewage into the river. The Twin Cities sewage treatment plant at Pig's Eye Island had been treating effluent since the late 1930s, but it was frequently overwhelmed by excess stormwater runoff. The solution was to separate stormwater drains from the sewer system, which Minneapolis actively pursued starting in 1986. In 2003 Minneapolis added Chapter 53 to its city code, prohibiting property owners from connecting stormwater drains (including roof drains) to the sanitary sewer system, and charging residents an additional stormwater management fee based on the amount of impermeable surface on their property. In response, Augsburg retrofitted its buildings so that roofs no longer drained into the sewer lines and installed a series of stormwater retention ponds adjacent to new buildings and parking lots. By 2007, as a result of the city's fifty-year sewer and stormwater separation effort, no untreated sewage was released into the river for the first time in the city's history.[34]

Water management in the Twin Cities continued to modernize, with increased demand for water and an ethos of abundance that can be traced back to the origins of the city's water plans in the early 1900s. Minneapolis planners built an industrial-scale water treatment plant to ensure economic growth, and water withdrawals from the river increased steadily during the twentieth century, peaking in 1988 when the city reported just under 30 billion gallons pumped from the river. Since then the total withdrawals have declined to an average of 21 billion gallons, another sign of the adequacy of local supply.[35] Aggregate supply, although not without its limits, was not the main problem. The Metropolitan Council projects that both Minneapolis and Saint Paul have adequate water supplies through 2050, unless "drought or contamination" limit the ability to draw from the river.[36] Reflecting concerns about contamination in the wake of the 1993 cryptosporidium contamination of Milwaukee's municipal water supply, Minneapolis built a multimillion-dollar ultrafiltration plant.

Currently Augsburg's total annual consumption of this ultra-filtered city water is around 31.7 million gallons, or eighty-seven thousand gallons a day, for which the college pays approximately $250,000. Though the water is inexpensive, it is not free, costing each student about $80 a year. Most of this water is used in the summer for irrigating lawns and for cooling buildings (ironically, the campus gets the bulk of its rain in July). Reflecting the increased costs, and increased disconnect from local rainfall patterns, campus grounds installed an automatic sprinkler system run by the aptly named Aqua Engineering, Inc.[37] When the system was first installed in 2000, the sprinklers could often be seen running even when it was raining; the facilities department has since adjusted the watering schedule and installed soil moisture sensors that regulate watering, a sign of the shift in campus water management practices.

The sources of drinking water on campus, reflecting broader social trends toward privatization, have also shifted from sole reliance on the municipal water supply to increasingly expensive, energy-intensive, and commodified sources. Starting in the 1990s, Pepsi and the Culligan Man joined the campus watershed. Bottled water, which was relatively unknown prior to the 1970s, has grown to a business on which Americans spent over $11 billion in 2011.[38] In a strange throwback to the days of mercantilism and state monopolies, many schools and college campuses now grant sole contracts to particular vendors. In 2000 Augsburg became a "Pepsi campus," signing its first contract giving that company exclusive rights to sell beverages on campus. The contract specifies that its terms be kept private, but it includes a provision for revenue sharing so that a certain percentage of the beverage sales goes to the college, giving the school an incentive to continue the arrangement. The campus has twenty-five beverage vending machines, which dispensed 9,250 bottles of water and 26,350 bottles of soda, juice, and energy drinks in 2009. At $1.50–$1.75 per bottle, that amounts to $54,000 spent on those beverages that year.[39] In addition to this direct cost, each vending machine uses approximately 6.2 kWh/day of electricity, which adds another $7,000 onto the campus's annual electrical bill. The bottles themselves take water and energy to be produced and transported to campus, part of the 100 million barrels of oil used annually for producing, bottling, transporting, and cooling bottled water containers globally.[40]

Despite the ready availability of ultra-filtered city water, there are also forty-five Oasis Culligan water dispensers scattered across campus. This purified water is actually just filtered and mineral-enhanced municipal drinking water from the Twin Cities suburb of Brooklyn Park. In 2011 the campus consumed on average 250 jugs a month, at $4.00 per jug, plus the cost of the electricity used to power those dispensers and the diesel fuel for the truck that hauls the water sixteen miles from the bottling plant. The electricity needed for these dispensers added another $4,000 to the annual cost of campus water. So for fifteen thousand gallons of filtered, cooled, and heated water, the college paid $16,000, or $1.07 per gallon. This was 150 times more expensive than the 0.7 cents per gallon for city water (a comparison of various campus water sources is summarized in Table 11.1).

Such energy-intensive conveniences are part of an increasingly expensive and technological campus culture, which now includes such amenities as battery-powered and motion-activated paper towel dispensers and "touchless" toilets and urinals. These represent choices about how money is allocated and the kind of collateral learning that takes place on campus. Bottled water sales, for instance, go to PepsiCo and its stockholders (which made $5.6 billion in profit in 2010). The gleaming bottled water dispensers and vending machines (some now sporting video screens) also reinforce the message that municipal

Table 11.1. A Comparison of Various Water Sources at Augsburg College

Water source	Gallons per year received/consumed	Cost/gallon	Contents, besides water
Rain	29–30 million	0	Trace elements from airborne pollutants
City water	31.7 million	$0.007	Chlorination byproducts, fluoride, trace metals from pipes & solder
Filtered city water	~35,000 from 10 on-campus filtered water dispensers	$.07	Traces of city contaminants from city water
Culligan (filtered city water)	~15,000	$0.80	Fluoride, leachates from PET bottles (unknown levels)
Pepsi (Aquafina)	~5,562	$9.60	High fructose corn syrup (HFCS), caffeine, flavoring

Source: All figures were gathered and analyzed by the author, using Augsburg College records.

water is unsafe and the battery-powered towel dispensers teach students that turning the crank on a paper towel dispenser is either dangerous or too much work or both. Today the college and the city are embracing new ecological worldviews and proactive environmental planning, but remain caught up in the larger socioeconomic dynamics that impel them to consume more. After several years of discussion on campus about bottled water consumption, in 2011 the campus installed several Elkay™ filtered water dispensers (at a cost of around $700 per unit) that are getting extensive use by students to fill their reusable water bottles. Campus water consumption practices are beginning to change, but are still firmly embedded in the larger modernist and capitalist context.

Changes in water use have been accompanied by shifts in the curriculum and pedagogy that promise to help counter some of these pressures. The political ferment and unrest prompted by the civil rights and antiwar campaigns of the late 1960s ushered in curricular reforms that included the introduction of interdisciplinary, critical, and environmental courses and pedagogies—reflecting a healthy degree of reform and experimentation that drew on the Lutheran concept of *semper reformanda* (constant reform). With its small size, minimal bureaucracy, and history as part of the democratic Lutheran Free Church, the institutional culture was amenable to new initiatives. The first of these was the American studies program and Crisis Colony (a residential program in the inner city) spurred by the race riots of the 1960s. Soon after, Augsburg helped create the experiential and interdisciplinary Higher Education Consortium on Urban Affairs (HECUA).

Conservation and forestry programs date back to the early work of Gifford Pinchot and Yale's School of Forestry (founded in 1900), but environmental studies programs as such did not begin to appear in the United States until the mid-1960s with Middlebury College's environmental studies program starting in 1965 and the University of Vermont's in 1972. The 1970s saw a rapid increase in the number of environmental studies programs, and with time Augsburg College joined this growing national environmental movement with new courses and programs that reflected those concerns. The 1970 college catalog provided evidence of a sea change on campus, with concerns around race relations and the environment finding a prominent place on campus for the first time. Interdisciplinary courses were now listed as a category, including the classes White Racism as a Pathogenic Agent, the Urban Encounter, and a biology class called Man and the Environment that looked at "the present situation in the natural world in light of the Christian ethic."[41] The 1971 summer "urban term" included a biology course, Problems of the Urban Environment, that focused on "problems of population growth and the human ecosystem, and . . . the impact of urban desity [sic] and territorial behavior, environmental pollution and genetic damage, and man's attempts to overcome his biological limitations."[42] There were a range of other courses that included the "environment" as part of their titles—Environmental Esthetics, Man and the Environment, and Economic Issues of the Urban Environment. A metro-urban studies program began in 1973, and other interdisciplinary programs, such as international relations, women's studies, and American Indian studies soon followed.

With the rise of environmental awareness in the wider culture, a small group of faculty and staff formed a campus Environmental Concerns Committee in the late 1990s. This committee grew slowly, gained a small budget and the endorsement of the college's president, and helped to shift campus culture and practices in various ways. Student groups such as the campus chapter of the Minnesota Public Interest Research Group (MPIRG) successfully advocated on campus for composting food waste, purchasing wind energy, banning chemical herbicides on lawns, and attempting to ban bottled water on campus.[43] The college inaugurated its environmental studies program in 2007. That same year, the college became a charter signatory to the American College & University Presidents' Climate Commitment (ACUPCC) and joined the newly formed American Association for Sustainability in Higher Education (AASHE). The college's environmental studies program focused particularly on local urban environmental issues that reflected both its urban location and theological roots. Now the college offers a range of environmental field study courses that study sustainability both locally and in international locations. Students in chemistry classes now walk to the river to collect water samples for testing, and Augsburg students directly experience their connection to the watershed and the Mississippi River itself with courses that travel the river by canoe. Reflecting the dramatic improvements in water quality, students in these courses can now fish and swim in the river.

The new connection among curriculum, facilities management, and the watershed was evident when in the fall of 2007 a group of Augsburg College students in a new Environmental Connections course left their classroom and began digging up the lawn next to one of the campus's office buildings. The purpose was to direct water away from that corner and into a rain garden. In that catchment they later created a small native wetland, planted with some of the species originally found in the prairie there—dogwood, turtlehead, bulrushes, milkweed, cardinal flower, joe-pye weed, echinacea, and asters. When completed, the rain garden project added two dozen native species to campus, effectively doubling the floral diversity on campus in one day, and for the first time in the college's history taking a small step toward restoring the landscape to its presettlement form. Although the students did not know it at the time, they were digging on the very spot where the Seminarium's first building had been located. With the installation of that rain garden, the students had begun to think of the campus as a watershed, joining the grounds crew in figuring out how to keep the water out of basement offices and storm drains. Instead, they used it to irrigate a garden, which used the water, filtered it, and created a microhabitat for local wildlife. It was the latest in a long series of developments on campus, reflecting an evolving praxis and potentially a new paradigm for its relationship with the physical environment.

RECONNECTING TO THE WATERSHED

The evolution of Augsburg College's relationship to its watershed reflects both a paradigm shift and a source of inspiration for cities in Minnesota and elsewhere hoping to successfully move toward greater sustainability. But this transition is far from complete. In Minnesota, a water-rich state, water supply generally exceeds demand, with the exception of some rapidly growing suburban communities that draw on aquifers. That said, the current water regime is expensive, wasteful, and reflects a mindset connected more to short-term economic imperatives than to the dynamics of the local ecosystem. Irrigated lawns, bottled water, and cooled and heated water are neither efficient nor necessary. Despite the abundance of water here, concerns remain about water quality, about the large number of impervious surfaces that contribute to flooding, and about the $350,000 spent on water at Augsburg each year.

In looking at our local water supply system, perhaps its most salient characteristic is its continued lack of congruence with local ecological realities and the lifeways and traditional ecological knowledge used by the original Indigenous inhabitants of this land. The city has worked long and hard to bring clean, safe, drinking water into our homes, but we have ended up with a large-scale system that provides billions of gallons of potable water, most of which gets used for things like flushing toilets, watering lawns, or washing cars. The rain that we do receive, which is naturally purified water, is channeled as quickly as possible over contaminated roads and parking lots, to the river—where it then causes flooding and pollution problems downstream. We do not harvest that water or design our grounds to match local rainfall patterns. Instead we plant lawns that require copious amounts of purchased water, fertilizer, and herbicides, and ship in expensive bottled water to drink.

Given the increasing cost, complexity, and vulnerability of modern urban waterworks, one path to a lower-cost, more resilient water supply may be found in a return to some of the more sustainable water management practices found at Augsburg in the nineteenth century, when water was free and locally harvested, and when land management matched rainfall patterns. Slowly the city and campus are rebuilding wetlands, catchment basins, and rainwater harvesting systems that were part of the presettlement ecosystem. These smaller-scale, decentralized ways of collecting and managing water require new, local infrastructure that will in turn need both significant investments and new ways of imagining our relationship to the land. But we should be mindful of the strange complexity and inefficiency of the system we currently have in place so that resources we currently spend on wasteful water consumption can instead be channeled to higher priorities. A water management and landscaping system designed with local ecological dynamics in mind could significantly lower on-campus water costs, reduce the need for bottled water on campus, reduce the labor and cost of our landscaping, and further reduce our impact on the Mississippi River. To move in that direction, we will need to continue to foster a frame of mind that is capable of seeing the problems with our current practices, and that can reconnect us to our ecological context. We need to start seeing our cities and campuses as watersheds.

PART III

Environmental Politics, Thought, and Justice

MONUMENTAL ENCOUNTERS

The Politics of History, Conservation, and the Reconstruction of Grand Portage, 1922–1958

Chantal Norrgard

IN 1963 TWIN CITIES ARTIST DEWEY ALBINSON remembered his visits to the Grand Portage reservation at the northeastern tip of Minnesota on Lake Superior and wrote about his appreciation for its natural beauty: "I find myself more and more in communion with nature, rocks, waves, shoals, and cedars hanging on the cliffs, their branches twisted by the winds and winter sleet."[1] Albinson became fascinated with the environment of Minnesota at an early age, painting the area around Lake Minnetonka where his family vacationed during the summers. In 1922 he traveled to Grand Portage because he was drawn to the romantic lure of the place and was fascinated by the Ojibwe people who lived there.[2] Albinson returned to the reservation throughout much of his life to paint its landscape and people (though Grand Portage Ojibwes did not always welcome his efforts to paint them).[3] Albinson's dramatic paintings of Grand Portage and its people were featured in exhibitions in Minneapolis, Washington, DC, and New York City. They popularized the rustic, regional imagery of northern Minnesota.[4]

Albinson also became an energetic advocate for the protection of the northern Minnesota environment. He railed against what he saw as the destructive progress of roads and development on Lake Superior and the disappearance of a wilderness inhabited by Indians and a few hardy immigrant families.[5] Like many Euro-Americans at the time, he believed that Ojibwe culture was disappearing with the encroachment of modernity.

Albinson's presence at Grand Portage reflected a growing number of Euro-Americans who traveled to northern Minnesota to enjoy the region's natural environment and heritage. It also reflected a growing interest among Twin Cities residents in the protection of Grand Portage. That same year, Solon Buck, the superintendent of the Minnesota Historical Society (MHS), became concerned that private development threatened the site of the eighteenth-century fur trade post at Grand Portage. In July 1922 the MHS sent Albinson and his friend, Alban Eastman, to explore the site.[6] The maps and photographs the men compiled established a valuable framework for future efforts to uncover Grand Portage's past. The interest in conserving the natural environment of the area and revitalizing its history that began in the 1920s eventually culminated in the designation of the Grand Portage National Monument in 1958. Today the national monument includes a reconstruction of the Northwest Company fur trade post that stood on the site in the late eighteenth century and the 8½-mile Grand Portage trail, the historical trade route connecting Grand Portage to the interior waterways of the region.

This chapter explores the history behind the creation of the Grand Portage monument from 1922 to 1958. While a good deal of attention has been devoted to Grand Portage's historical importance as a fur trade center, the early twentieth-century movement that led to the designation of the monument and its implications for the Ojibwe residents of Grand Portage are less well known.[7] The movement to preserve Grand Portage was significant in its own right—infused with political tensions over who held ownership over the history and landscape of Grand Portage and shaped by the complex interplay among tribal, local, state, and federal interests.

Tracing the history of the monument reveals the perspectives of multiple groups involved in the reconstruction project at different points in time and their diverse attachments to this place. Environmental conservationists and historians (many of whom were from the Twin Cities) were interested in reconstructing the fur trade post and carrying on its history as part of the region's heritage on both a state and national level. They linked efforts to protect Grand Portage with the aims of the American conservation movement, using the site as an extension of a natural wilderness in need of protection and revitalization. They also utilized preservation efforts to gain a foothold in the region and to expand institutional control. In contrast, Grand Portage Ojibwes saw Grand Portage as a living, breathing place. They valued the history of Grand Portage in connection to their cultural and community identity, but they also cared about their economic, political, and cultural survival in the early twentieth century.

They drew on the site for their livelihoods, making a living through tourism and hunting, fishing, and maple sugaring in the surrounding area.

Although many Grand Portage Band members participated in efforts to reconstruct the fur trade post, their contemporary ways of life sometimes came into conflict with the movement to preserve the history of the place. Even though the post lay fully within the boundaries of an Ojibwe reservation, non-Indians utilized the project as a vehicle to celebrate and lay claim to the early settler history of Minnesota. They increasingly saw Ojibwe ownership of the lands on which the post and trail were located as an obstacle to preservation efforts. Following World War II the Grand Portage Band's inability to support the upkeep of the site financially, combined with a new impetus among conservationists to preserve the post and the natural environment surrounding it, led to a largely unprecedented action—the cession of reservation lands by treaty to the federal government for the purpose of creating a national monument.

The contexts surrounding the movement to restore and protect Grand Portage bring to our attention the complex relationships among public history, environmental consciousness, and state and federal colonialism in Minnesota during the twentieth century. Examining this movement on a local scale allows us to understand how the unique political, economic, and environmental setting of the region shaped events. This history joins two distinct but interrelated movements in American history: the movement to conserve the natural environment and the movement to preserve and commemorate America's past. Though these movements nominally focused on environmental conservation and historical preservation, they also became instruments for their proponents to exercise political influence and to control land use and the actions of local Indigenous people.

There are several works in environmental history and memory studies that explore the politics of environmental conservation as well as public history.[8] However, scholarship focusing on the implications of these politics for American Indians illuminates how environmental conservation and public history were intertwined with the project of American colonialism. Karl Jacoby, Robert Keller, Michael Turek, and James Feldman have demonstrated how, in the process of revitalizing and protecting the environment, the American conservation movement became a vehicle for state and federal control over American Indians, which ultimately led to further dispossession of tribal rights, lands, and resources in the nineteenth and twentieth centuries.[9] Scholars, including Katrina Phillips, Jean O'Brien, Coll Thrush, Paige Raibmon, and Philip Deloria, have shown how non-Indian expectations about American Indian identity and commemoration of American Indian history have displaced American Indians themselves, undermining their connections to place and erasing their contemporary presence in American society.[10] I build on this literature to illustrate

how non-Indians intertwined environmental conservation and historical preservation in an effort to restore and lay claim to Grand Portage. As a result, the
movement to restore Grand Portage had important political implications for
Ojibwe connections to the site.

EARLY HISTORY

Indigenous people have occupied Grand Portage for thousands of years.
A number of Indigenous peoples hunted game in the region's boreal forests,
fished in the waters of Lake Superior and nearby lakes and rivers, and harvested
wild plants and natural produce, including maple sugar and wild rice. Among
them were the Ojibwes, who migrated to Lake Superior from the east and made
their home along its western and northern shores. The French began trading
with the Ojibwe in the area in the late seventeenth century. By the 1740s Ojibwes had established a village at Grand Portage, or *Gichi-onigaming*, and had
long since become active participants in the Great Lakes fur trade.[11]

Grand Portage became a prominent place of commerce in the late eighteenth century. In 1778 alone, trade there brought in £40,000 and employed
approximately five hundred workers.[12] Each summer, thousands of Europeans
assembled at Grand Portage during the rendezvous season to trade with Native
people and to conduct fur trade business and policy.[13] In the 1780s the British
Northwest Company constructed a fur trade post on Grand Portage Bay. The
post held eighteen wooden buildings, including store houses, dwelling houses,
shops, a counting house, a mess house, and an impressive great hall, and was
surrounded by palisades made of cedar posts.[14] A fur trade post in this location
gave Ojibwes a safe haven that provided an income and access to trade goods.[15]
In 1802 Northwest Company officials relocated their trading business to the
mouth of the Kaministiquia River (near present-day Thunder Bay, Ontario) due
to the increasing political influence of the newly formed United States. However, the fur trade continued to present lucrative economic opportunities for the
Ojibwes and Europeans at Grand Portage until the 1840s, when the trade began
to diminish in profitability.[16]

The economic and political climate surrounding Grand Portage shifted in
the mid-nineteenth century with US expansion and settler colonialism. With
an interest in obtaining the rich copper and iron ore deposits in the region, in
1854 the United States negotiated a treaty with Ojibwes living along the shores
of Lake Superior. The treaty led to the cession of Ojibwe lands on the north
shore of Lake Superior and the establishment of the Grand Portage reservation,
which encompassed the old fur trade post. It also recognized Ojibwe rights to
hunt, fish, and gather in territory ceded to the United States.[17] More broadly,
these treaties established an exclusive relationship between Ojibwes and the
federal government over which state and local governments held no legal juris-

diction. Though the federal government negotiated treaties in order to obtain title to and extend jurisdiction over Indians lands, those treaties represented nation-to-nation agreements between tribes and the United States; they recognized Ojibwe tribal sovereignty.[18]

Following the treaties, Grand Portage Ojibwe weathered a range of assimilation policies designed to eradicate their culture and traditional ways of life. They lost additional lands in the early twentieth century when the federal government allotted reservation lands into 24,191 individual land holdings for the purpose of turning Grand Portage people into farmers while opening an additional 16,041 acres to settlers.[19] Historian Carolyn Gilman writes, "What an Ojibway was to do with 160 acres of inaccessible rocky cutover land was a mystery."[20] The soil and climate simply could not sustain agriculture. As a result, many Grand Portage people sold their allotments or abandoned them. The combination of reservations, assimilation policies, and allotment made it difficult for Ojibwes to rely solely on traditional subsistence activities to survive. Many suffered extreme poverty. Like Ojibwes living in other communities, Grand Portage people scraped by through a combination of seasonal subsistence activities; commercial fishing, hunting, and trapping; and wage labor in the regional lumber, shipping, and railroad industries.[21] They also found ways to sustain their culture even under the difficult conditions of reservation life, practicing their beliefs and ceremonial traditions in resistance to the surveillance of the federal government. As the Ojibwe community at Grand Portage adapted to the dramatic transformations of the nineteenth and early twentieth centuries, some of the old fur trade post buildings were abandoned and decayed, while others were adapted for other purposes.

THE MOVEMENT TO PRESERVE GRAND PORTAGE

Renewed historical interest in the fur trade post began in 1922 when a local property owner placed a "road closed sign" on the Grand Portage trail.[22] Another landowner notified the director of the MHS, Solon Buck, that the trail was threatened. He was concerned that the trail had become endangered and inaccessible.[23] Buck brought attention to the site at the State Historical Convention held in Duluth during the same year by presenting a paper on the history of Grand Portage. Following Buck's presentation, members of the convention discussed the need to mark and preserve historic sites. They recommended establishing a state park "to include the Grand Portage from Lake Superior to Pigeon River—the first white man's road in Minnesota."[24] Originally Buck thought that the site could be promoted as Fort Charlotte State Park, but the complex web of private, federal, and Indigenous landownership made it nearly impossible to designate the land for such purposes. Therefore it remained in Ojibwe hands.[25]

It was hardly coincidental that interest in protecting Grand Portage surfaced at this time. The Progressive Era, with its focus on reform based on scientific and scholarly expertise as well as the creation of administrative infrastructure, ushered in a new era of historical preservation in Minnesota. Solon Buck was a scholar of the American West and social history. He held appointments as a professor at the University of Indiana, the University of Illinois, the University of Minnesota, and the University of Pittsburgh. In 1914 he became the superintendent of the Minnesota Historical Society. Buck's predecessor, Theodore Blegen, described him as a "scholar-administrator." In true Progressive Era fashion, Buck transformed the historical society from a "skull capped institution with a genteel genealogical fragrance into a modern, scientific historical society that took rank among the very best in the nation."[26] As part of his efforts, Buck moved the organization to a new building, "re-formed the staff," established a quarterly magazine, and acquired new materials for the society's collection.[27] He also expanded the reach of the society beyond the Twin Cities, encouraging the formation of county historical societies and designating and protecting historic sites important to Minnesota's heritage.[28]

Buck's interest in turning Grand Portage into a state park suggests the extent to which renewed interest in preserving and protecting historical sites was intertwined with the aims of the American conservation movement. The movement began in the mid-nineteenth century but peaked in the early twentieth. Conservationists stressed the importance of regulating and managing natural resources to protect them from industrial development. The movement was prominent in politically progressive states, such as Minnesota and Wisconsin. In reality, however, it was not only a movement focusing on the protection of the natural environment but one that extended state and federal power over rural populations, including American Indians, imposing new categories of legal and illegal land use.[29] Karl Jacoby writes that conservation "extended far beyond natural resource policy, not only setting the pattern for other Progressive era reforms, but also heralding the rise of the modern administrative state."[30] With an interest in promoting conservation, state and federal governments stepped up fish, game, and forestry regulation, and created state and national parks for the purpose of reinvigorating and protecting the wilderness. In Minnesota the movement to reconstruct Grand Portage offers an example of how the state's conservationists used the project of historical and environmental preservation to extend their political influence.

Conservationists and historical administrators associated the fur trade post and the portage itself with Minnesota's distinct wilderness past and with the early presence of Euro-Americans in the state.[31] In June 1922 Paul Bliss, a writer for the *Minneapolis Journal*, and Cecil W. Shirk, the Minnesota Historical Society field secretary, explored the Grand Portage trail. Bliss wrote a long article about their expedition in which he called for the preservation of Grand Portage. His writing reflected a fascination with the natural beauty of the place as well as its history: "Seldom do the trail of history and the path of beauty merge together down through the yesterdays into today in the white man's country, here in the center of the continent. Usually, beauty walks alone, waiting long to clasp hands with civilization. Yet here are two centuries of the white man in Minnesota, marked by a definite thoroughfare right through one of the most beautiful regions in all the beautiful North Star state, whose span of time as the white man measures it runs only a little ways into the past."[32] Bliss saw the beauty of the landscape surrounding Grand Portage as intimately connected to the portage trail. He identified the portage as a marker of Euro-American presence and expansion over the area. Bliss connected his desire to preserve the fur trade post and portage trail with his desire to protect the surrounding natural environment. It apparently never occurred to him or other conservationists that the fur trade had been a highly extractive industry that greatly altered the region's wildlife and environment.[33]

State conservation policy undermined Ojibwe sovereignty and control over resources. The Minnesota Game and Fish Commission stepped up regulations targeting Ojibwe people, violating their treaty rights and their exclusive relationship with the federal government. Ojibwes hunted, fished, and gathered both on and off the reservation based on their treaty rights. State officials argued that these rights enabled Ojibwes to obtain disproportionate amounts of fish and game, causing resources to decline (despite no evidence supporting this conclusion). State officials criminalized Ojibwes because treaty rights presented a barrier to the state government's ability to gain full control over all resources in the state. As Bruce White argues, Ojibwes became "a negative example against which non-Indians could unite in their desire to preserve fish and game."[34] Ojibwe struggles to exercise their rights to hunt and gather resulted in arrests, fines, and confiscated equipment, which exacerbated poverty in reservation communities.[35] While the Grand Portage reconstruction project did not spark conflicts as overt as the struggles to exercise treaty rights, it did present another example of how the state extended its hold over Ojibwe lands. Political tensions arose over the reconstruction project and land use surrounding the site. Although Grand Portage residents benefited economically from employment under the project and the tourist income it generated, their contemporary interests came into conflict with state efforts to rebuild the post and maintain it in pristine historical condition.

The MHS's interest in Grand Portage continued throughout the 1920s and early 1930s. In 1931 the MHS held its annual convention at the site and thousands of people gathered for events sponsored by the Cook County Historical Society.[36] However, work did not begin on the reconstruction project until 1936 due to a lack of funding. The Indian Division of the Civilian Conservation

Corps (CCC-ID) appealed to Washington for relief work funding on the reservation.[37] This request provided the opportunity for the MHS to promote the reconstruction project to the Grand Portage Band and to attain the funds needed for the project. In 1936 CCC-ID appropriated $6,200 for the reconstruction of the palisade at Grand Portage.[38] Soon after, the Indian Service and the MHS began work hiring Grand Portage men to assist MHS Museum curator Willoughby M. Babcock and archeologist Ralph D. Brown with the excavation of the site. That same year Ojibwe crews began excavation of and then rebuilt the post's palisade and Great Hall in 1938. In *Indians at Work*, an Indian Service publication, Brown described the parts of the post that surfaced in the excavation, as well as the artifacts that the workers unearthed.[39] He noted that the project drew a number of visitors from all over the United States, some of whom took part in sifting dirt with screens in order to search for artifacts.[40]

Brown took photographs that revealed the extent of the work the Ojibwe crew members performed.[41] At times the pictures are eerie and at others, humorous. In one picture an Ojibwe crew member stands in a hole that was once a well in order to demonstrate its depth. Only his head and shoulders are visible, and it looks as if he had been unearthed. In another picture a large group of workers stands in front of the excavation laughing, shovels in hand. The photographs recorded two different and simultaneous goals: the Ojibwe workers' desires to find employment during the Depression and the MHS's desire to uncover and preserve eighteenth-century history. They also recorded the substantial role that workers from the Grand Portage community played in the project.

Grand Portage Ojibwes saw the reconstruction of the fur trade post as an economic and educational opportunity for their community. In 2000 tribal elders discussed their memories of the reconstruction of the post as part of a series of oral histories published by the band. The elders' stories of the reconstruction project were brief, focusing on the labor and social interactions that took place at the site rather than the historical importance of the project. Jim Wipson stated that the fur trade post "wasn't any fine finished building; it was supposed to be like in the olden days. And we built it like that. It was very interesting, and it was good work."[42] Wipson helped to rebuild the stockade. He remembered learning carpentry from a man named Claude Johnson, who directed the stockade construction and taught the Ojibwe workers how to build the walls. Wipson explained, "We were young fellows ourselves, who didn't know Adam and Eve as far as building was concerned, but he showed us just what to do."[43] In another interview Betty Lou Hoffman recalled that the post "played a big role in the life of the Grand Portage community." She noted that a number of people from the community helped build the post and "many women worked there making crafts, which were on display there."[44] Like Wip-

Figure 12.1. Members of the Indian Civilian Conservation Corps crew pose for a photo at the end of their first day's work. In 1936 the Indian Division of the Civilian Conservation Corps (CCC-ID) appealed to Washington for relief work funding. The request's approval provided the opportunity for the Minnesota Historical Society to finally secure the reconstruction of the palisade at Grand Portage. Courtesy of the Minnesota Historical Society.

son, Hoffman described the post's central place in the contemporary community at Grand Portage because of the work that Grand Portage people put into the reconstruction project.

Despite the employment it provided, the reconstruction project also generated conflicts between the Grand Portage Band and the MHS. In January 1938 the Grand Portage tribal council called a special meeting because stockade construction threatened to block road access to the Grand Portage dock. The council adopted a resolution which barred blocking access to the road because of its importance to community fishermen and guides. In January 1938 tensions over the dock came to a head at a meeting between the Indian Service and members of the MHS. Mr. Balsam, a member of the Indian Service, outlined practical problems with the reconstruction for the Grand Portage community. While the Indian Service was "anxious to maintain historic tradition as nearly as possible" and that while the community was interested in this, they "were forced to consider the matter from the cold blooded economic standpoint."[45] Balsam stated that "there comes a time in research when it is imperative to make a decision as to whether one should adhere to an ideal or whether this ideal can be modified to meet the conditions and needs of living people."[46]

In response, Theodore Blegen relayed the MHS's view on the matter, stating, "It is the feeling of [the] Society that a historical reconstruction should be as nearly accurate as possible," explaining that an inaccurate stockade would open the society to severe criticism.[47] Although the society did not wish to deviate from its plan, if necessary, it would build only part of the stockade rather than build the entire stockade inaccurately.[48] As a result some of the stockade was left open and the dock remained in use until the 1950s. Work on the fur trade post continued until 1940 when the Great Hall was completed. The Grand Portage Band managed the site. On one side of the post's reconstructed Great Hall, they ran a coffee and souvenir shop, and on the other they set up museum cases displaying artifacts and items that community members had made for a Works Progress Administration Arts and Crafts project.

The conflict that arose over the stockade construction exposed differing views on the reconstruction of the fur trade post. The MHS sought to maintain the historical authenticity of the construction, viewing the site much as one might view a collection item. In contrast, the Grand Portage Band saw the development of the site from a practical standpoint. While the reconstruction of the post was an interesting historical endeavor that employed community members, it became a problem when it stood in the way of their economic survival. As they had historically, Grand Portage Ojibwes saw the post as a site that served their contemporary economic needs.

During World War II federal focus turned to national defense, which caused funds for the reconstruction project to dwindle. Carolyn Gilman writes that at this time "Grand Portage was 'literally abandoned' by all but the Ojibwe."[49] The MHS ended its official involvement in the reconstruction project. However, despite the lack of funding, the site continued to be a place where Minnesotans devoted attention to conservation as well as the historical heritage of the state. For example, in 1946 the Indian Service and several businessmen from Duluth sponsored the Boy Scouts North Star Council of Duluth in its efforts to clear the portage trail. Over the course of the next three years, they cleared the trail to make it accessible to hikers and canoeists.[50]

In 1949, the year of the Minnesota territorial centennial, the Cook County Historical Society held a pageant at the site called "Grand Portage the Great Carrying Place." Both Grand Portage Band members and non-Indians participated. Elmer Albinson's film footage of the pageant parade illuminates settler colonial ideas about the history of Grand Portage as portrayed in the pageant.[51] The parade began with Ojibwe people dressed in traditional tribal regalia including jingle dresses (dresses covered with metal cones that "jingle" when the person wearing them moves) and black wool or velvet clothing decorated with floral beadwork.[52] Then came characters dressed in clothing from the fur trade era. Euro-American settlers dressed in mid-nineteenth century garb followed, and finally came the pageant queens who wore formal dresses and rode in the back of a convertible.[53] The parade reflected the progression of history as the organizers saw it, beginning with Indians and the fur trade, continuing with Euro-American settlement, and ending with modern, mid-century America. While the parade included Ojibwes, it was largely a celebration of the appearance of Euro-Americans at Grand Portage and the advancement of their history; there were no Indians at the end of the parade as part of the modern era.

The script from the pageant also provides insight into how the organizers saw the history of Grand Portage as well as the place of Ojibwe people within that history. In the section of the script detailing the costumes of participants, the pageant organizers stated, "Indian male actors make especially good appearances if their costuming consists of wigs, small headdresses, breech clouts, moccasins; a few beaded or fur trimmings, and a limited amount of grease paint." The pageant organizers did not designate regalia specific to Ojibwes. Instead, they promoted conventional Western stereotypes of Indian dress.[54] Their descriptions of the costumes for non-Indians were more specific. The outfits of the "explorers," for instance, would "depend upon the countries they represent."[55] These levels of specificity and nonspecificity in the script indicate the extent to which pageant organizers wanted to view American Indians as a static, generic population, while promoting more nuanced images of non-Indians.

In some instances the organizers had to recognize the contemporary Ojibwe people who participated in the pageant, but translated their presence through romanticized imagery and language. In "Episode I—Early Indian Life," the narrator stated that the Indians living at Grand Portage were proud of their heritage and were trying to keep it alive "thru coming generations."[56] After further narration about historic Ojibwe life, Lawrence Connors, the president of the Minnesota Chippewa Tribe, emerged from a wigwam dressed in regalia.[57] At his appearance the narrator asked, "But who is this Indian in all his regalia: his feathered headdress, black-beaded breastplate, sleevelets, breachcloths, ceremonial bag, and feathered banner? None other than an Indian Chief whose duty it is to rule over his tribe! He is Lawrence Connors, the president of the Minnesota Chippewa Tribe. And now he is calling his men together for a council."[58]

Next, Marjorie Lewis (or "Princess Starflower" in the script) emerged as the narrator and announced, "Cook County has an Indian maid who is an Indian Princess." They explained that Lewis was the descendant of a "Big Chief," and was "the last in the line of the Carribou family."[59] They also noted that Lewis's uncle, Peter Stevens, had acted as a guide for a party exploring Labrador thirty-five years before.[60] The narrator went on to explain that "Princess Starflower's American name" was Marjorie Lewis and that she and her aunt were "in charge of an Indian Relic Store, seven miles west of Grand Portage on the highway."[61] Lewis's appearance as an Indian princess played into one of the stereotypes central to Euro-American historical imagination of Indians. The fact that Lewis's uncle worked as a guide for an exploration party added flavor to the pageant's fur trade theme.

Most striking of all were the ways that the pageant organizers emphasized Grand Portage's past to call for the protection of Minnesota's distinct heritage and natural landscape. At the conclusion of the pageant, the entire cast sang "Minnesota Hail to Thee," and the narrator stated, "Even though great progress has been made since 1849 it has not always been wisely made. Records show mistakes, flaws in our dealings with the Indians; flaws in our use of natural resources. Look about us and we can see evidence of forest destruction; fewer species of wildlife, and mineral depletion."[62] The narrator continued with a sense of hope and improvement: "We are, however, attempting to make amends for these mistakes of the past. Let us then, with full knowledge of our mistakes, and with just pride in our progress join in deep humility and thankfulness to show our appreciation of Nature's Gifts, and for the heritage of courage and industry left us by our valiant forefathers, the pioneers of Minnesota Territory."[63] The pageant organizers envisioned conservation as a means to atone for past wrongs in Euro-American dealings with Indian peoples and to address the destruction of the environment. They linked this message to the distinct history (and colonial settler past) that was the heritage of Minnesota.

After World War II Grand Portage once again became the focus of reconstruction efforts. The National Park Service (NPS) became involved with the fur trade post reconstruction project in the 1950s, beginning the actual political process through which the site became a national monument. Initially, both tribal interests and conservationists sought federal involvement. In June 1950 the Grand Portage Band invited the NPS to a tribal executive meeting to discuss whether the site could become a national historic site. Band members wanted to capitalize on the economic potential of Grand Portage and attain grants for the upkeep of the site under Indian ownership. NPS representatives attempted to secure a signed agreement to establish Grand Portage as a national historic site, which would be owned and operated by the Grand Portage Band and not the federal government.

However, in a series of events that resembled nineteenth-century treaty negotiations, the NPS presented the terms under which it was willing to work with the Grand Portage Band. The tribal council raised concerns and objections during the three days of discussion about these terms. NPS representatives wrote six different draft agreements before reaching agreement with the council.[64] In the final draft the secretary of the interior declared Grand Portage a national historic site. In turn the Grand Portage Band promised to operate the site for "the benefit of the American people" and to build housing and food facilities for tourists.[65] In 1951 the Minnesota Chippewa Tribe, the Grand Portage Band, and the secretary of the interior signed the agreement. The site remained under Ojibwe control and the NPS provided technical assistance and support for efforts to preserve the site.

Minnesota conservationists, such as the well-known writer and activist Sigurd Olson, pushed for this designation in hopes that it would lead to further conservation in the state. Olson was a prominent member of the Wilderness Society, a national organization established in 1935 to protect American wilderness.[66] According to Ron Cockrell, the Wilderness Society "pressed the Park Service to establish the national historic site in northern Minnesota in order to secure a foothold or anchor in the region." The society hoped that once the federal government became involved in the protection of Grand Portage, other historic or wilderness areas would fall under the umbrella of federal protection.[67] The connections conservationists drew between Grand Portage and the expansion of federal power illuminate the close relationship between environmental conservation and historical preservation. Conservationists saw designating Grand Portage as a national historic site as an opportunity to gain greater control over local land use.

The problem for local conservationists and the NPS was that Ojibwe ownership and management of the site stood in the way of larger initiatives to preserve the fur trade post as they wished. According to the NPS director, Conrad L. Wirth, effective administration would require that the NPS obtain title to the land. Since the Grand Portage Band refused to give up ownership of the site and title to the land, the NPS could play a limited role at Grand Portage.[68] The NPS stressed that under its control more federal money would be available for upkeep of the site, since President Truman has specified that no more than $2,200 could be spent by the Department of the Interior there. This limited the Grand Portage Band's financial ability to maintain the site.[69]

Ultimately, the lack of funding forced the Grand Portage Band to reconsider the NPS's proposal. In June 1953 the Grand Portage executive body recognized the financial challenges the band faced in managing the post. Members of the executive body suggested that it would be "highly desirable to create Grand Portage National Monument in lieu of the Grand Portage National Historic Site" in exchange for preferential employment and other benefits.[70] However, band members would be required to sell allotments to the federal government as part of a Congressional action, eliminating tribal control of the site. Many Grand Portage people found the terms of this proposal controversial because of the immense historical connotations involved. As Cockrell put it, "Ceding land back to the Federal Government once it is set aside as an Indian Reservation guaranteed by treaty is a rare occurrence."[71] The Grand Portage Reservation Business Committee and the Minnesota Chippewa Tribe's Executive Business

Committee vigorously debated the implications of creating the monument. "Bitter opposition to the measure" among some tribal members delayed any action on the matter.[72] However, a number of influential Indian and non-Indian communities lobbied the Reservation Business Committee and the Minnesota Chippewa Tribe's Executive Business Committee, including Cook County judge C. R. Magney, president of the Cook County Historical Society Effie McClean, conservationist Sigurd Olson, representative from Minnesota's Eighth District John Blatnik, and Grand Portage Band member and business agent Alton Bramer.[73] They succeeded in convincing the Grand Portage Band and Minnesota Chippewa Tribe to agree to the creation of a national monument. However, it is also likely that the lack of funding to maintain the fur trade post site and the choice of federal jurisdiction over state jurisdiction also influenced the Grand Portage Band's decision to create the monument.

On September 2, 1958, Congress approved the final agreement for the monument under Public-Law 85–910. The law stipulated that the Grand Portage Band agreed to relinquish title to the lands on which the monument was located to the federal government in exchange for a handful of benefits.[74] These included preferential employment for Grand Portage Band members at the monument, permission to sell hand-made items inside the boundaries of the monument, no interference from the government in the trade or business of the tribe outside of the monument, the right to traverse the monument grounds in pursuit of treaty rights (though the secretary of the interior could regulate the conditions under which the monument would be traversed), and continued use of the Grand Portage dock.[75] In 1960 the monument was officially established, and in the same year the first park superintendent arrived, charged with the responsibility of restoring and transforming the monument to its appearance 200 years ago."[76]

Amid these developments, there was still opposition to the changes that took place after the designation of the monument continued. Ironically, Dewey Albinson saw the intensive restoration of the site in the 1960s as problematic because it encouraged tourism and development rather than the revitalization of a remote wilderness. In 1963 he decried the environmental changes caused by new roads that increased tourist traffic and the disappearance of local community members' homes at Grand Portage. He exclaimed, "God, I can see what's coming next. I can see the public bathhouse, public toilet, and New National Park architecture."[77] Albinson suggested that the NPS should have also built structures for local Ojibwes rather than just focusing on the preservation of the fur trade post. "The 'damned' tourists coming with their cameras would have something to photograph," he wrote, concluding, "I could growl."[78]

COLONIAL LEGACIES OF ENVIRONMENTAL CONSERVATION
AND HISTORICAL PRESERVATION

While many more events occurred in the history of Grand Portage, the creation of the monument marked a major turning point in the site's history—the end of Ojibwe ownership. Grand Portage Ojibwes have retained economic benefit and access to the monument. The economic and political position of the band has also changed as a result of both the federal government and American Indian focus on tribal self-determination in the 1970s. In 1975, the Grand Portage Band opened a multimillion-dollar hotel and conference center funded by the federal government and the band's income.[79] In the 1990s they added a gaming facility to the hotel after passage of the federal Indian Gaming Regulatory Act.[80] Today the NPS and the Grand Portage Band work together on joint efforts, such as park maintenance, a summer mentoring program for Ojibwe youth, shared municipal services, and the annual Rendezvous Days event that celebrates the historical legacy of Grand Portage.[81] In 2007 the NPS opened the Grand Portage National Monument Heritage Center, a 16,600-square-foot, $4 million building, with the aim of "honoring the area's history, people, and culture."[82] The center overlooks the fur trade post and features exhibits about Ojibwe culture and the fur trade.[83]

Though circumstances surrounding Grand Portage have changed, the history behind the creation of the monument carries symbolic importance and highlights the politics of historical preservation and conservation in Minnesota in the twentieth century. The story of Grand Portage shows how public history and the conservation movement have had broad political implications for Indigenous people and is intimately connected with the extension of state and federal influence and power. As Karl Jacoby argues, "In the case of American conservation, memory formation and policy making evolved in tandem with one another, for in justifying their programs, many of the movement's leading proponents found it useful to offer a vision of the past to which conservation emerged as the only logical response."[84] Grand Portage's history as a fur trade center became a vehicle through which state conservationists, and later the federal government, invoked a particular association between the fur trade post and a wilderness in need of protection as part of state and national heritage. Grand Portage became a place where Euro-Americans projected and commemorated settler colonial history in Minnesota. Local conservationists and historians as well as state and federal officials representing these interests ultimately concluded that Indian ownership of the site could not accommodate this vision. In their view, the only logical way to expand their project of protecting the site and its natural setting was the cession of the tribal lands on which the fort stood. Thus the history surrounding the monument not only embodied efforts to protect Grand Portage's heritage, but also tensions surrounding ownership of the history and environment of northern Minnesota.

PITTSBURGH'S COLONY IN SAINT PAUL'S HINTERLAND

Tensions over Environmentalism in Northeastern Minnesota's Iron Range

Jeffrey T. Manuel

ENOX, SHORT FOR "ENVIRONMENTALLY OBNOXIOUS," is a slang word heard occasionally in the northeastern Minnesota town of Ely. Like "tree hugger," *enox* mocks the tourists who flock to Ely on their way to the vast Boundary Waters Canoe Area Wilderness (BWCA), a federal wilderness reserve famous for clear northern lakes free from motorized vehicles. Although the term's local origins are obscure, readers outside Ely likely first heard the word in 2007 when it appeared in a criminal complaint against six young men from Ely who rampaged through the BWCA on a summer night in August. The men used a boat with an outboard motor—prohibited in the wilderness area—to race around Basswood Lake while shooting fireworks and chugging beers. Their behavior grew more threatening as the night wore on. They began shooting pistols and an assault rifle and then menaced visitors camping around the lake. A sixty-four-year-old man camping with his two children reported that the men rummaged through their campsite and threatened them while the terrified family hid in the bushes nearby. "[Expletive] tourists . . . go home [expletive] 'enox' tree-huggers," the young men allegedly yelled at terrified campers. "You dumb [expletive] with your dehydrated food," one man allegedly taunted, "maybe if you make some smores for an hour we won't kill you." The men were arrested by local police several hours later and eventually were sentenced to jail time, required to pay fines, and were banned from the BWCA for up to three years.[1]

The threatening behavior of these young men was an extreme example of hostility between wilderness tourists and local residents in northeastern Minnesota. The rampage hardly represented the typical behavior of northeastern Minnesota residents. In an editorial, a local newspaper wrote that the young men's "actions should horrify everyone." Yet the editorial also hinted at an underlying tension between local residents and tourists who came to the BWCA. "There has long been a tendency in our area to paint youthful rebels who run afoul of Boundary Waters regulations as folk heroes," the editorial noted. Actions such as "motoring in a paddle-only lake, or a late night border run on snowmobile," although illegal, were tolerated and even celebrated among local residents.[2] A former iron ore mining town that relied on the tourist trade after the last iron ore mines closed in the 1960s, Ely was a flashpoint for such tensions, but it was hardly unique among northeastern Minnesota's small cities and towns. Throughout northeastern Minnesota and especially on the Mesabi Range, deep opposition to environmental regulations was woven into the culture of everyday life. In its extremity the 2007 BWCA rampage brought into sharp relief tensions over environmentalism, tourism, and the post-mining landscape that ran within northeastern Minnesota's and especially the Iron Range's culture in the late twentieth and early twenty-first centuries.

What explains the growing resentment and hostility toward wilderness tourism and the environmental movement in Minnesota's iron ore mining regions during the late twentieth and early twenty-first centuries? What drove the six young men to single out enox tourists in their rampage around Basswood Lake? Although attitudes about environmentalism among Iron Range residents were not monolithic, the region has a history rife with tension and hostility to the modern environmental movement. This chapter argues that tensions between Iron Range residents and environmentalism sprang from the conflict between the Iron Range's economic ties to industrial centers outside Minnesota and its political ties to Saint Paul and Minneapolis. Economically the Iron Range looked east to Chicago, Cleveland, and Pittsburgh, the centers of the United States steel industry. Yet state politicians in Saint Paul often had little appreciation for the Iron Range's unique economic and cultural connections outside the state. As one Duluth official told a reporter in 1982, the Iron Range "is a part of Minnesota only in name. It is really an outpost of the eastern industrial establishment—Pittsburgh's colony, if you will." According to this official, residents of northeastern Minnesota were "a lot more concerned about what's going on in Cleveland and Detroit and Pittsburgh than we are about what's happening in the Twin Cities."[3] Yet the Iron Range was nevertheless beholden to state laws, including many environmental regulations set by lawmakers in

Saint Paul who had few ties to the Iron Range. These tensions were heightened in the late twentieth century when other regions of Minnesota, especially the Twin Cities, began to emphasize service sector jobs and the experiential economy. "During the last half of the twentieth century," geographer David Lanegran argues, "Minnesota has gradually evolved from a landscape of work toward a landscape of play."[4] The Iron Range largely resisted this transformation and thus found itself economically and culturally out of step with state politicians by the late twentieth century.

Three major conflicts illustrate how the Iron Range's economic ties outside of Minnesota and its unique regional identity led to tensions over environmentalism. First and most important, conflicts over the economic value of iron ore mining versus that of wilderness preservation and tourism drove tensions between Iron Range residents and environmentalists. In many ways the Iron Range faced a classic jobs-versus-the-environment showdown in the second half of the twentieth century as new environmental restrictions challenged older mining practices. When the mining economy waned—for reasons that had little to do with environmental regulations—the region increasingly depended on tourism, which fed deepening resentment of tourists from outside the region, especially from large cities such as Minneapolis/Saint Paul and Chicago. Yet a simplistic jobs-versus-the-environment paradigm fails to adequately capture the evolving nature of the dispute, which began with tentative alliances between labor and environmentalists that then broke down in the late 1970s and 1980s.

Second, growing rifts within Minnesota's Democratic-Farmer-Labor (DFL) party in the second half of the twentieth century exacerbated tensions over the environmental movement. The Iron Range was one of Minnesota's most reliable DFL voting blocs, but this solidarity masked deep fault lines as the statewide DFL adopted more consumerist and proenvironmental policies in the late twentieth century. Emphasizing a producer ethos and backed by powerful labor unions, the Iron Range DFL was increasingly out of step with the statewide party.

Third, Iron Range residents clashed with the modern environmental movement because they interpreted the landscape—both on the Iron Range and statewide—very differently from environmental movement activists. When they looked out across the mine pits and overburden dump piles of their hometowns, many Iron Range residents beamed with pride at the physical transformations wrought by a century of hard work. In contrast, environmentalists saw the Iron Range's mining landscape as a scarred and battered version of the pristine North Woods that was preserved in wilderness areas like the BWCA. Iron Range residents feared that the mining landscape, which many saw as direct evidence of generations of hard work in a hostile environment, would be replaced with a sanitized landscape of North Woods tourism, cleaned and packaged for consumption by weekend visitors from urban regions to the south. These tensions, which often overlapped and intermingled with one another, ran beneath decades of minor conflicts and hostility between residents of the Iron Range and the modern environmental movement.

JOBS VERSUS THE ENVIRONMENT

By the early twenty-first century tensions between iron ore mining and the environmental movement on Minnesota's Iron Range appeared to be a classic jobs-versus-the-environment dispute. It pitted the mining companies and their employees, who valued employment above ecological consequences, against a coalition of environmental groups, mainly based outside the region, who argued that the economic value of mining jobs had to be weighed against the environmental harm that came from mining. Yet the sharp dichotomy of jobs versus the environment does not adequately capture the complexity of change on the Iron Range over the course of the twentieth century, especially residents' changing understandings of the environment and the economic value of industrial mining. A standoff between miners who wanted jobs and environmentalists who urged caution was indeed the final result of this process, but it took half a century of evolution to reach that point.

In the broadest sense, the divide between jobs and the environment is a false one. All jobs involve alteration and manipulation of the nonhuman world, whether directly in the case of mining or indirectly in the case of modern office workers whose computers rely on coal-fired power plants and mines that unearth rare minerals. "Work," environmental historian Richard White argues, "offers both a fundamental way of knowing nature and perhaps our deepest connection with the natural world."[5] Yet this sensibility—understanding work as connection to nature—is rarely encouraged in modern culture. Instead, jobs that involve direct manipulation and alteration of the nonhuman world, such as mining, are often seen as especially destructive of the natural order.

The Iron Range was not alone in its conflict between those employed in a resource extraction industry and the modern environmental movement. From disputes between loggers and protectors of the northern spotted owl in the Pacific Northwest to battles over offshore oil exploration, there were many local conflicts between jobs and the environment in the United States in the late twentieth century. Yet on the Iron Range and elsewhere conflicts between industry and environmental groups were only the final result of decades of developments. A longer historical lens reveals that over the second half of the twentieth century environmental politics moved from tentative alliances with the new environmental movement, largely via labor unions, to conflict over jobs during the 1980s and 1990s.

By the post–World War II period, labor unions in the United States had been fighting to protect workers' health for decades. Central to their struggle were efforts to protect workers from the harmful effects of industrial pollution, both on the job site and in nearby communities where many workers and their families lived. Fighting to protect workers from silicosis and other industrial diseases of the early twentieth century, labor unions were among the first institutions to sound the alarm over deadly industrial pollution. In the 1950s and 1960s the nation's largest unions, such as the United Auto Workers and the United Steelworkers of America, forced lawmakers to consider environmental protections well before the environmental movement was popular on the national scene. Reviewing this early history of labor unions' push for environmental regulations, historian Scott Dewey argues that the major unions were "proto-environmentalists" that spurred national consideration of environmental regulation. These unions, Dewey notes, "took [environmentalism] seriously, favored it consistently, and discussed it in terms and arguments sophisticated for the late 1950s and early 1960s." By the 1960s it seemed that a type of labor environmentalism was developing that would marry labor's organizing might with the concerns of the young environmental movement. Steelworkers pressed legislators for cleanup of pollution in their communities and were influential in passing the Clean Water Act and the Clean Air Act in the early 1970s. Later tensions between laborers, especially those organized in labor unions, and the environmental movement were not visible as late as the mid-1970s.[6]

Alliances between labor unions and the environmental movement broke down during the late 1970s. Due to economic shocks such as energy crises and stagflation, unions nationwide turned their focus to bread-and-butter economic issues that left little room for the concerns of environmentalists. The wider environmental movement, for its part, increasingly challenged the economic growth model that had guided the postwar American political economy. With a labor movement fighting to preserve the paychecks of workers and an environmental movement increasingly questioning whether industrial growth was compatible with their goals, the nascent labor environmentalism of the 1960s and early 1970s broke down.

The paths of the labor and environmental movements further diverged in the 1980s. Under assault from the Reagan administration and deindustrialization, the labor movement was increasingly desperate to preserve past gains. Many environmental groups adopted new consumer-focused positions that had little common ground with labor unions. Thus fifteen years from the end of the century the divide between the labor movement and the environmental movement had been sealed, both in northeastern Minnesota and nationwide.[7]

Minnesota's iron ore mining regions followed a similar path from possible alliances between labor and the environment to open hostility between the two by the end of the twentieth century. The Iron Range is an imprecise label for a string of small cities and towns strung along the Mesabi Range, which has a deposit of rich iron ore that runs southwest from Babbitt to Grand Rapids. Major cities on the Mesabi Range include Virginia, Hibbing, and Eveleth. Minnesota's iron ore mining districts are part of the larger Lake Superior iron ore mining district, which also includes the Marquette and Menominee Ranges in Michigan's Upper Peninsula and the Gogebic Range in northern Wisconsin. Iron ore mining in Minnesota was limited to the Mesabi Range by the 1960s, but two other ranges, the Vermillion Range near Ely and the Cuyuna Range near Crosby, had been mined in the early twentieth century.[8]

In the early decades of the twentieth century conservationists held up the Mesabi Range's mechanized open-pit mines as exemplars of modern, scientific conservation of natural resources. During the Progressive Era, conservationists emphasized rational management of the nation's finite resources, including iron ore, and hailed the Mesabi's open pits for their complete removal of ore. Whereas underground mines typically left 10 percent or more of the mine's ore in the ground, open-pit mining could quickly remove all the ore from any given deposit. From the conservationist viewpoint, the open-pit mines of the Mesabi Range were exemplars of efficient use of natural resources.[9]

The middle of the twentieth century brought both an increasing emphasis on efficient, technologically advanced mining practices and a growing desire among Iron Range residents for natural recreation areas. The major development in iron ore mining and milling in the middle of the century was the implementation of low-grade iron ore mining on the Mesabi Range. This form of mining, often referred to as taconite mining in reference to the flinty gray rock used for the process, took low-grade ore and then used extensive milling processes to create pellets with high iron content and uniform shape. The taconite process was credited with saving the Iron Range's mining industry at mid-century. Previous worries about depletion of the natural ore mines were alleviated since the Mesabi Range contained enormous quantities of low-grade taconite ore.[10]

Yet taconite mining and milling generated waste in a manner as enormous and efficient as the giant taconite plants. Since less than 30 percent of the taconite ore was iron, processers disposed of the remaining 70 percent of the rock in a slurry near the plants. The taconite process also used massive amounts of water and energy, generating both large holding ponds of wastewater and concerns about air pollution that bedeviled the Iron Range throughout the second half of the twentieth century.

Just as the new Mesabi Range taconite plants were coming online in the 1950s and 1960s, residents became concerned with maintaining clean and accessible natural areas for outdoor recreation. Like many blue-collar Americans

Figure 13.1. The major development in iron ore mining and milling at mid-century was the implementation of low-grade iron ore mining, also known as taconite mining, on the Mesabi Range. Since less than 30 percent of the taconite ore was iron, processors disposed of the remaining 70 percent of the rock in a slurry near the plants. Here the Reserve Mining Company's plant at Silver Bay discharges its waste on the shore of Lake Superior. *Source:* Library of Congress, Record Group 412: Records of Environmental Protection, EPA DOCUMERICA, 412-DA-9122, Donald Emmerich, photographer.

in the postwar era, Iron Range residents celebrated wilderness and outdoor recreation spaces. Union miners on the Iron Range, flush with postwar prosperity, flocked to the woods and lakes of northern Minnesota for hunting and fishing trips. Iron Range towns often had active chapters of Ducks Unlimited and offered firearms safety courses. With the promise of steady paychecks and union-backed vacations, many Iron Range workers grew increasingly concerned about maintaining the natural beauty and accessibility of the North Woods.[11]

Potential alliances between Iron Range miners and the young environmental movement, however, were upended in the 1960s and 1970s by a high-profile pollution lawsuit against one of the region's largest taconite producers. The lawsuit, *United States of America v. Reserve Mining Company*, pitted the Environmental Protection Agency (EPA) and the Minnesota Pollution Control Agency (MPCA) against the Reserve Mining Company in a conflict over dumping tailings into Lake Superior. Reserve Mining Company was one of the first producers of low-grade taconite ore on the Iron Range. It mined the ore on the eastern Mesabi Range and then shipped piano-sized chunks of ore by railroad to its giant mill in Silver Bay, which crushed and ground the taconite to the consistency of flour. Powerful magnets then separated the magnetite iron from siliceous gangue—only 20 to 30 percent of the original ore was iron—and finally the iron was rolled into marble-sized pellets for use in steelmakers' blast furnaces. The tailings were pumped out of the mill via two enormous tubes and dumped into Lake Superior.[12]

The Silver Bay mill produced an enormous quantity of ore and dumped an even more massive amount of tailings into the lake: an average of sixty-seven thousand tons each day. This amount, historian Thomas Huffman notes, was "more than two times the estimated solid waste garbage produced by New York City during the same period."[13]

Minnesota politicians and local residents both enthusiastically supported the Reserve Mining Company mill when it was first built in the late 1940s and 1950s. Northeastern Minnesota was hungry for jobs, and the plant promised a new era of high-tech iron ore production. By the late 1960s, however, attitudes about the environment, industry, and pollution had changed dramatically. Minnesotans statewide were more keenly aware of industrial pollution as a threat to their health and the state's environment than they had been just decades earlier. With this new concern, the Reserve Mining plant's enormous tailings flowing into Lake Superior now appeared to be an egregious example of prioritizing industrial jobs over the natural environment.

In 1969 the federal government filed a lawsuit against the Reserve Mining Company alleging that the company was violating several federal water pollution laws. According to complex federal regulations, the case first wound its way through numerous public forums and fact-finding conferences before arriving at the federal courthouse in Minneapolis. Public conferences in Duluth quickly became flashpoints for the emerging conflict between workers in the iron ore mining industry and Minnesotans concerned about pollution. While scientific experts and company officials presented dense testimony to the enforcement conferences, protestors picketed outside. Ultimately, however, the enforcement conferences were indecisive and the EPA sued the Reserve Mining Company to clean up the pollution.

In federal court the case was derailed by accusations that the tailings being pumped into Lake Superior by Reserve Mining Company were asbestos-like and potentially carcinogenic for the cities that drew their water supplies from the lake, including Duluth. The trial took place just as Americans were awakening to the health risks posed by airborne asbestos. Paul Brodeur's 1974 book, *Expendable Americans*, for instance, reported that thousands of Americans were exposed to asbestos and now faced life-threatening diseases.[14] The judge presiding over the case, Miles Lord, hyperbolically described Reserve

Mining's pollution as "potentially . . . the number one ecological disaster of our time."[15]

It took a decade for the lawsuit to wind to an ambiguous and unsatisfying conclusion. The federal judge hearing the case ordered Reserve Mining Company to stop dumping taconite tailings into Lake Superior on April 21, 1974. But the Eighth Circuit Court of Appeals immediately overturned this injunction. Evidence of widespread cancer risks from taconite tailings proved inconclusive, and the company grudgingly agreed to switch to on-land tailings deposits. Reserve Mining's threat of pollution became moot in the early 1980s when the company went bankrupt amid the steel crisis that shut down steel mills nationwide.

During the decade that the Reserve Mining trial wound through the federal courts, workers on the Mesabi Range and in Silver Bay followed the case with a mixture of worry and indignation. The first reaction of many workers, especially those who worked for Reserve Mining in Silver Bay, was outrage that they could lose their livelihood over the pollution lawsuit. Many workers assumed that the mill would face additional regulations or be forced to switch to on-land disposal, but few considered that the operation would be shuttered entirely by a federal judge. Yet closer analysis revealed that these workers were also concerned about their health and the health of their families. For example, one taconite plant worker met secretly with a reporter and confided that he was deeply worried about the health effects of working in the taconite mill.[16] The Reserve Mining lawsuit was only the most prominent of several conflicts between jobs and the environment that swept the Iron Range during the 1970s and 1980s. But the end result was the same: industrial miners saw themselves pitted against environmentalists in a zero-sum game.

What jobs were available for Iron Range workers if iron ore mining was no longer available, whether because of a decline in the industry or environmental regulations? One of the most frequently mentioned options to absorb former miners was tourism. Jobs in the tourist industry, it was imagined, would replace many of the jobs lost in the iron ore mines.

There was a long history of proposing tourism as a panacea for job loss on the Iron Range. Given the Iron Range's location in northeastern Minnesota, a region famous for pine forests, clear lakes, and what people used to call "good sleeping weather," it is not surprising that outsiders saw potential for a tourism industry. In a prescient observation, a sociologist visiting the Iron Range in the 1930s predicted that civic boosters in the region would begin promoting tourism by "advertising the hinterlands as a summer vacation center of unusual attractions" to visitors from nearby urban areas.[17] Statewide, tourism was a booming industry throughout much of the twentieth century. By the 1960s tourism was Minnesota's third-largest economic sector, fueled largely by the fishing vacation trade in the northland.[18] Yet the continued strength of the region's iron ore mining economy kept tourism in the background until the postwar years.

As many of Minnesota's iron ore mines slowed down dramatically after World War II, regional leaders turned to the tourism industry as a potential source of jobs for the struggling regional economy. A 1968 report, for example, noted that tourism could replace jobs lost in the declining iron ore mines. "An expansion of tourism," the report noted, "would help to diversify this area's source of income."[19] Tourism was thought to be one of the only feasible means to diversify the Iron Range's economy away from its one-industry reliance on the iron ore mines. Indeed, desperation for a more diversified economy fueled much of the interest in tourism during the postwar decades. A *New York Times* reporter visiting the Iron Range in 1964 described towns building ski resorts and touting their natural beauty, but warned that "the tourist industry is not growing fast enough to offset the loss of jobs and income caused by the decline of mining."[20] Tourism was ultimately a second-best option for the mining region as its traditional industrial base dried up in the postwar decades.

The need to diversify the Iron Range's economy led regional development agencies, especially the Iron Range Resources and Rehabilitation Board (IRRRB), to promote tourism during this era. Created in the 1940s as a state-level agency charged with economic development in Minnesota's iron ore mining regions, the IRRRB did not initially consider promoting tourism as part of its mandate. But in the 1960s tourism moved to the forefront of the agency's agenda due to the pressing need for economic diversification in a region that offered few options for new industries. From the 1960s tourism development was central to the IRRRB's mission.[21] Yet the agency was always conflicted about the best ways to promote tourism in an isolated open-pit mining region. It tried many different approaches over the years, from developing campgrounds near lakes to stocking fish in abandoned iron ore mines that had filled with water.[22] By the late 1970s tourism development efforts came to center on an ambitious heritage tourism center in Chisholm, the Iron Range Interpretative Center. The IRRRB had high hopes for the center, intending it to become "the most outstanding tourist draw in the Midwest."[23]

There is little evidence that Iron Range residents embraced tourism as a viable replacement for mining jobs, despite the official emphasis on tourism as a panacea. Iron Range residents were skeptical about jobs in tourism for many different reasons, but many emphasized tourism's negative economic position relative to mining. A 1982 business report found that tourism-sector jobs paid, on average, one-third the wage of jobs in taconite mining.[24] In addition to lower wages, northeastern Minnesota residents worried that a tourism economy would be less egalitarian. One laid-off miner explained that tourism would do little to alleviate unemployment in the region and might exacerbate inequality, since resort owners would benefit disproportionately from government programs to help the tourism sector.[25] Iron Range residents also struggled to imagine their mining-dependent hometowns as destinations that tourists would want to visit. Consultants researching possibilities for tourism on the Iron Range in 1980 wrote, "Many of the individuals on the Range interviewed . . . expressed skepticism about tourism development on the Range. They were concerned about the lack of lakes, and the lack of tourist accommodations. The lack of strong, vocal community support for tourism on the Iron Range may well be the most difficult of all problems to overcome."[26] For Iron Range residents who saw their hometowns as gritty but hardworking mining enclaves, it was difficult to imagine that they could—or should—be attractive to tourists from outside the region.

While Iron Range residents took jobs in the tourism sector—whether out of necessity or choice—many resented outsiders who, they believed, wanted to relegate northeastern Minnesota to a playground for recreational tourists. As Iron Range bartender Darryl Rice complained in 2010, "They pushed tourism down everybody's throats. They got rid of industry and replaced it with restaurants. Now what do we have to show for it?"[27] From the perspective of many Iron Range residents, tourism was a poor alternative to mining jobs.

FRACTURES WITHIN MINNESOTA LIBERALISM OVER ENVIRONMENTALISM

Another factor that contributed to tensions between Iron Range residents and the environmental movement was a growing rift within Minnesota's Democratic-Farmer-Labor (DFL) party over environmentalism's role within liberalism. For much of the mid-twentieth century Iron Range districts routinely voted for DFL candidates in statewide elections by large margins. Yet Iron Range support for the DFL was driven primarily by economic considerations. The DFL was the party of organized labor and blue-collar workers. When Minnesota's DFL and the national Democratic Party began taking liberal positions on social issues in the 1960s and 1970s, many Iron Range DFL supporters observed these developments with considerable trepidation. The DFL's adoption of strong support for environmental regulations and restrictions in the late twentieth century was a more direct blow against the economic liberalism of the Iron Range.

The DFL has long dominated Minnesota's liberal political landscape, but it has been internally divided many times during its history. During the early Cold War years, for instance, progressives aligned with Elmer Benson feuded with anticommunist liberals led by Hubert Humphrey. Iron Range miners were often among the most prominent and influential factions within the party's labor wing.[28]

Beginning in the late 1960s and continuing into the 1970s, New Left liberals within the DFL began to exert more influence and push a political platform that included environmental regulations. In many ways Minnesota's DFL followed the lead set by the national Democratic Party, which adopted environmental positions into its national party platforms in the late 1960s and early 1970s. The DFL split also played out along regional lines, with liberals from the Iron Range aligning with other rural DFL members against the interests of urban progressives from the Twin Cities.[29]

John Blatnik, the long-serving congressman for northeastern Minnesota's Eighth Congressional District, perfectly illustrated the growing tensions between environmentalism and industry within Minnesota's liberal political landscape. Born in 1911, Blatnik grew up on the Iron Range during the Great Depression, and his firsthand experience with searing poverty and unemployment in the mining region shaped his later political beliefs, especially his deep commitment to using federal funds to help regions stricken with long-term poverty. Blatnik first ran for Congress as a DFL candidate in 1946. He won the election and joined a cohort of new Democratic congressmen, including his friend John F. Kennedy, who were recently returned veterans committed to continuing the New Deal's legacy in the postwar era. Following his initial election in 1946, Blatnik was reelected thirteen times. His seniority led to a prominent position as chair of the House Committee on Public Works, a key legislative position that controlled federal spending on public roads, waterworks, and other internal spending.

Blatnik emphasized two main initiatives during his long career in Congress. The first was economic development for Minnesota's Eighth District, anchored by the beleaguered mining region that, Blatnik believed, needed federal aid to alleviate suffering. To this end Blatnik was a strong advocate for iron ore mining and other economic development projects on the Iron Range. Blatnik was especially interested in developing taconite mining. By the late 1940s he had become a strong supporter of the Reserve Mining Company, which he thought would bring much-needed industrial jobs to the Eighth District.

The second initiative that Blatnik spearheaded was clean water legislation. Blatnik authored several of the keystone clean water acts of the 1950s and 1960s, notably the 1956 amendments to the Water Pollution Control Act. Throughout the late 1950s and early 1960s, Blatnik was known as one of the strongest supporters of tough water pollution laws in Congress. He was honored by Hubert Humphrey—who described Blatnik as "the daddy of all these Water Pollution Control measures"—and by the Izaak Walton League.[30]

There was initially little conflict between the two priorities of Blatnik's legislative agenda. In fact, water pollution legislation began as a way to funnel federal money to localities for economic development. There is little indication that Blatnik was interested in cleaning up Minnesota's waterways. Historian Paul Milazzo describes Blatnik and his fellow congressmen, such as Edmund Muskie, who passed the initial clean water legislation as "unlikely environmentalists" since they cared little about the nascent environmental movement. Instead, Blatnik saw water pollution legislation as an excellent vehicle for pork-barrel spending. Transferring federal funds to small towns and counties throughout the nation, Blatnik's legislation continued the New Deal tradition of federally funded but state-led infrastructure building.[31]

Yet the growing environmental movement of the 1960s grew increasingly suspicious of Blatnik's legislative approach to pollution regulation. These groups criticized what they saw as a conservative approach taken by Blatnik and Congress's Public Works Committee. Blatnik championed local control over pollution regulation, which gave him powerful political leverage, but environmental groups argued that cities and towns did not strongly regulate industrial polluters, who often wielded enormous influence at the local level. As the Reserve Mining trial developed in the 1970s, Blatnik became a focus for many environmentalists angry at what they perceived as cozy ties between politicians and the mining industry. One leader of the opposition to Reserve Mining bitterly joked that the delta of taconite tailings spreading into Lake Superior outside Silver Bay should be named Blatnik State Park.[32] Battered by the trial and sensing that his brand of New Deal liberalism had run its course, Blatnik declined to run for reelection and left Congress in 1973.

Blatnik's career, and especially the growing rift between economic development and environmental legislation, illustrates the changing priorities of the environmental movement and the schisms that the movement created within the national Democratic Party and Minnesota's DFL in the last third of the twentieth century. Blatnik had combined economic development of his rural, industrial district and antipollution legislation as a seamless whole. More precisely, antipollution legislation was, for Blatnik and other mid-century liberals, another means to economic development and industrial growth. Blatnik was blind to the possibility that antipollution measures might challenge his larger mission of industrial and economic growth on the Iron Range. Yet this is precisely what happened by the late 1960s. By the end of his career in Congress, Blatnik was no longer a friend of the environmental movement. He was now portrayed as one of many liberal politicians who would sacrifice the nation's environmental health in the name of industrial growth.

By the early twenty-first century many Iron Range residents felt that Democratic support for environmentalism had gone so far that they were considering turning away from the DFL altogether. During the 2012 congressional election, for example, Ely's mayor, who generally supported DFL candidates, explained why he was now supporting a Republican for Congress: "The Democratic Party has gone too green for some of us in northeastern Minnesota. . . . They care more about the environment than jobs and people."[33] Yet dissatisfaction with DFL environmentalism did not lead all Iron Range residents to abandon the party. Instead, the tension largely played out within the DFL, opening a notable divide between an Iron Range caucus hostile to environmental regulations and the mainstream party dominated by the Twin Cities.

LANDSCAPE FOR PRODUCTION, NOT CONSUMPTION

A final factor that propelled Iron Rangers' hostility toward the modern environmental movement was a sharp contrast between the way in which Iron Range residents saw and valued the Iron Range's landscape versus the interpretations that outsiders, including environmental activists, drew from it. This landscape is disorienting for outsiders. It is difficult to know at first glance which hills and valleys are products of geology and which are the result of mining. Enormous mounds of overburden encircle many Iron Range towns. Many of these "hills" are close to a century old and have been reforested in the intervening years. Those searching for pristine wilderness will be disappointed by the Iron Range. Few areas of rural Minnesota have been more heavily affected by human activity in the past century. Yet individuals have interpreted this landscape very differently depending on their point of view.

When Iron Range residents looked out over the scarred and battered landscape of their home, they often saw one created by a century of hard labor. Through grueling and dangerous work, Iron Range residents transformed an isolated northern forest into one of the world's great mining districts. The enormous mine pits and dumps of overburden bore mute witness to this herculean labor. "The blood-red tablelike escarpments of Minnesota's Iron Ranges," geographer Richard Francaviglia notes, are "some of the grandest manmade topography on earth."[34]

In the early years of its development, observers from outside the Iron Range saw the region's mines as symbols of both productive industry and environmental conservation. The enormous pits were early adopters of several powerful new technologies, such as railroads that ran directly into the mine pits, and steam shovels that mechanized much of the labor-intensive work of moving ore.[35] As historians Steven High and David Lewis note, "Industrial landscapes were once the proud symbols of human progress and modernity."[36] Although historians typically associate this mindset with the Progressive Era, celebrations of industrial beauty continued well into the mid-twentieth century in northern Minnesota. In 1957, for instance, *Fortune* magazine published a series of watercolor illustrations of Minnesota's first taconite mills. The illustrations portrayed the giant mills as scenic additions to the landscape that combined industry, nature, and human labor.[37]

For other observers the open-pit mining landscape of the Iron Range has been interpreted as an ugly or even horrific destruction of the natural world. Writing in the late 1980s, geographer Thomas Baerwald described the Iron Range as "long, reddish gray gashes [that] marked the sites where miners had torn iron ore from the earth."[38] In recent decades, building on the premise that mining mars the natural landscape, concerted efforts to remediate or "reclaim" landscapes altered by mining have gained popularity. At the federal level, the Surface Mining Control and Reclamation Act of 1977 first brought federal oversight to the reclamation of abandoned mines. In Minnesota responsibility for administering mine land reclamation was split between the IRRRB and the Department of Natural Resources (DNR). Both on the Iron Range and nationally, mine-land reclamation efforts grew alongside the modern environmental movement. On the Iron Range such efforts have focused on planting trees and grasses on overburden piles to make them more attractive and prevent erosion. They have also focused on stocking abandoned pit mines—which are typically filled with water—with game fish. Geographers Peter Goin and Elizabeth Raymond note that Iron Range reclamation projects assume that the mining landscape is unnatural and in need of restoration. "The implicit premise of . . . environmental restoration . . . is that the Mesabi landscape, despite the massive upheaval to which it has been subjected, can and should be made to appear more 'natural.'"[39] Mine-land reclamation has been one of the most influential means by which the Iron Range landscape has been reimagined as unnatural and needing assistance from outside regulators.

Hostility to environmentalism was an important part of the culture of Minnesota's Iron Range region in the second half of the twentieth century. Yet as this chapter has argued, this hostility resulted from specific historical developments that shaped the Iron Range and similar industrial mining regions nationwide during this era. The fundamental source of this tension is the fact that the Iron Range's economic ties point east to the nation's industrial nerve centers, while the region is tied politically to Saint Paul. As a result, many Iron Range residents were frustrated that state politicians did not always understand their region's unique economic base and sought to impose unwieldy environmental regulations on iron ore mining. Conversely, many state politicians and outside observers believed that the Iron Range resisted change and stubbornly refused to accept modern environmental regulations on noxious industries.

Many Iron Range residents remain skeptical about the environmental movement and those who identify as environmentalists. Unfortunately, their hostility caricatures the nuance and complexity of environmentalism in the twenty-first century. Conversely, environmentalist sentiment often obscures the deep and direct connections that Iron Range residents have to the natural world. Yet there are also signs that the iron ore mining industry has accepted the importance of environmental considerations as part of its business practices. During a 1990 interview, one mine manager explained how awareness of mining's environmental impacts had changed during his career: "You never had to worry before about dumps as to where you put them and what kind of condition they were in as far as revegetation. I think nowadays it's watched a lot closer and rightfully so. When you get through mining why should it destroy the countryside? There should be something to do with it afterwards."[40] Environmental permitting was an integral part of the mining industry by the 1980s and continues into the twenty-first century. When U.S. Steel announced that its KeeTac facility in Keewatin would be expanded in 2008, for example, the company addressed environmental concerns immediately. An executive explained, "We want to bring environmental stakeholders into this process."[41]

For better or for worse, it is likely that tensions over environmentalism will continue on the Iron Range. Even if the iron ore mining industry has made a limited peace with new environmental regulations, proposed nonferrous mines—typically copper-nickel mines—are poised to split the region yet again in the twenty-first century. Since at least the 1970s geologists have recognized that northeastern Minnesota contains valuable copper and nickel deposits. Yet a nonferrous mining industry has been slow to develop, in part due to concerns that copper-nickel mining would have significant environmental consequences in the region. Unlike iron ore tailings waste, which is chemically inert, waste products from copper mining typically include large amounts of sulfuric acid and other noxious chemicals. These toxins are an unavoidable result of low-grade copper mining and milling, although the modern mining industry argues that they are much better prepared to handle this waste safely than in past mining operations.[42] The potential environmental risks from copper-nickel mining in northeastern Minnesota are high. As debate continues over how to balance the region's economic needs against the potential ecological harm of mining, it is likely the region will continue to divide along predictable lines.

A HOUSE DIVIDED

The Minnesota Experimental City and Competing Narratives of Conservation

Todd A. Wildermuth

SWATARA, MINNESOTA, IS A THREE-STREET TOWN—a six-street town, if you count the county roads that come together at its outer edges. The town sits 130 miles north of Minneapolis, near the northwestern corner of Aitkin County where the total population is about sixteen thousand people. Aitkin County contains plenty of such towns. One of them—Hill City, about 7 miles north of Swatara—advertises itself as a city. But according to the US Census Bureau, not one of Aitkin County's sixteen thousand residents is "urbanized." Hill City residents and others may protest, yet Aitkin County officially remains what it has long been: a city-less, rural county where named lakes outnumber named towns by a wide margin. Interstate highways and international airports have missed the county entirely, and a dozen or so townships remain legally "unorganized." Much within its 1,822 square miles looks as it did forty years ago.[1]

Had the dean of the University of Minnesota Institute of Technology, Athelstan Spilhaus, and the publisher of the *Minneapolis Star* and *Minneapolis Tribune* newspapers, Otto Silha, had their way, it all would be very different. Had their shared vision for Aitkin County become reality, Swatara would now be surrounded by pollution-free industry. Hill City would sit aside a worldwide air transportation hub. The two towns, the new industries, and the new airport would be connected by a "guideway network system"—a type of light-rail transit system. Up to 250,000 people would live within a ten-mile radius of either Hill City or Swatara, far greater than the 600 or so who reside there today. In the grandest visions of Spilhaus and Silha, all of this would have unfolded over the course of a mere ten years between the mid-1960s and the early 1970s. As the two men occasionally said, they dreamed of an "instant" city. Why such rapid development? So that all technology would be current, they said; so that, for once, a city and everything in it could come close to technology's cutting edge.

What Spilhaus and Silha wanted to do was build the greenest modern city ever. As they envisioned it, the new city would cap its overall population and keep residents in place with a strict urban growth boundary. Within that boundary, nature would have a strong say in what was built and where. Planners would deliberately site and group buildings to provide consistently pleasing views of nature. Carefully planned housing would yield widespread access to places where residents could play, swim, walk, or bike. City codes would exclude polluting industries and require complete recycling of fresh water. Indeed, using the latest "space-age technology," planners would largely erase the age-old divide between the "dirty" city and the "clean" countryside. Their guideway transportation network would eliminate cars, the first and key step in freeing the city from the internal combustion engine. If that goal proved elusive, they would at least eliminate combustion engines from the city's surface. Beneath the city would lie an elaborate underground tunnel system, big enough to house "fume sewers" (to pump air pollution away from the city), underground roads (for delivery trucks), and construction machines (for raising and removing buildings on the surface immediately above).

Technology in this new city, as Spilhaus and Silha envisioned it, would respond not just to known problems and needs but to future ones as well. Their city would be composed of modular parts, precast or preconstructed pieces, which would allow for energy-efficient assembly and disassembly. It would be, by design, a city in a constant state of flux; a city that need never grow old; a city that could always feature the latest industries with the best pollution controls. As Spilhaus and Silha said time and again, it would be an *experimental* city first and foremost: an international proving ground for city technologies that could, after testing in northern Minnesota, be exported worldwide, to old and new cities alike. They would call it the Minnesota Experimental City, or more typically the MXC.

Between 1966 and 1973 this green-tech urban vision was much more than a fantastic idea. Indeed, selective snapshots of the proposed city's progress would testify to an impressive trajectory: in July 1966 we would see Spilhaus and Silha sitting down to discuss the city with the Secretary of Housing and Urban Development as well as the administrator of NASA. In April 1967 we

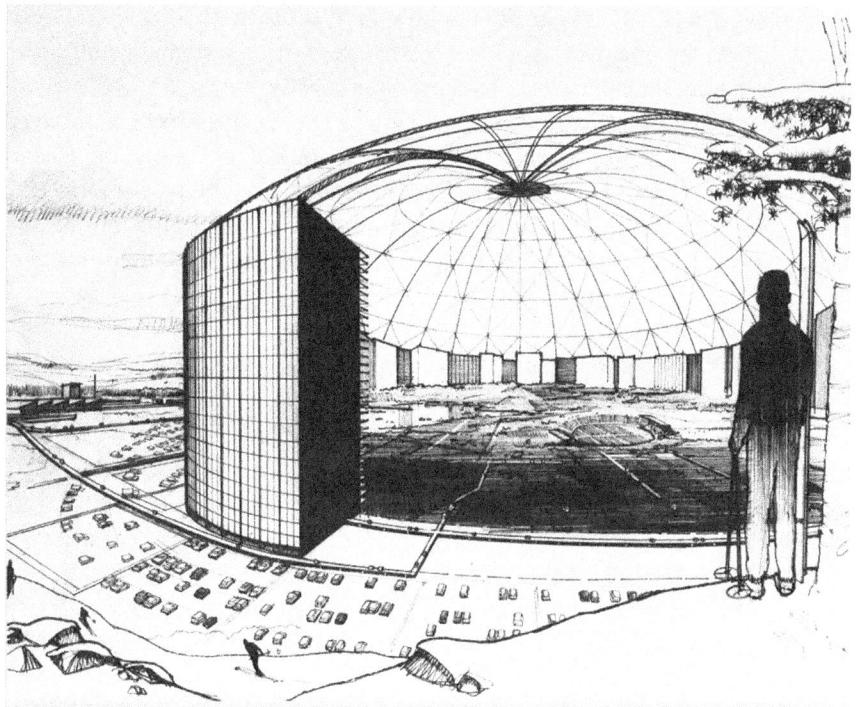

Figure 14.1. The Minnesota Experimental City (MXC) project aimed to build the greenest modern city ever, proposing an updated version of Ebenezer Howard's Garden City capped by a dramatic use of Buckminster Fuller's geodesic dome. For a project launched for the very purpose of eliminating urban pollution, however, the MXC had to answer surprisingly often to charges that it would cause environmental damage. *Courtesy of Chad Freidrichs.*

would see the two men sharing a banquet table as the nation's vice president, Hubert Humphrey of Minnesota, congratulates them on their vision and federal funding. In December 1967 we might catch Spilhaus from a different angle, this time on the dais himself, delivering a distinguished lecture on the Experimental City before the prestigious American Academy for the Advancement of Science. Come 1969, we would find Silha again sitting with a vice president, this time Spiro Agnew. With them are the Secretary of Commerce and other Washington insiders, with Silha once more receiving encouragement for his project. In 1971 we would catch Silha watching as Minnesota governor Wendell Anderson signs into law a bill to create a Minnesota Experimental City Authority (MXC Authority). And at various points throughout 1972, we would see Silha and other members of the Minnesota Experimental City Authority boarding a private plane, jetting about the state, and evaluating potential city sites. Flipping through that collection of snapshots—and perhaps tossing in one last one, of Silha in 1974 promoting the city to yet another vice president, Gerald Ford—we might wonder how a project with such well-positioned friends could fail. We might even double-check to make sure there really is no Minnesota Experimental City in Aitkin County. Yet Swatara and Hill City stand today much as they were then, with no Experimental City in sight. Conceived as an environmental supercity, the MXC ultimately foundered in the face of a determined opposition that effectively used environmental law—especially provisions governing public participation in environmental decision making and pollution prevention—to bring the elaborate vision crashing to the ground.

THE PROPER LOCUS OF POWER

The MXC got off to an auspicious start on June 4, 1971, when Minnesota governor Wendell Anderson signed into law legislation creating a new agency of the state, the Minnesota Experimental City Authority.[2] The result of more than two years of lobbying and testimony by Otto Silha and his allies, the MXC Authority made the idea of an experimental city a going concern of state government. The MXC Authority hired its own staff, worked up an official letterhead and, not least, had at their disposal a two-year appropriation of $140,000.

The state act gave the MXC Authority a specific, narrow public purpose: "to provide a vehicle for cooperative, coordinated effort by the state and federal governments, academic institutions, business and industry in partnership for planning and effectuating . . . the Minnesota experimental city or MXC."[3] The enabling legislation further specified the primary work of the MXC Authority as developing criteria for how to select a site for the MXC and then, based on those criteria, selecting a site to recommend to the governor and state legislature.[4]

After more than a year's work, the MXC Authority announced at its November 17, 1972, meeting that it had narrowed its consideration to two final sites—one in Aitkin County, where General Manager James Alcott had secured a county resolution "very strongly supporting" the MXC, and one in Douglas County, where support was less clear.[5] Most of the MXC Authority meetings had been small affairs to this point with only the board, a few conservation and civic groups, and some MXC staff in attendance. The November 1972 meeting, however, marked the beginning of open opposition to the MXC, as could be observed by the sudden interest of and appearance by local television and print media.

A board of commissioners from Pope County, located immediately south of the proposed Douglas County site, was the first to lodge a formal complaint against the MXC Authority and its plans. The proposed city was, according to the Pope County commissioners, only the latest in a long string of unwanted intrusions into the quiet counties of west-central Minnesota. Worse than the intrusion itself, however, was the sense among the commissioners that outsiders were making plans that affected Pope County without ever discussing them with the people of Pope County. "It is strongly suspected," the commissioners wrote in their resolution, "that even now there are proposals existing in public and private documents not known to the people [who] would be most affected," adding that "such procedures run contrary to customary and democratic process." The commissioners accused the MXC Authority of causing "the residents and Township Boards stress, confusion, anxiety and fear" by not giving any early warning and, in the view of local residents, springing the MXC on unsuspecting people with little time to react. Fed up with the MXC and other outside intrusions, the board resolved, with no little sense of outrage and desperation, that they be told "at the outset" of any plans affecting their county, that all communications be in plain language, and that anyone making a proposal affecting the county answer questions promptly. Finally, and most to the point, the commissioners resolved—with more hope than enforceable power—that the "future use of these townships rests with the desires of the residents and property owners." The commissioners of Pope County stood for self-determination by the most local form of government; the MXC appeared to threaten that spirit.[6]

Months passed and more concern arose, culminating in a public hearing on the MXC held on February 13, 1973, by the Minnesota Pollution Control Authority (MPCA) at the Minnesota Department of Health Building in Saint Paul.[7] After a series of official inquiries by state officials, the floor was opened to comments from the public. Dick Pemberton from Fergus Falls stepped to the microphone first. "I speak here today on behalf of an organization known as Park Region Organization to Ensure Citizen Togetherness [PROTECT]," he said. PROTECT had a broad appeal in the Douglas County area, Pemberton claimed, with members from the city councils of Fergus Falls and Alexandria, the Minnesota Township Officers Association, the Minnesota Taxpayers Association, and over fifteen sportsmen's clubs in the area. With that kind of membership, Pemberton demanded the MPCA's attention and consideration, saying, "I believe when I stand here this morning and speak to you that you hear not one voice but 10,000 voices or more and these voices say we challenge the concept of [an] experimental city." Douglas County had, by this time, been officially passed over by the MXC Authority in favor of the Aitkin County site, but Pemberton's and PROTECT's objections to the Experimental City—wherever it landed—were still hot. In one particularly fiery portion of his prepared statement, Pemberton declared:

> If growth does occur, we challenge that it must and should occur within the framework of spontaneously generated new cities. We have a social structure in rural Minnesota. We have a political structure. We have an economic structure. We have people in rural Minnesota that are capable of thinking for

themselves and capable of governing themselves. We ask that the people be allowed to speak, to say whether or not they want this magnificent gift that is being offered them, which some of us see to be a Trojan horse. We call for a referendum of the people. Let them speak. Let them vote.[8]

With this exposition, Pemberton moved the discussion beyond a mere matter of populism or local control of government. He was making, in a direct and plainspoken way, an appeal to a particular kind of traditional conservatism. Pemberton rooted his argument in a specific worldview that granted special privilege to the past: things in rural Minnesota were the way they were because those were the best ways to prove themselves over time; they were attributable, that is, to a slowly developed and emergent social logic. Such a logic or pattern came into being "spontaneously" and organically over time, had multiple layers (social, political, economic), could not be reproduced by any centrally planned project, and would be irrevocably displaced by the imposition of the MXC.

In this light, the MXC was twice damned for its attempted intrusion on the countryside. First, as shown by the Pope County commissioners, the MXC failed as a matter of courtesy and simple process; local governments and property owners were not asked to participate sufficiently in the decision making process. Second, the very nature of the MXC—centralized, logical, instant—threatened what Pemberton and others considered the essential human value of rural Minnesota: a voluntary and emergent community, shaped organically by some combination of space, time, and human will; an evolutionary feature of the landscape made possible by its distance from modern metropolitan influence.

DIVERGENT PERSPECTIVES ON PLANNING

The advocates for the MXC were driven by a concern that an increasingly urban America was wastefully expanding onto lands that could be used for agriculture and recreation, leaving behind underused infrastructure, and all the while never stopping long enough to figure out how to live somewhere without ruining it with pollution. The essential question they raised has been a recurring one in the American experience: Is there a collective, coordinated, positive vision of American settlement that can replace the individualistic, fragmented and catch-as-catch-can pattern that has characterized the country for most of its history?

Almost a hundred years before the MXC's time, John Wesley Powell in the late 1800s explored this same question—and all of its attendant tensions—as he effectively closed America's western frontier for a short, tumultuous time.[9] Thirty years later Lewis Cecil Gray and the national land-use planners of the New Deal did so again, mapping out the nation and parsing it into pieces that should be left alone, used for agriculture, or turned into cities.[10] Powell and Gray were representatives of an earlier, more agrarian America. The inspirations for their work differed markedly from those that would come later with urbanism: Powell worried that the arid western frontier would be inefficiently and undemocratically developed without coordinated planning; Gray feared that the same kinds of "submarginal" farming lands that fed the Dust Bowl would continue to come under the plow unless the nation designated them for protective use.

In the years between the mid-1960s through at least the late 1970s, Silha and his colleagues believed there could be such a rational, coordinated vision. They mostly talked about the MXC itself, but that was only the starting point for a bigger project. Silha and Spilhaus founded the whole idea of the MXC on a larger plan, and in service of a national demographic need. America, they said, would grow by a "Tulsa a month" for the next thirty years, and it needed a place to put those people.[11] The MXC was only one of those Tulsas. It was to be one part of a national program of planned spreading-out, a new American city for a new American age. After the MXC they hoped that others elsewhere would want a new city too, that everyone would see the merits of planned national settlement. Instead of an America of ever-expanding megalopolises— unplanned, shapeless developments on a map—the MXC would bring in an America of well-spaced, discrete dots on a map. After the MXC, all new cities would have a center and an edge, an urban center that eased out into lower-density areas mixed in with protected agricultural land, with even more open space—recreational or agricultural—providing a permanent buffer between one city and the next.

The basic decentralist idea was hardly new to America, having been imported in the form of Ebenezer Howard's Garden City by Lewis Mumford and the Regional Planning Association of America in the 1920s.[12] But Silha and Spilhaus gave it modern twists. They added Buckminster Fuller's geodesic dome to Ebenezer Howard's Garden City, mixed in projections about future population growth, combined a growing suburban critique with concerns about the city left behind, tossed it all together with first-generation pollution management technology, and aimed to drop the amalgam into the mixed pinelands of northern Minnesota. Their advice to the rest of the nation: mix and repeat, all across the landscape.[13]

For all of its emphasis on modernity and high technology, the MXC was in many ways an American throwback—an echo through time of a call for greater order in the long-term project of settling a growing nation. As others had done before, it tried once again to square a hope for efficient and orderly expansion with a strong current of American culture running against large-scale planning.

If Spilhaus and Silha were aware that this had failed before—and there is

no particular evidence that they were—they might have been excused for trying again. The conditions in the Twin Cities in the late 1960s and early 1970s
appeared especially promising for their experiment. The seven-county Metropolitan Council had been created by state legislation in May 1967 to coordinate
regional planning and public service delivery; it stood as one of the earliest
and few examples of regional metropolitan planning that was granted actual regulatory authority instead of merely advisory influence. Also, the Twin
Cities metropolitan area had joined a national "new town" trend in launching
two ambitious developments with federal funding and support: the Cedar-
Riverside high-rise complex, an example of the popular "new town in town"
model, and representing a different point along the new town spectrum, the
"satellite new town" development of Jonathan, located thirty miles southwest
of Minneapolis.[14] Minnesota was also particularly active in the national trend
toward active state management of pollution control, symbolized most readily
by the creation of the Minnesota Pollution Control Agency in 1967. For many
sessions of the US Congress in the early 1970s, a national land-use planning
bill had been actively debated as a complement to the nation's Clean Water
Act, Clean Air Act, and Endangered Species Act.[15] In that context the extensive
planning associated with the MXC can be seen as less a cultural outlier and
more a comfortable fit with a spirit of environmental thinking and positive attitudes toward planning in those years. The Twin Cities, in short, offered Silha
and Spilhaus encouragement—an impression that innovation of this sort was
being welcomed locally in recognition of a national zeitgeist.

That encouragement went only so far. In late November 1972, during the
peak of debate, proponents traveled to critical regions in rural Minnesota to
promote the MXC. Otto Silha and Eugene O'Brien, chairman of the state-
funded MXC Authority, presented their case to a crowd of over 150 at the
Alexandria Rotary Club in Douglas County. Their remarks were broadcast on
KXRA, the local radio station.[16] Other high-level staff of the MXC Authority met on December 6 with about 60 Douglas County residents in Evansville,
taking questions for over two hours.[17] They did so again on December 20 at
Brandon High School, packed with approximately 350 concerned residents; at
a town hall meeting at the Hill City High School with about 225 people in attendance, and in Millerville where another 150 area residents attended.[18] Overall the responses to the MXC were distinctly negative: in Evansville "lots of
hostility got vented"; in Brandon the attendees where "overwhelmingly against"
the MXC; only in Hill City were people "about evenly divided on the merits of
MXC."[19]

After a particularly disappointing series of meetings in January, James Alcott, the MXC Authority's general manager of the project, tried to set out on
paper the thinking of opponents. At the root of the MXC's troubles, Alcott

remarked, was a difference in perspective on the essential logic of planning. On this front there appeared to be an insurmountable difference in perspectives between "outstate" opponents and the mainly metropolitan members of the MXC Authority. Perhaps based more on frustration than reflection, Alcott described the key differences he saw between the two perspectives. Proponents of the MXC looked at a statewide, and even nationwide, set of trends—namely the growing population, the extraordinary growth of suburbs, and the emptying out of both urban cores and small, rural towns. For Alcott and his colleagues, this pattern was harmful—from the rural community all the way up to the national level—and its resolution the raison d'être of the MXC. As best as he could tell, many of the MXC opponents simply did not recognize this pattern as a problem, or at least not a problem that had much bearing on their lives. The opponents, he wrote, did not care to recognize the larger patterns of growth and population imbalance that he believed he saw all around. Mostly the opponents were "isolated and like it that way." Or, in other words, they "do not perceive much relationship between what happens in the Twin Cities metro area and in their area." Over and again, he heard them saying that the city and the country were simply separate things in separate places, a view that he and the MXC Authority fundamentally disputed.[20]

Even at the earliest stages of the MXC, when Spilhaus was still giving it shape among a few friends, there was always an implicit presumption that the world was interconnected. In the nature that Spilhaus studied—the upper atmospheric winds, the waters of the oceans, the continental air masses—interconnection was a scientific truth. In human spheres—transportation, economies, ideas, communications—global interconnection had not quite reached the status of established truth. But to the globe-trotting Spilhaus, as to the jet-setting Silha and Alcott, that truth was well on its way. The MXC, Alcott was realizing, had always been a cosmopolitan idea, an idea of people in metropolitan areas who saw the world as an interchanging network of cities and information and materials. For cosmopolites like himself, this view of globally interconnected cities and economies was just the way things actually were. This was a macro-scale view of the landscape, a view that considered the state, the nation, the continent, or the globe as a relevant and real operating unit of society. Once someone adopted the macro-scale worldview, it became a simple truth that neither Alexandria nor Swatara, neither Fergus Falls nor Hill City, could be thought of as having a fate separate from that of Minneapolis or Saint Paul. Once that proposition was accepted, the idea of the MXC as a solution to the unplanned growth of Minnesota was, as Alcott saw it, a simple and logical step. Opponents fought the MXC because they refused to see this logic, because they operated more on emotion than reason. Or, as Alcott put his frustrations in a confidential memorandum: "Because of the heavy emotional content of the outstate feeling in looking at MXC, I don't find the logical approach too convincing, but I can't find a nonlogical approach that works very well either."[21]

TECHNO-ENVIRONMENTAL SKEPTICISM

For a project launched for the very purpose of eliminating urban pollution, the MXC had to answer surprisingly often to charges that it would *cause* environmental damage. On multiple occasions throughout late 1972 and early 1973, the proponents of the MXC found themselves returning time and again to disputes over the city's green bona fides.

Responding to the increasing concerns of their constituents, elected officials in the Minnesota legislature began to draw environmental concerns into the debate. State representative Glen Sherwood of the Aitkin County area lodged a formal statement of opposition to the MXC, citing his concern that it could not "drop 250,000 people into our beautiful region without a devastating environmental impact."[22] State representative John Johnson, the assistant majority leader in the state House who introduced the original bill that established the MXC Authority, wrote to Otto Silha in January to express similar frustrations. "When I first started explaining the MXC," Johnson wrote, "I made the point that every effort would be made to put it on land that was marginal for use in agricultural or recreational purposes. Now I am getting heat from people who say the sites being recommended are excellent recreational lands, as well as agricultural to some extent."[23] Popular, and therefore elected, sentiment seemed to wonder quite directly how "prime" nature could benefit in any way from having a city built atop it.

Some experts clearly wondered the same. By statute the MXC Authority had to submit its site recommendations to the Minnesota Department of Natural Resources (DNR) and the Pollution Control Agency for review. The DNR's evaluation of two early proposed sites, one in Houston County and another in Cass County, reflected the standard environmental challenges that would continue to hound the MXC throughout its life. The Houston County site definitely needed to go, the DNR claimed, for a number of ecological reasons, including the cavernous limestone underlying the entire site that rendered it "particularly susceptible to pollution from percolating pollutants." "[A] suitable waste disposal system," they added, "would present major, if not unsolvable, problems." For completely different reasons, the DNR argued that the Cass County finalist also needed to be eliminated. The DNR's case against Cass County was easily summarized: "This site represents the largest roadless area in this portion of the state, encompassing a diverse resource base, richly blessed in lakes, forests, and wildlife. This is a truly unique section of Minnesota not suitable for massive urban development."[24]

In the case of each recommendation, the DNR directly challenged the MXC

Authority on one of the foundational premises of the proposed city. In rejecting the Houston County site, the DNR doubted the MXC's claim that a pollution-free city could be built; in rejecting the Cass County site, the DNR questioned just how realistic it would be to expect that a city of 250,000 people could fit seamlessly into a wilderness that would, outside the firm borders of the city proper, continue to function as it did before the city would be built. Not surprisingly, both the Houston County and Cass County sites were dropped from further consideration to satisfy the DNR.

But once the Aitkin County site was selected, the MPCA picked up where the DNR left off. At the February 1973 MPCA public hearing, Commissioner Milton Fellowes took particular aim at the MXC's purported ability to force new technological solutions to the well-known pollution problems of modern living. "Are the systems within the experimental city," he asked, "compatible with all new legislation like for instance the 1972 water quality act?" Alcott admitted a lack of familiarity with the new Clean Water Act, but he assured Fellowes that the MXC would be well ahead of the curve by reducing water pollution "in essence to zero by 1985." What about the other new set of federal regulations, those of the Clean Air Act, Fellowes wondered. He referred to recent lawsuits that established anti-degradation standards and asked, "[Do] you anticipate you can build a city of 250,000 without degrading the air quality of the air beyond reasonable limits?" Yes, Alcott responded, we do believe we can nearly eliminate traditional air pollutants. In the MXC, he argued, cars with internal combustion engines would be excluded or significantly replaced by rail transit, and energy would be generated from alternative technologies.[25]

The MPCA's executive director, Grant Merritt, remained unconvinced. He cited with concern a comment that had been published in the Congressional Record in 1969, when MXC proponents were lobbying Washington for federal funding. In the record, Merritt said, "was a reference and analogy to the taconite development of Minnesota—how the Experimental City would be akin to the development of taconite in this state." This concerned Merritt because he and the MPCA had been battling the Reserve Mining Company for years over the deposit of taconite tailings directly into the waters of Lake Superior. "Now having spent a good share of my career involved with the problems of taconite tailings. . . . I am very much concerned about this analogy . . . and I guess that I am wondering if we are going to end up with the same kind of situation." To bring the analogy to a finer point, Merritt continued, "What I am concerned about . . . is that mistakes were made by reasonable people in the rush to develop taconite." A similar mistake appeared to be in the making with the MXC, he proposed. Based on his own firsthand experience, Merritt said he felt that "the rush to technology before adequate reasonable consideration has been given to the consequences" was one of the "keys to the environmental crisis."[26]

Proponents of the MXC found themselves in a bind. It seemed that they were asking Greater Minnesota for the policy equivalent of an unsecured loan. Even if the city made good sense as part of a larger statewide or national plan, what guarantee could the MXC Authority give that such a larger plan was on its way? Very little, it answered; it was admittedly an experiment. What evidence did the MXC Authority have that the MXC would not become just the latest point of entry of the city into the country, and that a polluting urban expansion would not follow in the its wake as it had in the past? None. Building a city used up rural land directly and would cause serious land-use alterations all around it, even with greenbelt legislation. Many people in Greater Minnesota enjoyed the land in its rural condition. What promise were they being given that the MXC could successfully "sacrifice" some rural land in order to protect a larger amount of rural land that would otherwise be developed? The MXC Authority could make no such promise.

There was, strictly speaking, no larger plan, and there were no larger commitments, either to protect rural land or to make sure any standing cities were rebuilt to further reduce the spread of cities into rural areas. On occasion MXC proponents insisted that they were seeking zoning power over seventy thousand or more acres, only a few square miles of which would be developed for a city and the rest zoned for strictly rural use. Yet they had no firm commitment that the legislature would guarantee that arrangement.

ANATOMY OF A FAILURE

Throughout the very first years of the 1970s, Minnesotans—or at least their elected representatives—were initially very open to the idea of the MXC. But as it moved from idea toward reality, deep suspicions emerged, redirecting the MXC discussion and bringing the entire experiment to a sudden end.

MXC opponents' suspicions were many and intertwined. They remain difficult to separate completely. Three themes dominated, however, and can be readily identified during the period of most active debate over the MXC, spanning from around November 1972 to April 1973. First, MXC opponents were, at the most visceral level, consistently concerned over the proper locus of public power. In many instances it mattered little what the specific proposals entailed. "Outstate" residents of Greater Minnesota were concerned that decisions affecting their lives were being made elsewhere. They resented the intrusion on local autonomy and on the presumed right of self-determination; they further resented that a single, centralized solution was being proposed in response to a dispersed problem. That the MXC proposal emerged specifically from the Twin Cities and from cosmopolitan elites only further shaded the opponents' views.

Closely related, the second theme of opposition revolved around the nature of land-use planning itself. The MXC proponents took for granted that land uses should be considered at some level greater than the most local level. At a minimum their model presumed a measurable degree of county control of land use. The "M" in MXC gave the accurate impression that proponents favored some degree of state-level land planning. Often MXC advocates even spoke of the need for America as a whole to consider the idea of planning land use with the entire nation in mind. For MXC opponents, formal land-use planning in the countryside was a suspect enterprise. Planning implied planners, of course, and the only planners participating in the MXC were those based in the Twin Cities; this ran right back into the objection that the MXC intruded on self-government. But opponents' concerns were more particular than that. Even if willing to cede some degree of local control, they frequently contested the chosen form of landscape-scale planning presumed by the MXC model.

Third, MXC opponents were greatly concerned with what might be characterized as the model's technological environmentalism. What proponents thought of as an environmental supercity was not universally taken to be environmental at all. Spilhaus, above all, imbued the MXC with such fantastic technological optimism that proponents took as given that engineered solutions would be found to nearly all pollution. Later this notion would, in environmental circles, come to be described as "technology-forcing" policy: if one simply outlawed pollution, the thinking went, human cleverness would provide a solution. Few MXC opponents shared this faith. Coupled with this technological skepticism, opponents were troubled by even the simplest matter: How could one call it "environmental" at all to build a city for a quarter-million people where there used to be trees and deer and bald eagles? Spilhaus was, in this sense, seen as less a scientist than an alchemist—and no amount of clean technology could convince opponents that a lead lump of city could be turned into a gold nugget of nature.

Many factors were at work when the MXC died in the summer of 1973, not the least of those being opposition from the public and from state agencies. A great many anti-MXC letters to the governor, for example, included plenty of short-sighted, self-interested, and discriminatory language. Many writers were quick to share that they lived in the country to stay away from the ghettos and crime of the city. They celebrated the sameness of class and heritage around them and noted that they felt most comfortable around people who felt the same way. While no writer felt it proper in a letter to the governor to say that he or she did not want minorities around, plenty of them openly said that they disliked "city" people and problems; whether taken literally or as a euphemism, a simple and uncritical disdain for all city people hardly made a socially reasoned case against the MXC. [27]

Similarly, expert state agencies, particularly the MPCA, held an open bias against the MXC that aided its demise. Oddly, there was nothing about the MXC fundamentally at odds with the MPCA's mission. The MXC would almost certainly increase pollution wherever it was sited, but it was intended to reduce pollution *overall* within the state and *over time*; this goal was entirely consistent with the mission of the MPCA. The agency, however, never evaluated what pollution the MXC could have saved the state in many different places; it focused exclusively on the pollution the city would cause in one place. At the most practical level MXC proponents mismanaged public relations and failed to convince their legislative backers that the project was worthy of continued votes in the face of organized grassroots and state agency opposition. Yet it would be more than incomplete to ascribe the fall of the MXC only to those sources.

At a more enduring level, the MXC suffered because its backers wrongly presumed that a coherent set of new cultural beliefs had emerged in the late 1960s and early 1970s regarding the limits to growth and the need for widespread land planning. They allowed themselves to believe that their own perceptions of the world were more widespread than they really were.

Silha had once declared boldly that the times were "rotten ripe" for change. [28] And certainly new cultural forces were at work. Minnesotans of the late 1960s and early 1970s, like most Americans, were adjusting to the firm realization of certain modernities: a known and predictable "baby boom" bulge in the domestic population that would require more houses, roads, offices, and malls; expanding new technologies such as never-before-seen chemicals and space-age electronics; a well-established trend toward metropolitan areas and away from farms; a growing sense that all such developments were creating a greater demand for pollution control and protected natural areas.

These dynamics were changing attitudes across the country. In Minnesota the shift suggested a dependable regional openness to innovation and experimentation. And yet openness was not the same as willingness. The old, worn paths of unplanned settlement ran in deeper tracks than any of the new dynamics or understandings could overcome. In discussions of the MXC, the new era presumed by Silha and Spilhaus translated only into a new vocabulary and a forum for the fresh expression of old concerns; the outcome remained the same.

DISSECTING A
NATION-LEADING LEGACY

The Minnesota
Acid Rain Story

Gregory C. Pratt

IN 1982 THE MINNESOTA LEGISLATURE PASSED, and Governor Al Quie signed, the Minnesota Acid Deposition Control Act. This was the first legislative action on "acid rain" passed in the United States, addressing what was arguably the hottest environmental issue at the time. The law directed the Minnesota Pollution Control Agency (MPCA) to develop an acid deposition standard and a control plan to ensure that the standard was attained and maintained. In 1986, despite resistance from coal and other energy interests, the MPCA carried out its directive and promulgated rules comprising a standard and control plan.

In 1990, eight years after Minnesota passed its acid deposition law, the US Congress enacted amendments to the Clean Air Act that included Title IV on Acid Deposition Control, which called for a 50 percent reduction in sulfur dioxide (SO_2) emissions. The reductions were based on what was thought to be technically and politically feasible, not on emissions or deposition levels that would necessarily protect sensitive resources. The federal law did not promulgate a standard for acid deposition, leaving Minnesota's measure as the only acid deposition standard in the nation.

Minnesota's unique experience on acid rain begs several questions than will serve as the focal points of this essay: How and why did Minnesota pass the Acid Deposition Control Act? What sort of political, social, and cultural factors facilitated the act's passage? And what can we learn from this experience? In attempting to answer these questions, I argue that several factors converged in the early 1980s that led Minnesota to take a leadership position on the acid deposition issue, including the strong environmental ethic found among its citizens, a committed and mobilized coalition that had recently been energized by the battle to preserve the Boundary Waters Canoe Area Wilderness (BWCA), a perceived threat to the state's resources, a progressive legislature, and the relatively inexpensive solution of switching to low-sulfur western coal at a time when western coal was working to expand its market share.[1]

MINNESOTA'S ENVIRONMENTAL LEADERSHIP

Minnesota is often recognized for its progressive stance on environmental matters. Perhaps the rich and varied landscape of the state, or the large population of conservation-minded hunters and fishers, has fostered an attitude of respect and preservation. In addition, the traditions of Native Americans living in the state and the many Scandinavians and northern Europeans who immigrated here drew on heritages that included environmental protection. Whatever the cause, the result is evident in the laws and history of the state. The MPCA was established in 1967, three years before the federal Environmental Protection Agency (EPA). Minnesota is one of about twenty states to adopt a state environmental policy act that requires review of the potential for significant environmental impacts of proposed projects, similar to the National Environmental Policy Act (NEPA). The list of actions Minnesota has taken at the forefront of environmentalism include the state Clean Indoor Air Act, chlorofluorocarbon product bans, the 1970s fight over the dumping of taconite tailings in Lake Superior, the establishment of the BWCA, and the Clean Water, Land, and Legacy Constitutional Amendment approved by Minnesota voters in 2008, among many others.

Not least among the highlights of Minnesota's environmental accomplishments is the state's response to the problem of acid deposition. In 1982 the Minnesota legislature passed the Acid Deposition Control Act directing the MPCA to:

1. Identify areas of the state with resources sensitive to acid deposition;

2. Adopt an acid deposition standard that applies to the sensitive areas;

3. Establish a control plan, addressing both in-state and out-of-state sources, to attain and maintain the standard;

4. Ensure that all Minnesota sources subject to the control plan are in compliance by 1990.[2]

This legislation and ensuing activities placed Minnesota at the forefront of the acid deposition issue. Attention was focused on the scientific and legal arguments occurring in the state as foreshadowing the inevitable coming national debate.

SETTING THE SCENE

Environmental consciousness was on the upswing in the 1970s, a trend heightened by the convergence of a number of political and societal factors. Oil and energy shortages drew attention to the idea that natural resources are finite, and environmental cataclysms like the Cuyahoga River bursting into flames and the partial nuclear meltdown at Pennsylvania's Three Mile Island reactor awakened awareness that human activities were egregiously damaging ecosystems that we rely on for sustenance. The perception was growing that something was amiss in humanity's relationship to the earth. The changing zeitgeist of this era was presaged in 1948 by British astronomer Sir Fred Hoyle, who predicted, "Once a photograph of the Earth, taken from outside, is available . . . a new idea as powerful as any other in history will be let loose."[3] This prediction became a reality on August 23, 1966, when NASA's Lunar Orbiter 1 took the first photo of Earth from the moon's orbit, and many observers have noted the impact of this image on the human psyche. The activism spawned by this growing awareness culminated in the celebration of the first Earth Day (1970) and the passage of pivotal legislation, including National Environmental Policy Act (1970), the Clean Air Act (1970), and the Clean Water Act (1972). Interestingly, it was Republican Richard Nixon who signed NEPA and the Clean Air Act, although as Mark Hertsgaard has noted, "Nixon did all this . . . not because he was a closet tree-hugger—the poor man wore wingtips to walk on the beach—but because he was a calculating politician . . . keenly aware that 20 million Americans—roughly 10 percent of the population in 1970—took some kind of civic action that first Earth Day."[4]

Several factors came together that provided fertile ground for the passage of these new environmental regulations: a rising demand for environmental amenities, increasing affluence to pay for them, blatant and obvious pollution, 1960s social and antiwar movements that gave rise to "a new questioning and politically active generation," a growing view that "corporations were getting away with murder," and a business community that was caught off guard.[5] The mindset embodied in these landmark pieces of legislation was that environmental problems were predominantly local issues that could be solved by controlling big polluters. The mechanisms outlined in the law came to be called "command and control" and were mainly comprised of directives to industries to reduce their emissions. The recognition that environmental impacts could cross political boundaries, that effects could be subtle and move among environmental media, and that solving these complex problems might require multi-jurisdictional and even multinational efforts was growing, but still a hazy glimmer on the horizon.

Acidic deposition has existed since coal burning became widespread during the Industrial Revolution.[6] The term *acid rain* was coined in 1872 by Robert Angus Smith, who first observed the acidic nature of precipitation in 1852 in Manchester, England, and later in other areas influenced by the combustion of coal. Modern awareness of acid rain dates to 1955 and is attributed to Earl Barrett and Gunnar Brodin, working in Sweden and Eville Gorham in the rural English Lake District.[7] Gorham and Gordon described acidic deposition in North America from metal smelting in Sudbury, Ontario, in 1960.[8] The issue gained national prominence in the United States in the 1970s with publications by Gene Likens and others describing the acidification and loss of fish populations in New England lakes.[9] The main cause of the problem was widely believed to be emissions from distant coal-burning power plants, but addressing this type of scenario did not fit the extant regulatory framework. In addition, the scientific work on long-range atmospheric transport and deposition of air pollution was a new field, and the definitive cause-and-effect relationship between sulfur dioxide emissions and acid rain was not yet established to the satisfaction of sufficient numbers of political decision makers to enact new legislation.

The election of Ronald Reagan in 1980 signaled a change in the national approach to environmental regulation. His environmental positions can be characterized, in part, by the following quotes: "A tree's a tree. How many more do you need to look at?"[10] "Approximately 80 percent of our air pollution stems from hydrocarbons released by vegetation, so let's not go overboard in setting and enforcing tough emission standards from man-made sources."[11] Reagan's appointees to head the EPA and the Interior Department (Anne Gorsuch Burford and James Watt, respectively) actively opposed regulation and worked to undo existing rules.[12] With powerful interests opposing action on acid deposition, the more politically palatable approach of "further study" was routinely offered. In 1978 the National Atmospheric Deposition Program (NADP) began its monitoring program, and in 1980 the National Acid Precipitation Assessment Program (NAPAP) began studying the causes of acidification. These efforts ramped up during the 1980s, consolidating knowledge and understanding of the acid rain phenomenon. Also, in 1980 lake acidification in New England made national news. Over time the issue captured public attention, prompting environmentalists to call for action on the matter. After more than ten years of these national studies, with the state of the science maturing and after leadership changes in Washington, Congress finally acted in 1990 when the reauthorization of the Clean Air Act included Title IV, which called for an approximately 50 percent reduction in domestic SO_2 emissions.[13] This mandate, however, was based on what was politically achievable at the time, not on a critical load analysis that defined the reduction in acidic deposition needed to protect lakes and forests, and no national standard was set. Later analyses found that further reductions were required to protect sensitive resources.

Despite the national political turn away from environmental regulation in the 1980s, Minnesota went in a different direction. During the years leading up to the passage of Minnesota's Acid Deposition Control Act, progressives were in the midst of an era of political dominance. Democratic-Farmer-Labor (DFL) governors held office for twenty-six of the thirty-six years between 1955 and 1991, although Republican Al Quie—who signed the acid deposition legislation—was governor from 1979 to 1983. The DFL had long held majorities in the state legislature, and by the late 1970s the party held as many as 49 out of 67 seats in the state Senate.[14] The party makeup of the state House of Representatives was more volatile than the Senate. The DFL held as many as 104 out of 134 House seats in the late 1970s, but by the early 1980s the body was more evenly divided.[15] While DFL dominance of the legislative and executive branches of government sometimes facilitated passage of environmental legislation, there was broad support for environmental causes across both parties. The House authors of the Acid Deposition Control Act included DFL representatives Arlene Lehto, Willard Munger, and Lee Greenfield and Republicans Gary Laidig and William Dean, and the bill passed the House with only four opposing votes. The Senate author was Gerald Willet (DFL), who later served as commissioner of the Minnesota Pollution Control Agency from 1987 to 1991. The final Senate vote tally was 56 in favor and 3 opposed. The law appropriated $81,455 from the general fund to conduct the mandated work in the first year (1983) and established an assessment on electric utilities for future funding.

Interestingly, the legislature did not enact a statute calling for specific controls, but instead delegated authority to the MPCA to study the issue and promulgate rules. This approach reflected the lack of specific knowledge about the sensitivity of resources in Minnesota. Although numerous acidified lakes had been identified in New England and Canada, and a handful as near as Michigan and Wisconsin, no acidified lakes had been found in Minnesota. The approach also reflected a level of respect between the branches of government that has become less prevalent over time. The view on both sides of the aisle was that the MPCA could be trusted to conduct a serious analysis and propose a reasonable solution.

DRAWING BATTLE LINES

Preliminary work by the MPCA in the early 1980s suggested that acid rain threatened some 1,000 lakes in the state, and an additional 3,500 could be threatened if emissions continued unchecked. There was also speculation that some of the thin, humic soils and coniferous forests of northeastern Minnesota might be vulnerable to acidic inputs. University of Minnesota professor Eville Gorham testified before a US House subcommittee in 1980 that "in Minnesota, the data we now have suggests that rain is approaching the acidity levels which have caused distinct damage in Scandinavia. The loadings of sulfate are similarly approaching the levels which have been shown to cause damage."[16]

These threats came on the heels of the recently fought battle to preserve the Boundary Waters Canoe Area Wilderness. Congress passed the act establishing the wilderness in 1978, and President Jimmy Carter signed it following hotly contested debates.[17] The National Association of Property Owners, the state of Minnesota, and other property owners in and near the area challenged it in court.[18] The Friends of the Boundary Waters Wilderness, the Audubon Society, the Izaak Walton League, the League of Women Voters, Minnesota Environmental Control Citizens Association, Minnesota Rovers, Minnesota Ornithologists Union, the Wilderness Society, Wilderness Inquiry, and the Sierra Club all worked to preserve the wilderness designation, which the US Supreme Court upheld in an 8–1 decision on March 8, 1982. The Friends and fellow preservationists constituted a highly mobilized and motivated coalition that was very sensitive to threats to the newly created wilderness. At the same time that these threats percolated in the public consciousness, the Canadian utility, Ontario Hydro, proposed to build a large, unscrubbed coal-burning power plant in Atikokan, Ontario, just a few miles from the Quetico and BWCA wilderness areas. Environmentalists and their allies were outraged and demanded that something be done.

These events proved very powerful in the political arena in Minnesota, and convinced lawmakers to proceed in the face of reluctance at the national level. House author Lehto noted that Minnesota should take steps to limit its own sources of acid rain before it tried to force other states to reduce emissions. "I believe it will lay a foundation for us to go after the other states who are causing 70 percent of our problem," Lehto said.[19] These arguments handily carried the day in Minnesota despite opposition from electric utilities, the taconite mining industry, and out-of-state coal-mining interests. They argued that Minnesota should not act alone because it would put the state's industries at a competitive disadvantage. Furthermore, emissions reductions ahead of national action might not be recognized and credited under a later national plan, and Minnesota's industries might have to make reductions again in the future.

In other states and nationally, industry arguments prevailed for nearly another decade. A major factor in the controversy was that the states with vulnerable natural resources were not the states with power plants and other industrial sources that emitted the bulk of the sulfur dioxide responsible for most of the acidification, nor were they the states producing the coal to fuel the power plants. Different states and regions were pitted against one another. The northeastern states with the best documented damages were united in calling for a national plan, but support was tepid in much of the rest of the country. The Ohio Valley states, with many large power plants burning high-sulfur eastern coal, adamantly opposed limitations on their emissions. The eastern mining states, with notable congressional leaders like Senator Robert Byrd of West Virginia, were also staunch opponents to regulation. The "Rust Belt" states (including Michigan, Pennsylvania, Ohio, and Indiana) were seeing a decline in the automobile, steel, and rubber industries that had long under-pinned their economies. Addressing acid rain would increase electricity costs and further damage those industries, and so these states also opposed action. Minnesota stood alone in taking the moral high ground by choosing to limit its own emissions, even though the bulk of the deposition in the state originated elsewhere. Even New England states with serious deposition problems did not act independently, but rather chose to support a national approach.

Following passage of the state's acid deposition legislation, the MPCA established a Technical Review Committee in 1983, made up of representatives from the forest, tourist, utility, and mining industries, as well as environmental advocates and various state agencies. The purpose was to discuss issues and possibly resolve some of them prior to the formal rulemaking process. The meetings were open to the public and attracted a far larger number of people than the formal list of representatives would suggest. Although these meetings narrowed the focus to the most controversial issues, they generated little agreement about how to resolve them. During 1984 and 1985, numerous additional meetings gathered public input and planned for the administrative hearings on the proposed rule. In October 1984 the MPCA published a Notice of Intent to Solicit Outside Opinion. The following month the MPCA Citizens Board authorized a formal rulemaking proceeding, and the agency filed documents with the chief administration law judge (ALJ) indicating the intent to hold hearings on the matter. MPCA director Thomas Kalitowski issued a Notice of Hearing in December, and eight days later Northern States Power Company (NSP)[20] filed a procedural objection to the proposed rules.[21] Despite a number of procedural filings in the succeeding weeks, the process moved forward.

PROMULGATION

Under the Minnesota Administrative Procedures Act (APA), rule promulgation requires a *Statement of Need and Reasonableness* (SONAR) that describes the statutory authority for promulgation, the reasoning for the rule, and an explanation of why the specific details of the rule are a reasonable means to accomplish the goal.[22] The intent to promulgate must be published in the State Register and otherwise advertised, defining a thirty-day public comment period. If twenty-five or more persons request it, an administrative hearing must be held in which testimony is taken before an ALJ. Minnesota has an unusual structure for rules review that includes legal review of proposed administrative rules by judicial offices. In addition, the attorney general's (AG) staff plays a dual role in rulemaking: they act as legal counsel to agencies, helping them write rules, and they review the over 80 percent of rules that do not require a public hearing. The Office of Administrative Hearings' ALJs also have two rulemaking roles: they preside at public hearings, and they review the approximately 20 percent of rules that require a public hearing. Both offices review rules to ensure that agencies have statutory authority to adopt a rule and that they comply with the APA's due process requirements. However, with respect to whether a rule is needed and reasonable, deference is given to the executive branch department or agency. Both the AG and ALJ apply a standard that requires an agency to demonstrate a rational basis for a rule, but it does not require the agency to show that it is the "best" rule. This is the same standard applied by the courts if and when a rule is subjected to judicial review.[23]

The MPCA prepared a very lengthy and technically detailed SONAR.[24] The process was somewhat unusual, because the MPCA conducted original research, and contracted with university scientists for additional research in developing the evidence presented in the SONAR rather than relying solely on preexisting information. One of the first steps was to determine whether there were resources in Minnesota that were sensitive to acidic inputs. The MPCA sampled 1,842 lakes and found that 474 of them had alkalinity values less than two hundred micro-equivalents per liter (ueq/l), the generally accepted upper limit for a sensitive lake. The sensitive lakes were mostly small in size (averaging about 240 acres), but many were valuable fishing and recreational lakes. Extrapolating from this sample, the MPCA estimated that there were as many as 2,200 sensitive lakes in Minnesota. Based on data from Canada, Scandinavia, and the northeastern United States, the most sensitive lakes were thought to have alkalinity values less than forty ueq/l and would be the first to undergo acidification. The MPCA estimated that approximately 200 of these very sensitive lakes existed in the state, primarily in Itasca, Lake, Pine, and Saint Louis Counties. MPCA scientists also evaluated soils and forests for their sensitivity to acidic inputs, but determined that the very low-alkalinity lakes were more sensitive and focused standard development on protecting these resources. If the standard protected the most sensitive resources, it would also protect less sensitive resources. The MPCA was not able to identify any lakes in Minnesota where the chemistry of the lake had been acidified to the point where it adversely affected aquatic life. Such lakes were known in neighboring Wisconsin, Michigan, and Ontario, however. A report and a listing of the areas of Minne-

Figure 15.1. Preliminary studies failed to identify any lakes in Minnesota where the chemistry of the lake had been acidified to the point where it adversely affected aquatic life. Such lakes were known in neighboring Wisconsin, Michigan, and Ontario, however, as demonstrated in this image comparing a healthy fish (above) to a starving fish (below) following the acidification of a lake in northwestern Ontario, near its border with Minnesota. *Courtesy of IISD Experimental Lakes Area.*

sota with sensitive resources were approved by the MPCA Citizens Board in April 1983.

Given the level of alkalinity in Minnesota's sensitive lakes, MPCA staff and university researchers determined that rainfall with a pH of 4.7 or lower (i.e., more acidic) would acidify Minnesota's most sensitive lakes. A critical link in the analysis was demonstrating that the pH of 4.7 corresponded with a wet sulfate deposition of eleven kilograms per hectare per year (kg/ha/y). Several chemical and statistical analyses confirmed this link, which became a major issue in the administrative hearings. Based on this analysis, the MPCA proposed a standard allowing no more than 11 kg/ha/y of wet sulfate deposition per year to keep the pH of precipitation at or above 4.7 and thereby protect sensitive lakes. The standard applied only in areas of the state with sensitive resources. The link between sulfate and pH was critical because air pollution sources do not emit acidity per se; rather, they emit substances like sulfur dioxide and nitrogen oxides that undergo chemical reactions in the atmosphere in which acids and acidifying compounds form. In 1985 the standard of eleven kg/ha/y was being met at all of the MPCA monitoring stations with the exception of the Sandstone site. The standard did not include dry deposition of sulfate because the available research most closely linked lake acidification with wet deposition and because dry deposition was more difficult to measure and model.[25]

In addition to proposing a standard, the MPCA simultaneously proposed a control plan, using long-range transport air dispersion models to determine which sources were responsible for sulfate deposition in the state. Based on the modeling, the agency determined that Minnesota emissions contributed about 7.6 percent of the sulfate deposition, and only two Minnesota sources (the Sherco and Alan S. King power plants) exceeded 1.5 percent of the total sulfate deposition at Sandstone. Of the 92.4 percent contributed from outside Minnesota, sources in only one state, Texas, exceeded 10 percent. The finding that Texas was a major contributor to wet deposition in Minnesota seemed surprising since it is so distant from the state. However, a closer examination made meteorological sense. It was well known that the Upper Midwest received the majority of its moisture in precipitation from the Gulf of Mexico. The same air masses that brought rain and snow to the state also brought pollutants that were added during their northward movement. Furthermore, Texas was a major polluter, emitting 1.16 million metric tons of SO_2 in 1980, mainly from its petrochemical industry, but also with a significant electric utility component.

The significant pollution from distant sources made developing a control plan a difficult undertaking. State and federal regulatory mechanisms at the time were ineffective in addressing acid deposition because they did not account for long-range interstate transport of pollutants; rather they were designed to address ground level pollutant concentrations near individual sources. The modeling data showed that reducing Minnesota's SO_2 emissions to zero would not result in attainment of the eleven kg/ha/y standard at Sandstone. Minnesota by itself could not control the bulk of the emissions coming from other states and provinces. Within Minnesota the largest SO_2 source, the NSP Sherco power plant, already operated with highly efficient SO_2 removal equipment. The MPCA concluded that a national approach to emissions reductions was necessary to attain the 11 kg/ha/y standard. Given this finding, MPCA proposed a control plan with provisions that would accomplish Minnesota's reductions if a national plan were enacted. The provisions of the control plan included:

1. A cap on SO_2 emissions from Minnesota's two largest utilities (Northern States Power and Minnesota Power);

2. A limit on statewide emissions of 194,000 tons per year (tpy), representing a reduction of 60,000 tpy from 1980 levels;

3. A requirement for Reasonably Available Control Technology (RACT) on Minnesota's two largest emitters; Allen S. King would be limited to 1.2 pounds of SO_2 per million British Thermal Units of heat input (lb/MMBTU), a 50 percent reduction; and Clay Boswell would be limited to 0.75 lb/MMBTU, a 20 percent reduction;

4. A two-year study period from 1990 to 1992 would reevaluate whether a second round of emissions reductions would be required.

The emissions caps and RACT requirements were achievable mainly through switching from high-sulfur eastern coal to low-sulfur western coal as fuel for Minnesota's coal-burning power plants. In fact, western coal, especially from the Powder River Basin in Wyoming, was ramping up production in the 1980s and 1990s. Low-sulfur western coal was inexpensive and nonunion, with low-cost transportation available on existing rail lines. Between 1980 and 2010 coal production from states east of the Mississippi fell from 600 million to 450 million tons per year while western states' production grew from 250 million to 650 million tons per year.

Regulators estimated the costs of the control plan at $40.9 million per year for NSP customers and $4.7 million per year for Minnesota Power customers. Benefits of the control plan were difficult to estimate because most of the socioeconomic value of natural resources, such as environmental amenities associated with pristine aquatic ecosystems, is external to the market system. The MPCA hired an economic consultant to conduct a contingent valuation survey of Minnesota citizens to estimate their willingness to pay for protecting sensitive lakes. The method entailed substantial uncertainties but estimated benefits between $1.0 and $89.4 million dollars per year, the midpoint of which exceeded the estimated cost. The economic study concluded that Minnesotans placed a high value on their aquatic resources and were willing to pay substantial amounts to protect them.

After publishing the SONAR and proposed rule, the MPCA prepared for administrative hearings. The Agency tapped the expertise of eleven state scientists at MPCA and the Department of Natural Resources and fourteen more scientists from the University of Minnesota, the University of Wisconsin, Bemidji State University, the US Environmental Protection Agency, the US Forest Service, the Wisconsin Department of Natural Resources, the Ontario Ministry of the Environment, and others. A coalition of environmental groups also participated, hiring an attorney and calling their own witnesses. Lining up to oppose the rulemaking were the state electrical utilities—Northern States Power and Minnesota Power—whose plants would be subject to controls. These utilities received various kinds of support from coal (particularly eastern coal) and taconite mining interests, as well as from the Electric Power Research Institute. They also contracted with consulting firms and scientists from the University of Minnesota, the University of Calgary, and others. Both in Minnesota and nationally, the electric utilities and other industries that opposed the regulations were not well organized, and their public image was at a low ebb.[26] The sense of human responsibility for nature commanded a moral high ground that may have been a factor in leading Minnesotans to take altruistic actions for the environment. On a more practical note, the $45 million cost was a relatively small amount to the affected utilities.

Hearings began on January 22, 1986, in Roseville before Administrative Law Judge Allan Klein, and concluded on May 1, 1986.[27] The judge took testimony from seventy-five witnesses over a total of thirty-five days. Approximately 965 exhibits were entered into the hearing record, as well as more than eight hundred letters from members of the public, and petitions bearing more than eight thousand signatures indicating support for acid rain standards. Evening hearings in Roseville, Duluth, Hibbing, and Rochester facilitated public participation. The hearing record was arguably the most complete and exhaustive examination of acid deposition assembled to that date. In addition to MPCA staff, major participants in the hearing included the Minnesota Department of Natural Resources, North American Water Office, Minnesota Power, North ern States Power, and a coalition of environmental organizations that included the Sierra Club, the Friends of the Boundary Waters Wilderness, the National Audubon Society, and the Izaak Walton League of America. Public testimony and exhibits overwhelmingly favored adoption of the rule. Most persons acknowledged that the rule would result in higher electric bills, but expressed their support for its adoption nonetheless.

Expert testimony sponsored by the major participants was compiled in thirty volumes. The testimony generated frequent controversy, with witnesses cross-examined at length by opposing parties.[28] Rather than accepting the work of the MPCA, NSP funded its own long-range transport air dispersion modeling, lake and watershed modeling, and statistical analyses of the data. The utilities presented their results as superior to those of the MPCA and suggested that their witnesses had more impressive credentials than MPCA experts. Many days of hearings turned into battles among experts, often on the minutiae of scientific technicalities.

In June 1986 Judge Klein released a fifty-six-page report which included 177 findings of fact.[29] In summary Judge Klein offered three key findings. First, he found that the MPCA fulfilled all relevant substantive and procedural requirements of law in regard to the rules, proposed standard and control plan, and the hearings. Second, he determined that the MPCA documented its statutory authority to adopt the proposed rules. And third, he found that the MPCA demonstrated the need for and reasonableness of the proposed rules. Judge

Klein recommended that the proposed rules be adopted consistent with his findings and conclusions. The MPCA Citizens Board adopted the rules in July 1986 and published them in the State Record in August. At the conclusion of the process, the Minnesota Acid Deposition Standard and Control Plan became law. The standard remains the only one on acid deposition ever promulgated in the United States. The opponents could have chosen to further dispute the rule promulgation in the courts, but elected not to do so, perhaps because of the relatively low costs of the program and the fact that those costs could be factored into their rate base.

Acid deposition control eventually became law at the national level. In 1990 the US Congress amended the Clean Air Act to include provisions (Title IV) for regulating acid rain. These amendments called for the reduction of 10 million tons of SO_2 emissions per year below the 1980 levels, to be achieved by the year 2000. Prior to this amendment, projections showed the emissions of SO_2 continually increasing up to the year 2000 and beyond. The emissions reductions needed to protect sensitive resources were hotly debated in Congress, which heard a great deal of expert testimony. The 10-million-ton reduction eventually adopted by Congress represented an approximate 50 percent reduction in emissions. National coal companies and utilities reluctantly accepted the deal as likely to be the best they could get. They worked to have the required reductions specified in statute rather than authorizing EPA rulemaking on the issue, fearing that the rulemaking process might lead to even more stringent reductions.[30]

WHAT CAN WE LEARN?

The Minnesota effort to control acid rain succeeded because several factors converged. A progressive DFL dominated the Minnesota legislature and was able to strongly influence the terms of the debate. An organized and committed constituency fresh from the BWCA battle stood ready to protect the state's new wilderness area (acid rain seemed to threaten the BWCA and other important fishing lakes). Commitment to environmental protection was on the upswing in public consciousness. Coal and utility interests were not well organized in their opposition. The solutions (scrubbers and switching to low-sulfur coal) were relatively inexpensive and played into the hands of the expanding western coal concerns. In addition, Minnesota had the sensitive resources, but did not have the coal resources responsible for the problem.

Beyond these factors, one intriguing aspect of Minnesota's acid rain story is the state's willingness and ability to act alone on an issue that demanded national and international action. Stalwart legislators, public servants, and ordinary citizens stood up and said, "Yes, we need a national plan to solve this problem, but we feel strongly enough about this issue that we are willing to do our part now, ahead of the rest, even though it may hurt our pocketbooks." All the more remarkable, they did so at a time when national leaders (e.g., Reagan) were putting the brakes on environmental regulation. However, Reagan's turn away from environmentalism did not reflect national opinion, which continued to favor pollution controls. By 1988 Republican presidential candidate George H. W. Bush proclaimed himself an environmentalist and campaigned at polluted Boston Harbor to show up the environmental shortcomings of his Democratic opponent, Massachusetts governor Michael Dukakis. The broadly held sense of discord between humanity and nature that led to the legislation of the early 1970s persisted through the 1980s, Reagan notwithstanding, and the majority of citizens remained eager to address the problems.[31] There was also a developing value placed on taking precautions in our interactions with nature, although the term *precautionary principle* had not yet come into the environmental lexicon. It seemed like common sense to try to head off problems before they got out of hand.

A second intriguing feature of the Minnesota acid deposition process is that it marked a watershed point in environmental regulation. Politicians could no longer simply lay problems at the doorstep of local polluters. Issues had become national and international in scope, and solutions required a wide net and multiple players. This trend has only accelerated over time; we now recognize that some of our most intransigent problems are caused by "nonpoint" pollution coming from the integrated and cumulative effects of the small actions of whole populations. Solutions to such problems seem to require, at least in part, the same type of altruistic actions that characterized Minnesota's acid deposition history.

The years since this story unfolded have seen changes in the standing of environmental concerns in the public arena. There is increasing tension between viewing our social interactions exclusively through the lens of the marketplace and acting as environmental stewards. In addition, antigovernment, antitaxation, and antiregulation actors have become well-funded and well-organized forces in the political arena.[32] The amount of money spent to influence opinion and undermine action on environmental issues like climate change has grown exponentially, and the money on the table in electoral politics has skyrocketed since the 2010 Supreme Court decision in *Citizens United v. Federal Election Commission*, in which the court held that the First Amendment prohibits the government from restricting political independent expenditures by corporations, associations, or labor unions. There is a sense of unease in our ecological relationship to the earth, compounded by a jaded cynicism stemming from the drumbeat of messages in the media, advertising, and product placement promoting the idea that purchasing products will lead to fulfillment, success, and happiness. Together these currents nurture an ambience of environmental futility.[33] As Philip Shabecoff has noted in the foreword to the second edition (2003) of *A Fierce Green Fire*: "Because of . . . developments since the first edition of this book appeared [in 1993], I have been unhappily forced to modify its previously optimistic tone in this updated and expanded edition. While I still believe that the environmental movement has the latent strength to change our society for the better and to keep us from descending into ecological darkness, I can no longer forecast that such progress is inevitable. Unless there is a profound shift in our current political, economic, and diplomatic trajectory, I believe that we, and especially our children and grandchildren, will live in a darker, drearier, and more dangerous world."[34]

Despite the trends that have undermined environmental progress, SO_2 emissions have continued to decline. Nationally, emissions fell from 26 million tons in 1980 to 16 million tons in 2000 and 6 million tons in 2012. Minnesota SO_2 emissions were estimated at 142,000 tons per year in 1994, falling to approximately 100,000 tons in 2000, and 21,000 tons in 2010. The continuing Minnesota reductions are largely due to the Metropolitan Emissions Reduction Project (MERP), in which three coal-burning power plants were converted to natural gas, and advanced SO_2 scrubbing was installed at a fourth plant.[35] These investments allowed the utility to upgrade aging equipment at rate-payers' expense while at the same time reducing SO_2 emissions. Improving their environmental performance nationally has helped the electrical utility industry fend off efforts to address climate change.

In the years since this story of acid rain unfolded in Minnesota, arguments about climate change, arguably the most pressing environmental problem of our day, have been surprisingly similar to those of the acid rain issue. National (and international) action is required to solve the problem. Any actions by a single state like Minnesota, or even a single country, cannot solve the problem. Yet the challenge before us is much greater than what we faced with acid rain, and the solutions will be more difficult. Many of us who came of age during the second wave of environmentalism in the 1960s and 1970s saw the destructiveness wrought by modern society and endeavored to do better than our parents in protecting nature and turning over a healthy planet to coming generations.[36] In large measure we have failed to accomplish that goal, and we hand off immense, complicated problems to our successors. Will we humans present and future face our environmental problems and the unsustainability of our culture and take actions to reduce our impacts? Or will we continue business as usual and suffer (or let our children and grandchildren suffer) the consequences?

THE URBAN ROOTS OF
MILITANT INDIAN PROTEST

AIM's Origins in the
Twin Cities, 1968–1973

William C. Barnett

NATIVE AMERICAN ACTIVISTS MADE HEADLINES between 1968 and 1973 with a series of dramatic occupations of symbolic sites, and these protests are part of the standard portrayal of 1960s political unrest in history textbooks. The American Indian Movement (AIM) was the critical group. Formed in 1968, AIM quickly became a national organization, achieving widespread recognition with its takeover of the Bureau of Indian Affairs (BIA) in Washington, DC, in 1972 and its lengthy occupation of Wounded Knee, South Dakota, in 1973. AIM activists also led a series of lesser-known land seizures in rural and urban locations, and their efforts inspired pan-Indian unity and pride.[1] While AIM became famous for its militant protests, the group's origin story is often overlooked. AIM developed in a surprising context, rising up as a challenge to degraded environmental and social conditions in industrial cities. Its defiant brand of protest swept across the rural West, but the group originally formed to protest urban environmental problems in the slums of Minneapolis.[2]

Indian slums were a new part of urban America created by the marked shift in Indian policy beginning in 1953. The federal government set out to interact with Native Americans as individuals rather than as members of tribal nations, and this termination policy reconfigured rural and urban communities. Over one hundred tribes ceased to officially exist, and whites bought up rural lands once owned by tribes. At the same time relocation policies moved large numbers of Indians into industrial cities during the 1950s and 1960s. Relocation was intended to help Native Americans find work, but in practice it moved people with profoundly rural orientations into grim urban slums.[3] Rural migrants to industrial cities experienced culture shock and dislocation even when they were white, as seen in the struggles of white Appalachian coal miners in Midwest cities during the same era.[4] Being members of a minority group compounded these problems, and Indian ghettos were bleak landscapes inhabited by struggling people. The activists who joined together in Minneapolis and then Saint Paul to create AIM were responding to local environmental conditions such as inferior housing, police brutality, alcoholism, and other problems resulting from poverty and discrimination.[5] From this base AIM expanded to confront unacceptable living conditions that Indians faced in both urban and rural environments, sparking a new determination for Native Americans to gain control of their local communities.

AIM'S ORIGINS AND THE QUEST FOR ENVIRONMENTAL JUSTICE

AIM confronted local injustices with militant direct action, uniting Indians who moved to the Twin Cities from Great Lakes and northern plains reservations around the common cause of improving slum conditions. Several defining characteristics of these campaigns helped AIM become a regional and then national movement. Ongoing communication and migration between the Twin Cities and reservations in rural Minnesota, Wisconsin, and the Dakotas gave AIM a strategic location at a hub connecting a large Indian hinterland. The group's pan-Indian structure and militant approach appealed to both urban and rural Indians. Additionally, the socioeconomic problems that AIM fought in Minneapolis, particularly deplorable housing and police brutality, resonated for Indians across the nation.

For generations reservations had been the focal point of Indian politics, but that changed after 1953 due to the campaign to dissolve the relationship between the federal government and the tribes and move reservation residents to cities. An unintended consequence of relocation was that the new, shared experiences of Indians thrust together in urban slums enabled them to overcome tribal divisions. It was in this context that a new generation of urban Indians developed the attention-getting tactic of seizing public lands—usually abandoned federal facilities—and claiming them as the rightful property of Indian peoples. They asserted this claim based on an 1868 Sioux treaty that gave Indians the right to unused federal lands, but the legal basis of the claim mattered less than its popularity.[6] These protests attracted significant media coverage, bringing the challenges facing Indians to national attention for the first time in decades, and soon gained strong support on reservations. In just six years AIM activists sparked a wave of protest that swept from the Twin Cities across Indian country.

Seen from a national perspective, the American Indian Movement, along with the Black Panthers and the Weathermen, fits with the popular view of the rapid rise and violent collapse of radical 1960s movements. AIM was indeed involved in dramatic confrontations like the BIA and Wounded Knee occupations, which the federal government ended with force. But when AIM is examined more closely, with attention to community organizing and not media coverage, a different narrative emerges. Accounts that emphasize violent clashes with authorities fail to see long-term changes in communities like the Twin Cities that AIM created by bringing Indians together across tribal divides, awakening pride, and building institutions such as alternative schools and American Indian centers.[7] Overemphasis on the violence linked to occupations also obscures the ways these activists prompted improvements in federal policies, including the end of termination, the expansion of tribal authority over federal antipoverty programs, and the growth of tribal schools and colleges. Analyzing AIM's origins provides an opportunity to understand the local problems that led Indians to organize, including environmental justice issues, and reveals the web of connections that bound far-flung Indian communities together. Ideas about place are crucial to this story, as profoundly rural Indians rejected urban slum conditions, and then brought militant protest to the reservations, where they asserted a far-reaching argument about the substandard conditions in which Indians lived.

Environmental justice is not typically linked to Native American activism in the 1960s, but this concept, which gained recognition in the 1980s and 1990s, offers valuable insight into this era. Sociologist Robert Bullard summed up the core idea of environmental justice with a simple statement: "People of color in all regions of the country bear a disproportionate share of the nation's environmental problems." Although Bullard's path-breaking research examines African American communities, he wrote, "This analysis could have easily focused on Latino Americans in the Southwest or Native Americans in the West."[8] Some authors have analyzed environmental justice issues on reservations, most notably exposing the devastating impact of uranium mining on the Navajo, but environmental problems facing urban Indians have received little attention.[9]

Militant Indian activists began by protesting the environmental conditions in industrial slums, including housing, sanitation, and health problems, and they linked these injustices back to impoverished reservation lands. The parallels between urban and rural poverty were noted in the 1969 Alcatraz Island takeover, which was the protest that started the wave of land occupations. The young Indians occupying the empty federal prison issued a proclamation that drew attention to the grim environmental conditions on rural Indian lands. Their sarcastic proclamation analyzed the bleak prison, reporting, "this place resembles most Indian reservations in that:

1. It is isolated from modern facilities, and without adequate means of transportation.
2. It has no fresh running water.
3. It has inadequate sanitation facilities.
4. There are no oil or mineral rights.
5. There is no industry, and so unemployment is very great.
6. There are no health care facilities.
7. The soil is rocky and unproductive; and the land does not support game."[10]

The activists occupying Alcatraz were angry because Native Americans were relegated to the nation's most environmentally impoverished lands. When read today, their list discusses environmental justice, but that idea was not in use in 1969. The term does not appear in the autobiographies by AIM leaders Dennis Banks and Russell Means, but a significant body of recent scholarship analyzes the environmental side of racial inequality, and this research provides valuable insights.[11] Native American activists of the 1960s began by expressing their outrage at the squalid living conditions in Indian slums, and as the statement from Alcatraz demonstrates, they immediately connected urban problems to those on reservations. Their protests highlighted the tragic irony that the Native Americans who were pushed onto the most degraded urban and rural environments once owned all the nation's land.

It is useful to link Indian activism from 1968 to 1973 with the idea of environmental justice because the widespread belief that the problems of people of color are about poverty, but not environment, has sharply limited the way Americans understand both environmentalism and racial inequality. Robert Gottlieb has analyzed this problem, stating, "The issues for the African-American, Chicano, and Native American groups tended to be defined less in environmental than social justice terms. By establishing a distinction between environmental and social justice themes, these struggles have further reinforced the prevailing assumption that environmentalism continues to be a white movement."[12] The broader definition of the environmental movement that Gottlieb and others call for directly addresses race, class, and ethnicity, and focuses attention on urban conditions and health issues in addition to wilderness protection. It is significant that the militant Indians who sparked a resurgence in Native American pride initially focused on inner-city problems like inadequate housing and poor access to environmental resources. Viewing their movement for self-determination as linked to environmental justice can help bridge a gap between Indian activism and environmentalism.

AIM'S FORMATION IN MINNEAPOLIS

By the late 1960s relocation had created large numbers of urban Indians, but their living conditions in cities like Minneapolis were grim.[13] In 1967, a year before AIM formed, Clyde Warrior, a young Ponca from Oklahoma, testified at a federal hearing on poverty: "If there is one thing that characterizes Indian life today it is poverty of the spirit. . . . We are not free. We do not make choices. Our choices are made for us; we are the poor." He continued, "For those of us who live on reservations these choices and decisions are made by federal administrators, bureaucrats, and their 'yes men,' euphemistically called tribal governments. Those of us who live in non-reservation areas have our lives controlled by local white power elites."[14] Warrior's comments on oppression and spiritual poverty in both rural and urban areas reveal a sense of crisis. In a 1966 interview he was even more blunt, asking, "How long will Indians tolerate this? Negroes, Mexican Americans, and Puerto Ricans could only take colonialism, exploitation, and abuse for so long, then they did something about it." He asked, "Will American Indians wait until their reservations and lands are eroded away, and they are forced into urban ghettoes, before they start raising hell with their oppressors?"[15] Warrior predicted a new Indian activism, identifying African American protests as a model and urban Indians as the likely leaders.

Clyde Warrior's 1966 question was answered in 1968 when Indians in Minneapolis came together to fight the unacceptable conditions in its Indian ghetto. Dennis Banks, recently released from Stillwater Prison, and George Mitchell, his friend from an Indian boarding school, organized the July 1968 meeting that launched the American Indian Movement. In his autobiography, Banks discussed 1960s social movements and his desire to organize Native Americans: "I started to educate myself while in solitary and found there was a lot of social and political unrest happening on the outside. I began to follow the anti-war movement, the marches and protests, the Students for a Democratic Society, the Weathermen, and the Black Panthers." Banks continued, "I desperately wanted to be part of a movement for Indian people. . . . there was no movement specifically addressing the police brutality that was an everyday fact for Indian people or the discrimination in housing and employment in Minneapolis."[16] George Mitchell described similar motivations: "I was fueled by anger and rage at the poor condition of the lives of Indians in the community." "We had no attorneys to defend us in court," Mitchell explained. "We had no way of knowing how to apply for a job. Indians were finding it difficult to find adequate housing."[17] These comments from two AIM founders reveal that they envisioned a civil rights campaign to address inner-city problems such as poor housing in order to remake a hostile urban environment.

Over two hundred people attended the initial meeting, and the first issue the group took up was the mistreatment of Indians by Minneapolis police outside Franklin Avenue bars. This effort quickly involved AIM in a range of legal and health issues. The organization began by using the Black Panthers as a model. Banks recalled, "We immediately established an Indian patrol to prevent the police from further harassing our people. We patterned it after the patrol created by the Black Panthers in Oakland." The activists, who adopted the name AIM a month later, walked the streets in red jackets and berets, gathered outside bars at closing time, and used three red cars to drive intoxicated Indians home so they would not be jailed. They acquired walkie-talkies for communication, radios to monitor police calls, and cameras to record police misconduct. Banks said they told the police chief, "We will bring you films and photographs that show your people using excessive force all the time and acting in a sadistic way."[18] In addition to trying to reduce arrests, the group offered legal services and taught Indians to stop pleading guilty to charges. Vernon Bellecourt, a founding AIM member along with his brother Clyde, said they asked problem drinkers to join the Indian patrol, which required participants to be sober, explaining, "So it was really an alcoholic rehabilitation program at the same time."[19]

Although AIM's bold actions created a backlash from Minneapolis police, they instilled pride in the activists. Clyde Bellecourt testified in 1969, "We became known as radicals right away—militants, red militants—and part of a power movement. All kinds of other names."[20] "The local cops hated and feared AIM, of course," recalled Russell Means, a key AIM leader but not one of its founders. "The embarrassed Minneapolis police went after Clyde with a vengeance. In twelve months during 1969 and 1970, they arrested him more than fifty times for nearly every alleged minor infraction imaginable. . . . Clyde was a former convict, a repeat criminal, so the cops always had an excuse to watch him—or work him over. Clyde was belligerent and would never knuckle under. He kept his head up even when they hit him across the face with blackjacks."[21] From the start AIM leaders defiantly resisted white authority.

The shared experiences of Dennis Banks, the Bellecourt brothers, and George Mitchell explain their fierce opposition to white control. AIM's founders were Ojibwe (or "Chippewa") men with roots on rural reservations. These men had moved between reservations, Indian boarding schools, Minnesota's Stillwater Prison, and urban slums, so they had many direct experiences with enduring, and attempting to resist, white authority. AIM's leaders shared the jarring experience of having been moved off their reservations, so they had inhabited two different worlds and had faced brutal treatment from white authorities. Russell Means described boarding schools run by the BIA and by missionaries as prisons for children: "Harsh, military-style discipline was used to break the children's spirits. Each student was assigned a number . . . just as in prison . . . [and] kids were often called by their numbers instead of their names."[22] He described physical, psychological, and sexual abuses experienced by family members and friends, noting that school administrators protected sadistic abusers.[23] The mistreatment of Indians by police that these men challenged was part of a long pattern.

AIM's founders had spent more time in prison than in college, but they had developed a sophisticated view of community organizing. Testifying in 1969 at a Public Forum of the National Council on Indian Opportunity, Bellecourt spoke about empowering everyday people: "Some of our great white Indian experts . . . feel that you have to have a college degree to take care of your problems and to really understand what the system is. We maintain the people who are going to make the changes are the people who have faced the conditions. I know because I am one of them." He also stated, "We made moves immediately that night for our organization to become a self-determination, self-concept, self-help group."[24] Bellecourt's philosophy of grassroots activism and participatory democracy paralleled the ideas of the Student Nonviolent Coordinating Committee and the Black Panthers. The patrol had an immediate impact in reducing arrests of Indians, and Russell Means noted a sharp drop in Indians in jail, calling the campaign "one of AIM's first recognized successes."[25]

Confronting the police was a powerful experience, and that defiance became a crucial part of AIM's identity as members began to conceive of themselves as warriors. Vernon Bellecourt described the effect of participating in the patrol: "I watched what they were doing, and I could see the pride in these young men and women. A new dignity, a new awareness, a new power, a new strength." The experience changed him, and he abandoned efforts to live a middle-class life and joined his brother as an activist. "I looked at myself, I was making money and living in White suburbia. . . . Finally I got so involved I started letting my hair grow long, and I stopped wearing a tie and started to sort of deprogram myself to become just a simple person, a simple man. More humble."[26] The long hair and distinctive clothing that AIM members adopted was central to their image, combining an aggressive urban style with traditional Plains Indian symbols. Leonard Crow Dog of South Dakota described the impact of this new look: "When the traditional Lakota and the city militants got together, that was the moment AIM took off. Suddenly men wore their hair long or in braids. They threw away their neckties. Everybody started wearing bead or bone chokers. They began wearing ribbon shirts. They wore Levi jackets with AIM patches and buttons reading, INDIAN POWER or INDIAN AND PROUD. They had eagle feathers tied to their hair or stuck into their hatbands.

We became warriors again."[27] This warrior image combining rural and urban styles was important to AIM's appeal, and added to the dramatic impact when caravans of activists arrived for a protest.

AIM'S COMMUNITY PROGRAMS

AIM's warrior image was important, and it echoed the Black Panther image, but it obscured community organizing efforts. Both groups developed a hypermasculine image and were famous for confronting white authorities, overshadowing their day-to-day work in building programs such as AIM survival schools and Black Panther breakfast programs. From the beginning, AIM had male and female members, although men held the leadership roles, and the group worked on programs relating to children and families as well as the Indian patrol. One early member stated, "God, it was a really exciting time. Not only did issues of police brutality get brought to the forefront, but so many others. I was just overwhelmed by the problems in the community. They talked about unemployment, alcoholism, juvenile delinquency, and welfare issues. . . . The primary focus was on the urban area and the kinds of things we could do for our own people living in the ghetto."[28] Dennis Banks also noted AIM's breadth of activities in the Twin Cities, explaining, "We began to do a lot more than just confront police brutality. We had an attorney from the Legal Aid Society available every Thursday evening. . . . We established a program to improve Indian housing . . . [with] a hauling trailer to help Indian families move." Banks also described newsletters, a radio program, and the Little Red School House, but gave no details about the alternative school.[29]

Female members staffed the Little Red School House that AIM started in Saint Paul. Pat Bellenger, a woman who was important in AIM from its beginning, described her goals in 1968: "I was young and angry and full of energy. I think the whole thing was that this all happened at a time when Indian people didn't have anything in the Cities. There were no programs to support families." In describing her anger, Bellenger sounded like AIM's men, but her discussion of families revealed a different focus. Steven Couture interviewed her in the 1990s and summed her up as "a lifelong activist." Bellenger described the Little Red School House's origins: "We started out with 35 students and basically a group of volunteers. Everyone worked for free. The school district gave us a hard time and tried to shut us down. They eventually backed off because the Mayor of Saint Paul, at that time Cohen, joined forces with us."[30] This cooperation with the mayor shows that AIM was able to negotiate with white leaders and was beginning to assert some control over life in the Indian slums.

AIM activists openly defied the police, but they also formed alliances with white politicians and with churches, universities, foundations, and corporations, bringing in considerable funding. In addition to donations from Minnesota churches and companies, AIM programs received money from the Office of Economic Opportunity (OEO), an antipoverty program begun by the Johnson administration. This was a key victory for urban Indians, because OEO funding was originally available only for reservations. It is significant that Russell Means worked for an OEO Community Action Program on the Rosebud Reservation before joining AIM, because he saw the impact of federal aid.[31] AIM also gathered state and local government funding, and gained support from Minneapolis public schools and the University of Minnesota to send juvenile offenders to its alternative schools. By 1972 AIM operated two schools for Indian youth, the Little Red School House Survival School in Saint Paul and the Heart of the Earth Survival School in Minneapolis.[32] AIM had successfully created two physical locations in the Twin Cities where Indian children could learn from teachers who respected their cultural heritage, providing a safe place in a hostile urban environment.

In addition to these efforts to start schools, AIM focused on improving the built environment in Indian slums. From its very start, AIM worked to address the substandard housing for Indians in the Twin Cities, a severe problem that drew the attention of varied observers. Patrick Lussier, an Ojibwe working for the State Employment Service, reported, "The main problem I find, in working with the Indian population, is housing. Finding the Indians a job isn't the hardest thing, it's housing." A 1969 University of Minnesota study repeated this diagnosis: "The gravest threat to Indians' health and welfare is found in the terrible housing situation which confronts newcomers when they arrive in the city[;] . . . without safe, hygienic, and comfortable housing there can be no satisfactory solution to the health problems (mental and physical) of the Indian in our midst."[33] A 1968 University of Minnesota study also referred to a singular Indian: "He occupies the worst housing in the worst neighborhoods in the city, where he is likely to become 'host' to an extended family, creating overcrowding." To make matters worse, the fact that Indians moved back and forth between city and reservation made it difficult to meet the one-year residency requirement to qualify for public housing. In addition, many landlords refused to rent to Indians because they feared large extended families crowding into small apartments.[34]

One of the few surviving issues of the *American Indian Movement News* from this era reveals that the Twin Cities housing crisis was atop the priority list. The April 1970 newsletter declared that AIM's first long-range objective was "to generate unification within the Indian peoples," and that its primary short-range objective was "to establish a program to better the Indian housing problems." AIM also called for "legislation to revise the residence requirements for public housing to allow greater benefits for urban Indians." And the newsletter voiced AIM's support for the joint resolution by the American Indian

Federation of the Twin Cities and the Twin Cities Tribal Council, calling on the federal government "to update the Bureau of Indian Affairs so that equal services may be given to American Indians in off-reservation areas."[35] This resolution shows that AIM could cooperate with older Indian organizations, and the presence of multiple AIM leaders on an eighteen-person delegation to the BIA commissioner in 1970 is evidence that the two-year-old group was recognized as a significant voice for Indians.[36]

AIM's efforts on behalf of urban Indians allowed its leaders to sit alongside delegates from established groups like the National Indian Youth Council, and AIM's leaders did not sit quietly. When BIA commissioner Louis Bruce refused to meet with the entire delegation, perhaps because of its militant members, Clyde Bellecourt responded aggressively, inviting television, radio, and newspaper reporters into the BIA for a press conference. Mary Arpan of the Minnesota Council of Churches summed up this confrontation: "Bellecourt directed questions to the Bureau officials and asked for immediate answers. The B.I.A. resented the news media present and asked them to leave and we demanded that they remain. After two initial pertinent questions, the B.I.A. officials walked out of the press conference." Arpan criticized the federal officials, not Bellecourt, concluding, "Our questions, unfortunately have still not been answered."[37] Commissioner Bruce later said AIM's attitude did not represent "the Indian way," and AIM's newsletter mocked this claim, replying, "Bullshit! . . . people can state that passivity is the Indian way if they wish, AIM will state otherwise!"[38] Bellecourt's actions at this 1970 hearing made it clear that AIM would use its confrontational tactics with federal officials. But AIM was not yet at the forefront of Indian activism, as its BIA occupation was two years away.

THE OCCUPATION OF THE TWIN CITIES NAVAL STATION

The lengthy occupation of Alcatraz Island from November 1969 to June 1971 inspired AIM to take bolder steps in the Twin Cities. Native American students attending varied California universities organized the Alcatraz occupation and gained widespread media attention, but they lacked strong ties to any specific Indian community. AIM, however, had a solid base in the Twin Cities and borrowed the tactic of seizing federal property to address local problems. On May 16, 1971, AIM occupied the abandoned Twin Cities Naval Air Station, citing the housing crisis in Minneapolis as a key reason for claiming a vacant federal facility.[39] The initial *Minneapolis Star* article explained these goals with a headline reading, "Indians Seize Air Station; School, Housing Demanded."[40] The protest was a response to inadequate local housing, but its tactics also fit with national radical protests. Other *Star* headlines that week included reports on Black Panther tactics in Oakland and on a large Minneapolis rally against the Vietnam War featuring Eugene McCarthy and John Kerry.[41]

Figure 16.1. On May 16, 1971, AIM occupied the abandoned Twin Cities Naval Air Station, citing the housing crisis in Minneapolis as a key reason for claiming a vacant federal facility. The protest—and the Nixon administration's ensuing crackdown on protesters—highlighted the ways that AIM responded to local environmental conditions such as inferior housing, police brutality, alcoholism, and other problems resulting from poverty and discrimination. *Photograph by Dick Bancroft.*

The link to California's Indian activists was more direct, as Dennis Banks displayed an endorsement from Alcatraz's occupiers, and the front-page photograph of Indians climbing the naval station fence showed "Alcatraz" painted across one man's jacket. This protest had national ties but local origins, and the *Star*'s summary of Indian demands emphasized Twin Cities issues, quoting AIM leaders as saying the empty barracks "could help solve the deplorable housing situation found here in Minneapolis."[42]

During the occupation's first days, the *Minneapolis Star* and the *Minneapolis Tribune* presented Indian viewpoints and depicted white officials as unlikely to evict the protestors. A sympathetic article about relocation's problems titled "Indians Lose Identity in City" ran in the *Star*, stating, "But those who go to the city should not simply be dumped there and left to founder. They deserve a reasonable chance."[43] A *Tribune* editorial said officials had no plans to press charges, quoting Mayor Charles Stenvig as saying that "he understood the plight of the Indians," and called on the BIA to allocate funds to urban Indians.[44] Navy and county officials initially treated the protestors with respect, as the navy supplied portable toilets and fire extinguishers, and the district attorney praised the activists for "the lack of threats, violence, and destruction."[45]

These accounts suggest that AIM had won some support from journalists and politicians in progressive Minneapolis.

Nixon administration officials, on the other hand, responded with force. The Naval Air Station occupation ended abruptly on its fifth day when eighty-five federal marshals and Hennepin County sheriff's deputies arrived before dawn. AIM leaders expected to meet with Senators Hubert Humphrey and Walter Mondale that day, and both the navy and the senators were preparing for that meeting, unaware that the Justice Department was launching a raid. The tactical squad entered swinging riot clubs. Sixteen Indian men were arrested and charged with trespass, including Dennis Banks, Clyde Bellecourt, and Russell Means, while a dozen women and children were released. A few protestors fought back with baseball bats, two officers sustained injuries, and two Indian men were charged with felony assault.[46]

After appearing in court on trespassing charges, Banks and Bellecourt criticized the government's use of force. They told reporters "the troops were sadistic and brutal" and accused marshals of "feeling up the girls they arrested."[47] Recalling the excessive use of force by police against Indians that led to the 1968 creation of the Indian patrol, AIM's leaders concluded that law enforcement still used violence to subdue Native Americans. Local and federal officials ended AIM's most militant action in Minnesota without resolving the underlying issues that had prompted the occupation. The seizure of the Naval Air Station did not become national news like Alcatraz, the BIA, and Wounded Knee, but Minnesotans remembered it. This failed occupation became part of the experiences of AIM's leaders, and it increased their militancy.

HEIGHTENED AWARENESS OF URBAN-ENVIRONMENTAL PROBLEMS

In 1972, during a series of AIM protests, the *New York Times* ran a detailed front-page report on problems facing urban Indians titled, "Militancy of Urban Indians Spurs Hope for Change." Reporter Homer Bigart cited a pattern of Twin Cities Indians accepting terrible housing, sanitation, and health conditions before AIM's formation. He reported that "uncollected garbage, mice, cockroaches, exposed wiring, and debris piled in the yards of old houses plagued Indian tenants, but the tenants do not generally complain." "The militants have been trying to change this," Bigart wrote, citing Dennis Banks in reporting that after years in which no Indians filed complaints against landlords, "60 complaints were filed with the Minneapolis Civil Rights Department, resulting in 12 convictions." Banks stated, "There has been a very slight improvement. Landlords are trying to live up to minimal standards." Bigart saw additional progress resulting from the support of the Minnesota Council of Churches. He reported

that their efforts to improve Indian housing in the Twin Cities won the approval of "the first federally-funded urban housing development primarily for
Indians," describing a $5.4 million project containing 216 units. At the same
time, however, Banks and a church official depicted growing economic hardships due to factory layoffs. "There are 8,000 Indians here, mainly Chippewa
and Sioux," the church official explained, "and they are in a worse position than
ever in housing, jobs, and health. Unemployment has gone up to 60 per cent."[48]

Analyzing these inner-city conditions through an environmental justice lens
helps build a picture of the urban world that gave birth to AIM. Most Native
Americans lived in the Phillips and Ventura Village neighborhoods of South
Minneapolis, where old, poorly maintained buildings contained high levels of
lead that put children's health at risk.[49] Franklin Avenue was the main commercial artery, and the 1971 city directory reveals a bleak neighborhood dominated
by liquor stores, bars, thrift stores, and social service providers. AIM's office
was on the same block as Goodwill and Catholic Welfare Services, and other
neighbors included Lars Liquor Supply, Corral Tavern, Nelson Liquor Store,
Addison's Bar, and Bliss Liquor Store.[50] This dense neighborhood was cut off
from adjacent residential areas when the interstate highways were built, with
I-35 running just to the west and I-94 to the north, bringing increased exposure
to environmental pollutants, including lead particulates and asthma-inducing
compounds from vehicle emissions. The neighborhood's east edge had a third
rigid boundary, Hiawatha Avenue, a heavily trafficked artery lined with railroads and heavy industry, including a pesticide plant that used large quantities
of arsenic and later became a controversial Superfund cleanup site.[51] The result
was an isolated and heavily polluted area of concentrated poverty known as
"one of the oldest areas in the city" with "some of the poorest housing."[52]

These problems drew researchers to the Phillips neighborhood, and their
inquiries revealed high levels of frustration among Indians who knew they
lived in a pocket of stark poverty inside a wealthy city and nation. In 1968 University of Minnesota researchers noted that over two-thirds of respondents felt
that Indian neighborhoods faced sanitation problems, and they reported in
1969 that 72 percent of Indian housing was substandard as a result of problems
such as "broken or inoperative doors" (36%) and "broken plaster, light fixtures
inoperative, and broken steps inside and out" (75%).[53] The 1968 report also said
Indian neighborhoods lacked green space: "The Minneapolis Indian tends to
live where parks and recreational facilities are inadequate," so there was little
respite for a newly urbanized population raised on rural reservations.[54] The
negative impact of crowded slum housing on these newcomers is not surprising, and the 1969 Minnesota researchers reported, "Over half of the Indian
people interviewed expressed disappointment, anger, bitterness and anxiety in
various degrees over their present housing situation."[55] While the term *environ-*

mental justice was not yet in use, these reports reveal a basic truth about environmental risks in industrial America: poor people and people of color bear an increased burden of risk from pollution. These oppressive environmental conditions contributed to the emergence of militant Indian activism.

The urban Indians who formed AIM in 1968 knew that they experienced oppression in the Twin Cities slums. In response, they came together and launched wide-ranging efforts to demand equal treatment and to improve living conditions. AIM members were not only patrolling the streets by night to combat police brutality; they were grappling with slum housing and inadequate schools during the day. Their efforts to address the unacceptable conditions facing urban Indians have direct links to environmental justice. These militant Indians were claiming the right to improve the deplorable urban environments where whites had chosen to house relocated Indians. In time AIM would build bridges from the Twin Cities to rural communities in a wider campaign for Native American control of their local environments.

A TWIN CITIES ORGANIZATION BECOMES A NATIONAL MOVEMENT

In 1972, the American Indian Movement was a strong and expanding organization, with charismatic leaders in Minneapolis and growing support from Indians in other cities. AIM had chapters in Cleveland and Milwaukee that revealed its evolution into a national organization. Russell Means gave AIM its foothold in Cleveland, while Herb Powless founded a Milwaukee chapter that occupied an empty Coast Guard station on Lake Michigan in 1971 and won permission to make it an Indian school.[56] Connecting with Indians in cities beyond the Twin Cities was significant, but AIM needed to forge alliances on reservations before it could claim to speak for all Native Americans.

Building partnerships with reservation communities and demonstrating a willingness to fight injustices there were critical to AIM's transformation into a national movement fighting for all Indian peoples. Russell Means played a key part in forging the alliance between urban activists and South Dakota's reservations because he was a Lakota Sioux born on Pine Ridge Reservation. AIM's founders were primarily Ojibwe men from the Great Lakes, but Means supplied a link to the most celebrated Plains tribe of buffalo-hunting warriors. In the late 1960s many rural Sioux were seeking allies in two struggles: against the violence and racism they experienced in reservation border towns and against corrupt tribal governments that mistreated traditional Indians. AIM willingly joined both campaigns, and the resulting urban-rural alliance launched the movement to national prominence.[57]

When AIM brought its defiant style of protest to impoverished South Dakota reservations, it won support across Indian country. Traditional Indians initially viewed AIM's urban focus as having little relevance to their lives, but many came to view the urban militants as trusted allies. Mary Crow Dog, wife of Lakota spiritual leader Leonard Crow Dog, revealed this shift: "In the beginning, AIM was mainly confined to St. Paul and Minneapolis. The early AIM people were mostly ghetto Indians, often from tribes which had lost much of their language, traditions, and ceremonies." Dennis Banks recognized that Leonard Crow Dog's traditional spiritual practices offered something vitally important to AIM and sought him out as a spiritual teacher. In exchange, AIM provided defiant militancy and a willingness to challenge tribal officials. Mary Crow Dog wrote of this urban-rural alliance: "AIM opened a window for us through which the wind of the 1960s and early '70s could blow, and it was no gentle breeze but a hurricane that whirled us around. It was after the traditional reservation Indians and the ghetto kids had gotten together that AIM became a force nationwide."[58]

Dennis Banks also analyzed the special bond between urban militants and traditional Lakota: "Spirituality is the heart and soul of Indian life, but we AIM people had been raised in white boarding schools, had lived in the Indian ghettos of big cities, had done time in prison. We did not know what we should believe in or how we could find sacredness."[59] With these words, Banks traced his life's journey through a series of environments where whites exerted brutal control over Indians. AIM's founders forged friendships in these harsh conditions while challenging white authority. In the Twin Cities they continued to reject white control, especially police mistreatment. First in Minneapolis, and then in rural locations, AIM organized Indian communities to remake the hostile environments they inhabited.

Many Native Americans remember the American Indian Movement protests from 1968 to 1973 as a time of remarkable passion, energy, and pride. In six years Indian activists seized the nation's attention and focused it on some of America's poorest communities. AIM's founders began in Minneapolis by demanding the power to improve the grim conditions in Indian slums. From this urban base AIM built bridges to the reservations, developing a national movement that asserted a shared Indian identity across tribal and geographic boundaries and challenged centuries of mistreatment. The shift from the dark days of termination in the 1950s and early 1960s to the cultural revival from 1968 to 1973 was rapid and dramatic. Rick Williams, an Oglala Sioux, said, "My grandmother died in 1969. . . . She went to her grave believing that Indian people had lost their culture. Had she lived another five years, she would have changed her mind. . . . My grandmother didn't know about Alcatraz or AIM or self-determination; all that would follow. She would have been absolutely shocked to see that the Sun Dance has returned . . . shocked that her great-grandchildren speak Lakota."[60] In just six years AIM's protests brought national attention to Native American poverty and sparked a major cultural revival.

At the heart of the American Indian Movement's protests was the message that Native Americans had been dispossessed of their lands and relegated to the nation's worst urban and rural environments, and that this injustice needed to be reversed. AIM's occupations did exactly that, seizing control of lands held by the federal government. Most gains were short-lived, as authorities responded with force, but both Indians and whites grasped the message that Native Americans wanted their stolen lands returned. AIM was created in the Twin Cities in 1968 to fight deplorable urban conditions, and the militant group grew from its efforts to improve one troubled urban environment to inspire campaigns for land rights and tribal sovereignty that continue to resonate across Indian country.

RADIOACTIVE WASTE, PUBLIC DEBATE, AND ENVIRONMENTAL JUSTICE AT PRAIRIE ISLAN

James W. Feldman

Prairie Island has a stunningly illogical geography. The island lies off the western bank of the Mississippi River, thirty miles southeast of Minneapolis and Saint Paul. The river is quiet here, restrained by Lock & Dam No. 3, making it a popular spot for boating, fishing, and waterskiing. The northern half of the island provides a home to the Prairie Island Indian Community, a Mdewakanton Sioux reservation with a population of just over two hundred people. The small commercial area consists of a community center, a medical clinic, a school, and the tribal administrative offices. Not far from the commercial center, the Buffalo Project pastures over forty head of bison, part of an effort to restore traditional Mdewakanton relationships to the animal and to revitalize tribal culture. The tribe operates the Treasure Island Resort and Casino, complete with a hotel, a golf course, and a marina, which attracts thousands of visitors each year from the Twin Cities area. The southern half of the island also has social, economic, and environmental connections to Minneapolis and Saint Paul, but in a radically different way. The twin pressurized water reactors of the Prairie Island Nuclear Generating Station stand less than a mile from the casino, generating 1,076 megawatts of power that the Xcel Energy Corporation sends to its largest market. The tribal homes, recreational facilities, casino, and nuclear plant that share Prairie Island present stark contrasts of modern and traditional, urban and rural, work and play. These contrasts are rooted in the island's peculiar history—and particularly in its nuclear history.

Public debate about the Prairie Island plant started even before construction of the facility began in 1969. For over forty years different constituencies have voiced their concerns about the risk of accidents, the environmental hazards of standard operating procedures, the reliance on nuclear energy, and a host of other issues. The storage of radioactive wastes—spent nuclear fuel—at the reactor has generated the most sustained and significant controversy. In the early 1990s, opponents of the Prairie Island plant secured an important victory by forcing the decision about waste storage out of the regulatory arena and into the legislative one—ensuring that the Minnesota legislature would take a contentious vote on the subject in 1994. Industry insiders, federal regulators, and antinuclear protestors from Minnesota and around the nation wrestled with the vexing questions of radioactive waste: Where to put it? How to store it, and for how long? Who should face the risk—if any existed—of living near it? All parties involved in the controversy tried to balance concerns about economics, public and environmental health, and social justice in a way that led to the greatest public good—even as they disagreed on how to define that public good. All realized that the future of nuclear power depended on the resolution of these questions. Both supporters and opponents of nuclear power from across the country watched the situation closely; the plant had become a national test case. Opponents of nuclear power hoped that Prairie Island pointed to new arguments and protest strategies that would shut down such power plants elsewhere, while supporters of nuclear power expected that a successful resolution of the controversy at Prairie Island would move the industry forward in the absence of a permanent waste storage solution.

As the controversy developed, the antinuclear coalition—local residents, environmentalists, and the Prairie Island Indian community—offered increasingly sophisticated arguments that they hoped would both prevent the storage of waste at the plant and shut down nuclear power production at the site for good. The idea of environmental justice lay at the center of this campaign— the recognition that neither the risks of environmental harm nor the power to make environmental decisions are equally distributed across social divides. Environmental justice emerged in the late 1980s as both a grassroots movement and as an analytical critique of mainstream environmentalism and entrenched social and economic structures. The antinuclear movement prefigured many of the central components of this discourse, such as the focus on procedural and participatory democracy, critiques of corporate and governmental power, and a concern with public health. The debates over radioactive waste at Prairie Island serve as an important case study of the fusion of social and environmental concern that has transformed modern environmental rhetoric. The controversy over radioactive waste storage at Prairie Island was part of a reformulation of environmental issues that put social analysis squarely at the center of environmental questions. This reformulation marked a significant turning point in both nuclear history and in the history of American environmentalism.

PROMOTING AND PROTESTING NUCLEAR POWER IN THE TWIN CITIES

When Northern States Power Company (NSP) executives first proposed the construction of a nuclear generating plant at Prairie Island in the early 1960s, they did not expect much opposition. The nuclear industry remained in a honeymoon phase, with experts predicting a rapid expansion of generating capacity in response to spiking demand for energy. The Atomic Energy Commission (AEC) suggested the possibility of one thousand reactors generating more than half the nation's energy supply by the year 2000. In Minnesota, NSP positioned itself on the cutting edge of this trend. One NSP official explained the power plant as a response to concerns about the sulfur dioxide pollution produced by coal plants and suggested that the goal of environmental protection motivated the switch to nuclear generation "even though the economics were not positive" when compared to coal. NSP started acquiring land on Prairie Island in 1960 and began construction of twin 520,000-megawatt pressurized water reactors in 1968. Construction ran slightly behind schedule, but workers completed Unit 1 in September 1973 and Unit 2 early the following year.[1]

As construction proceeded, an antinuclear movement based in Minneapolis–Saint Paul coalesced. The trigger was not Prairie Island, but another NSP reactor about to come online at Monticello, thirty-eight miles to the northwest. The plant had sailed through the federal permitting process with little controversy, and NSP officials hoped to begin power generation at Monticello in 1970. But everything changed as the Minnesota Pollution Control Agency (MPCA) completed the permitting process that allowed the discharge of water into the Mississippi River. At the MPCA's February 1968 public hearing, a group of scientists from the University of Minnesota raised concerns about the impact of the thermal and radioactive discharges on the river's marine life and potential implications for the water supplies of Minneapolis and Saint Paul, as both cities drew their drinking water from the Mississippi downstream from the Monticello plant. Mayor Arthur Naftalin told the MPCA that the issue was "a matter of the gravest concern to the people of Minneapolis." He presented the issue as a question of balancing costs with public health. "Perhaps we are imposing standards which may turn out to be unnecessary but I believe the public is willing to pay for adequate health protection and I do not think we can be too careful at any time where public health is involved." Even as the MPCA debated the Monticello permit, NSP applied for a construction permit from the AEC for two reactors at Prairie Island. The controversy grew to include permits for both plants and the role of states in regulating the entire nuclear power industry. In June 1969 MPCA issued a permit that allowed the discharges, but at levels of radioactivity approximately one-third of that allowed by the AEC. The MPCA drew immediate opposition from both those promoting nuclear power and those calling for stricter regulation. The controversy thrust Minnesota into the national eye as a test case for the regulation of radiation in the environment.[2]

NSP officials believed that the standards set by the AEC permit both protected public safety and allowed for the efficient operation of the plant, and also that *only* the AEC, and not the state of Minnesota, had the right to regulate radioactive discharges. Company officers believed that the MPCA had set permissible releases of radioactivity at levels "so low that they cannot be measured by presently developed technology." Remaining in compliance with the permit would force the Monticello plant to run far below capacity and would "eliminate it as a dependable source of electric power." The economic consequences of an unreliable power supply would be far higher than the negligible impacts of radioactive releases. NSP had been charged with supplying reliable power, and a "failure to adequately discharge this responsibility would have consequences more severe in their import and effect upon the economy of the area and the public health and safety than many of the speculative and ill-founded fears voiced . . . by opponents of the plant." Disregarding the state permit while it filed a federal lawsuit questioning the MPCA's regulatory authority, NSP began operating the Monticello reactor in 1970. The case worked its way to the US Supreme Court, which ruled in 1971 that the state of Minnesota did not have the authority to insist on its own radiation standards—a precedent-setting decision that paved the way for the expansion of commercial nuclear power around the country.[3]

Steve Gadler, a member of MPCA's policy-setting citizens board, emerged as the most consistent and vocal critic of NSP's plans for nuclear power. Gadler objected to the use of the river water for cooling the plant, to the planned radioactive discharges, and to what he perceived as lax regulation by the AEC. "Presently the NSP Nuclear generating facility at Monticello is being operated under the extremely lax AEC standards," complained Gadler, "which allows the dumping of radioactive wastes . . . regardless of the damage done to the environment and endangerment of the Metropolitan water supplies." Gadler also expressed his concern that the AEC's role as both promoter and regulator of the nuclear power industry compromised its ability to protect public health—a frequent complaint of the national antinuclear movement. One of his chief objections lay with the process for permitting and siting nuclear facilities. "[The] elaborate structure of administrative and legal mumbo jumbo interposed through layers of governmental bureaucracy should be changed to insure [sic] the rights of citizens," Gadler complained at one hearing. He labeled the system "a charade, a sham, a façade built and architected by the nuclear establishment and orchestrated by the AEC with Madison Avenue type publicity" that not only prevented public participation, but also deliberately prevented citizens from receiving necessary information. In highlighting the importance of participatory and procedural democracy, Gadler articulated ideas that would later became key components of the environmental justice discourse over spent fuel storage at Prairie Island. He became a regular at the state and federal hearings in the permitting process for both Monticello and Prairie Island, and his colorful language made him a favorite source in media coverage of the issue.[4]

Concern about the Monticello Plant continued to spread, and the Twin Cities remained the center of opposition to NSP's nuclear power initiatives. The Minnesota Environmental Control Citizens' Association (MECCA) formed in 1968 and became one of the most significant local environmental groups in the Twin Cities. Opposition to the discharge of radioactive wastes from the NSP plants emerged as one of the group's key concerns. In August 1970 MECCA delivered a petition signed by ten thousand Twin Cities residents, asking the state to prevent the fueling of the Monticello plant. "This petition is one more example of how the citizens of this community have repeatedly and strongly objected to the operation of a nuclear reactor . . . that will release radioactive isotopes into the air we breathe, the water we drink and the soil in which our food is grown," explained one MECCA officer. Local media coverage intensified after the Minneapolis newspapers reported in July 1971 that an unplanned discharge of radioactive water at Monticello had passed through the city's drinking water plant before the intake pipes could be closed. City health officials expected no adverse consequences, but the issue heightened concerns about the safety of nuclear power. The media coverage of the releases and the Supreme Court ruling that denied the MPCA regulatory authority raised awareness of the issue and led to the expansion of the antinuclear movement.[5]

The growing coalition of activists opposed to nuclear power in Minnesota drew on an expanding national movement. The initial protests in Minnesota predated the larger, headline-grabbing controversies over the construction of plants at Seabrook Station in New Hampshire in 1975 and Diablo Canyon in California in 1977. These protests gathered thousands of people using direct-action techniques—protesters chaining themselves to gates at reactor sites or lying in the road to disrupt construction. Although it never coalesced into a nationally organized campaign, the antinuclear movement attracted visible, charismatic leaders, such as Ralph Nader and Barry Commoner, and developed a group of experts who could effectively challenge pronuclear positions on scientific grounds. Protesters in Minnesota corresponded with others around the country as they increased the pressure on NSP to move away from nuclear power.[6]

The emerging antinuclear movement—both in the Twin Cities and around the country—developed differently from other strands of environmental thought. Antinuclear activists saw themselves as fighting for citizen's rights and participatory democracy against the deeply entrenched power structures of corporate boardrooms and secretive government bureaucracies. Many of the environmental reforms sought by activists in the 1960s and 1970s hinged on using the power of government regulation, enabling state and federal governments to curb air and water pollution or to protect wilderness. But to antinuclear activists, the AEC *was* the problem—an agency more concerned with promoting nuclear power than regulating it. These charges led to the dissolution of the AEC and the creation of the Nuclear Regulatory Commission (NRC) in 1974—but this did little to allay the perception that the government remained vitally interested in promoting nuclear power. The secrecy that shrouded the development of nuclear technology during the Cold War further exacerbated concerns about the government's inability to protect citizens from nuclear dangers. Although there was overlap in membership and interest, these anticorporate, antigovernment stances and the emphasis on participatory democracy and direct citizen action set the antinuclear movement apart from mainstream environmental groups. These tendencies also foreshadowed the environmental justice movement, which adopted many of the same principles in the 1980s.[7]

SPENT NUCLEAR FUEL AT PRAIRIE ISLAND

As the antinuclear movement grew more sophisticated, the focus of the debate shifted from radiation exposure and accidents—issues that protesters had used to raise public awareness about the risks of nuclear power—to more technical questions, such as the viability of commercial reprocessing and the storage of spent fuel. When engineers designed the first generation of commercial reactors, they planned to store spent fuel at the reactors in cooling pools for only a few years, after which the fuel would be transported to centralized locations for commercial reprocessing. While reprocessing allowed for a more efficient use of nuclear fuel, it also separated plutonium—the key ingredient for nuclear weapons—from uranium, raising concerns about the spread of nuclear weapons to other nations. Reprocessing also created liquid rather than solid wastes, which were much more difficult to store and transport. Furthermore, the commercial viability of reprocessing remained unclear. All of these concerns—although primarily fear of proliferation—caused President Jimmy Carter to issue a moratorium on commercial reprocessing in 1977.[8]

These national-level policy issues had on-the-ground consequences at Prairie Island and at reactors around the country. In its initial design, Prairie Island had space for the storage of 210 spent fuel assemblies. Used fuel would be covered by circulating water until it had cooled to a point to allow transportation to a reprocessing facility. In 1976 NSP requested and received permission from the NRC to expand its spent fuel pool at Prairie Island to 687 assemblies, arguing that it needed the additional space until commercial reprocessing became viable. The MPCA unsuccessfully challenged the decision to allow this expansion without the preparation of an environmental impact statement (EIS).[9] Just two years after completing its initial expansion, NSP again requested permission to reengineer the spent fuel pool, this time to hold a total of 1,582 fuel assemblies.[10]

This second NSP request led to another round of regulatory hearings. Although it continued to deny the right of the state of Minnesota to regulate radioactive waste, NSP was still required to apply to the Minnesota Energy Agency for a "certificate of need," and also for an amendment to its operating license with the NRC. The company justified the expansion on the grounds that the economic benefits of nuclear power production far outweighed the negligible risks to public health and the environment. As spent fuel accumulated at the reactor, and without the availability of commercial reprocessing or away-from-reactor storage, NSP suggested that it had only two options: shut down the Prairie Island plant or modify the spent fuel pools. The reactors provided 30 percent of NSP's electricity; closing the facility would lead to an interruption of electricity service, a crippling of the energy infrastructure for the Twin Cities, and an estimated cost of $160 million to replace the lost capacity with purchases from other utilities. Without an expansion of the spent fuel pool, NSP would be forced to shut down the reactor by 1983. Weighing the combined economic, social, and environmental impacts of their request, NSP officials concluded that "the real 'socially beneficial use' of the proposed modification is that it allows continued operation of the Prairie Island Plant."[11]

The antinuclear movement attempted to use the regulatory proceeding to prevent the expansion of the spent fuel pool, and possibly to shut down the plant entirely. Opponents of the plan contended that the expansion of the spent fuel pool constituted the creation of a semipermanent nuclear waste site at the reactor. MECCA representatives shifted the focus away from economics and onto the health and environmental issues at stake. "As dangerous high level radioactive wastes continue to pile up at NSP's Prairie Island nuclear plant, a new debate is taking place: Should NSP be allowed to continue to accumulate radioactive wastes when the wastes represent an unprecedented threat to public health and safety? . . . when there is no solution for their disposal? . . . when the Prairie Island nuclear plants containing the wastes may become external dumpsites for them? . . . when the waste will be a public burden for thousands of years?" The plant had not been engineered for permanent storage, and environmental assessments had not been conducted with permanence in mind. Steve Gadler, still one of the most vocal opponents of nuclear power, compared the issue to one of the most high-profile environmental controversies of the time: "The Love Canals continue . . . now NSP proposes a permanent radioactive Love Canal for permanent storage of radioactive waste." The MPCA appealed to the NRC to require an EIS for the fuel expansion, but without success.[12]

Building on these concerns, a group called the Prairie Island Intervention Project formed to pressure the state to prepare an EIS, pleading its case in public hearings before the Minnesota Environmental Quality Board (EQB).[13] The coalition consisted of eight organizations, including avowedly antinuclear groups like MECCA, but also several grassroots organizations that stressed the need for broad public input. The American Indian Movement (AIM) joined the coalition, as well, and Native American opposition to the Prairie Island reactor would become increasingly prominent in the next decade. The coalition represented a broadening of concern about the impact of the Prairie Island reactors, and dozens of residents of the Red Wing area testified in front of the EQB. The anticorporate, antigovernment tenor of the opposition surfaced once again. One resident, for example, objected to the company's repeated requests for expansion of its storage capacity: "NSP will say one thing in order to get the agencies to go along with it. . . . They say that they will have something to take care of the spent fuel, but then that fills up. . . . It's despicable as well as harmful to the public health and safety, the safety of the people of this State, but the utilities have the agencies right around its [sic] finger." In public hearings in Red Wing and Saint Paul, almost all who testified as citizens—as opposed to representatives or experts of NSP or the Prairie Island Project—opposed the expansion of fuel storage at Prairie Island, or at least demanded the completion of a full EIS on the issue.[14]

Local concern about the Prairie Island reactor peaked just as NSP submitted its request for expanded storage, and this changed the tenor of the debate. In October 1979—just one month after NSP filed its certificate of need with the state and six months after the crisis at Three Mile Island in Pennsylvania—a ruptured steam tube triggered a radiation alarm at Prairie Island, shutting down the reactor and sending the plant's workers streaming out of the facility. Only a minimal amount of radiation escaped, but the political fallout from the incident was much more significant. Local residents did not hear of the radiation alarm and plant evacuation from NSP or from the government, but rather from a public radio station in Red Wing. Residents of the Mdewakanton Sioux community, located next to the reactor, did not hear about the scare from anyone; they simply noticed the cars leaving the plant well before the end of the shift. The confused response by the company and the government sparked an immediate reaction. Three hundred protesters rallied at the NSP headquarters in Minneapolis and fifty more marched on the Red Wing city hall, demanding the shutdown of the plant. Clyde Bellecourt, executive director of AIM, charged that reservation land had been "totally contaminated by NSP." While this was an exaggeration, the AIM leader's comments highlighted the perceived threat that the reactor posed to the Indian community.[15]

Despite public pressure the EQB determined in September 1980 that the expansion of the spent fuel pool "did not have the potential for significant environmental effects" and did not require an EIS. The NRC came to the same conclusion. The Minnesota Energy Agency subsequently granted the certificate of need, largely on the basis of economic arguments. The agency's director commented that his role involved only the narrow question of the need for an EIS and that only the legislature could address the broader public concerns about nuclear power. And that is just where the controversy was headed.[16]

PERMANENCE AND ENVIRONMENTAL JUSTICE AT PRAIRIE ISLAND

The continuing failure of the federal government to determine a national policy for the storage of high-level radioactive waste ensured that NSP, and nuclear reactors around the country, would continue to face the spent fuel storage dilemma. In 1982 Congress attempted to resolve the issue with the Nuclear Waste Policy Act (NWPA), which required the Department of Energy (DOE) to identify sites for a federal repository and stipulated that the government should be ready to accept commercial waste by 1998. The bill mandated the selection of one site in the eastern United States—where the vast majority of nuclear waste was generated—and one in the West. The act included provisions to ensure equity and local participation in decisions about siting the repository. But the NWPA did not function as intended. The DOE ceased its search for an eastern repository in the mid-1980s, and Congress amended the act in 1987, selecting Yucca Mountain—against the strident objections of the state of Nevada—as the site of a single facility that would accept all of the nation's high-level radioactive waste. Nevada's refusal to cooperate and other delays ensured that the DOE did not initiate its study of the site until 1992. By this point it was clear that the government would default on its 1998 deadline, and that utilities around the country would need to pursue other solutions for storing spent fuel.[17]

In 1988 NSP announced that it could no longer wait for a federal repository. The company proposed to transfer fuel that had resided in the cooling pools for over ten years to an independent spent fuel storage installation (ISFSI). The ISFSI consisted of "dry casks"—forty-eight massive, pressurized, concrete-and-steel containers that would reside outside of the reactor on concrete pads. NSP's announcement once again thrust Prairie Island into the national eye. Although it would not be the first reactor to construct a dry-cask ISFSI—three others already used the technology—industry watchers expected the Minnesota case to become the most controversial. As one commentator explained, the looming regulatory battle would "reverberate far beyond Minnesota . . . other utilities and regulators are watching. And antinuclear forces here and elsewhere believe that waste storage is the industry's Achilles' heel, the best way to roll back nuclear power and eventually strangle the nation's 111 nuclear plants."[18]

Hoping to avoid some of the regulatory acrimony that had marked its previous requests, NSP officials asked the EQB to conduct an EIS on the proposed facility in 1989. With an EIS finding no significant impact as a part of the record, NSP submitted a formal application for a certificate of need.[19] Once again, the company's justifications rested on economic grounds. Without the ISFSI Prairie Island's two reactors would close by 1995, leading to an economic disaster. Nuclear power represented the company's cheapest source of energy, and NSP relied heavily on the two reactors at Prairie Island, which then accounted for 20 percent of the state's power supply. Company officials estimated the cost of replacing the plant's generating capacity and building a new power plant at $1 billion and cited millions of dollars of lost local and state tax revenues. Dry-cask storage represented the best possible solution to the spent fuel dilemma. The technology protected public and occupational health, had minimal environmental impact, and could be installed without impairing the reactors' operation.[20]

NSP officials also made the case that the Prairie Island plant had served as a world leader in safety and efficiency. One industry journal had rated Prairie Island Unit 2 as the second most efficient reactor in the world and Unit 1 as seventh; no other American reactors had placed in the top ten. Prairie Island's workplace radiation exposure numbers measured less than a quarter of the national average. The reactors had won awards from the government as well as from several industry groups. The Prairie Island plant's strong record raised the stakes in the looming controversy over the future of nuclear power; if antinuclear protesters could use the radioactive waste issue to shut down a model plant like Prairie Island, what would this signal for commercial reactors elsewhere?[21]

Before the Public Utilities Commission (PUC) would make a decision on NSP's application, however, state policy sent the issue to administrative law court for a review of the legal questions at hand and a nonbinding recommendation. At hearings in Red Wing, the Prairie Island Reservation, and Saint Paul in December 1991, Judge Allan W. Klein heard expert testimony from both sides, as well as the comments of 162 citizens, and he also received several thousand written statements. After ruling in favor of NSP on several complicated legal questions, Klein determined that the ISFSI should be considered a permanent repository and therefore required legislative authorization, in accordance with a state law passed in 1977 to prevent the construction of a facility like Yucca Mountain in Minnesota. Klein explained his rationale: "Past delays in federal siting raise questions about whether the dry cask storage will be temporary or will end up being permanent. . . . Once the casks are in place, the path of least resistance is to leave them there indefinitely." Klein's ruling radically changed the nature of the debate over radioactive waste storage at Prairie Island.[22]

This was exactly the kind of ruling that antinuclear forces had worked for more than two decades to achieve. It ensured that the entire public would engage in the debate over the future of nuclear power in Minnesota. Previous discussions had never left the shadowy spaces of the regulatory process; even public hearings saw only limited citizen participation. One member of the Mdewakanton community called the ruling "a great victory and a great day for democracy . . . the citizens of the state will at last have a voice, through their elected officials, in the critical decision of whether to store high-level nuclear waste on the banks of the Mississippi River." After one more round of legal appeals, the controversy headed to the state legislature.[23]

Now squarely in the public eye, the controversy became, in the opinion of one local commentator, "the hottest and most divisive environmental, energy, and jobs issue seen in Minnesota in decades." In February 1994, after seven months of hearings and deliberation, Democratic senator Steve Novak introduced a bill that would allow NSP to store seventeen dry casks at Prairie Island—fewer than the forty-eight requested by the company, but enough to keep the reactor running until the federal government could assume responsibility for the spent fuel. Four hundred supporters of the bill traveled to the state capitol for a rally on the day that a Senate committee voted in favor of the bill. Protests against the bill were more frequent and often larger. Singers Bonnie Raitt and Dave Pirner sang at a protest concert at the University of Minnesota; marches in downtown Minneapolis blocked rush hour traffic and led to two dozen arrests; a rally at Prairie Island gathered eight hundred opponents of the storage plan and resulted in more arrests. The debates inside the legislature were no less acrimonious. Senate majority leader Roger Moe, a twenty-four-year veteran of the legislature, called it the most divisive issue of his tenure. "It has members at one another. . . . It has evolved into a very, very nasty issue." At least two state senators received death threats, and the state police limited admission to the public galleries of the legislature to minimize disruptions.[24]

The controversy split traditional alliances and created strange political bedfellows. Democrats controlled the state house, and Steve Novak, a Democratic senator from suburban Minneapolis, took the lead in advocating for legislative authorization. But most of the opposition to the bill came from Democrats, as well. Most, but not all, Republican legislators supported dry cask storage. In addition to NSP, the Minnesota AFL-CIO supported the storage bill because of

ACT NOW TO
"DUMP THE DUMP"
AT PRAIRIE ISLAND

On October 12th, the Corporate-based Minnesota
Environmental Coalition (MEC) will be meeting at the
Historical Society. MEC parades under the facade of
being "environmental", but there is not one environmental
group in their Coalition. The focus of their meeting at the
Historical Society is to insure NSP's nuclear waste dump
at Prairie Island is authorized by the state legislature and built.

PICKET
TUESDAY, OCTOBER 12
INTERNATIONAL DAY OF SOLIDARITY WITH INDIGENOUS PEOPLE
10:00 - 11:30 am
MINNESOTA HISTORICAL SOCIETY
345 Kellogg Blvd. West
SAINT PAUL

TOGETHER, WE CAN CHANGE MINNESOTA HISTORY.

501 years of oppression is enough! Stop environmental racism.
Let's invest in Minnesota's clean energy future: safe energy,
jobs and dollars for Minnesota now.

For more information: Ken Pentel, Greenpeace 333-5807
Prairie Island Coalition Against
Nuclear Storage Hotline: 920-5943

Figure 17.1. After an administrative law court ruling in 1992 found that the storage casks
on Prairie Island should be considered a permanent repository and therefore required
legislative authorization, the nature of the debate over radioactive waste storage at Prairie
Island radically changed. Discussions that had never left the dark rooms of the regulatory
process suddenly became, in the eyes of one local commentator, "the hottest and most
divisive environmental, energy, and jobs issue seen in Minnesota in decades." This "Dump
the Dump" picket flier promoted a rally to protest the ongoing storage of spent fuel at
Prairie Island. *Courtesy of the Minnesota Historical Society.*

the impact of plant closure on local jobs, creating an unlikely alliance between organized labor and a large corporation. On the other hand, the controversy brought the American Indian Movement and the state's environmental organizations together—two groups that did not always work toward the same ends.[25]

Indeed, the opposition to dry cask storage at Prairie Island hinged on a coalition of Indian and environmental groups that used the increasingly sophisticated rhetoric of environmental justice to make their case to the broader public. Environmental justice emerged in the 1980s as a response to the recognition that minority communities faced a disproportionate risk of environmental harm while having little chance to participate effectively in environmental decision making. The siting of incinerators, toxic waste dumps, and landfills near these communities served as a particularly powerful example of environmental racism. The movement blended the racial critique of the civil rights movement with the anticorporate commentary that emerged after the Love Canal disaster in the 1970s. Environmental justice advocates adopted a much more inclusive view of nature than the mainstream environmental groups, defining the environment as their places of work, play, and worship, rather than as pristine wilderness areas or national parks. A central component of the environmental justice movement derived from the belief that mainstream environmental groups did not effectively represent the concerns of communities of color. All of these elements were present in the controversy at Prairie Island, an event that fused the older concerns of the antinuclear movement with the new rhetoric of environmental justice.[26]

As the debate intensified, opponents of dry cask storage increasingly relied on the rhetoric of environmental justice. As had been the case in 1979, a radiation scare peaked concern just as the issue became controversial. In 1990 surveys discovered traces of the radioactive isotope tritium in tribal well water, likely from a leak in the plant's discharge system. Although the releases remained well within NRC safety limits, the presence of radiation raised the fears of those living closest to the plant. Tribal vice chair Darelynn Lehto described the fundamental reason for her community's opposition to the project: "Would you want this dumped next door to your homes, your day care center, your business?" As the dry cask bill worked its way through the legislature, Indian leaders held a press conference with the state chapter of the NAACP. The groups labeled the plan a classic example of "environmental racism" and yet another case of the risk of environmental harm falling disproportionately on people of color. "Would any other community be expected to make room for a radioactive waste dump?" Lehto asked. "We do not want to live in fear of radiation sickness, we do not want to live in fear of a nuclear accident and we do not want to be forced off our land for the convenience and profit of a utility company."[27]

A key point of the environmental justice movement was the demonstration that these incidents were not isolated, but part of a systematic pattern that cut across time, geography, and racial group. The Mdewakanton saw the Prairie Island proposal in just this way. "The Community finds itself in a vulnerable position, a position in which most groups of Indian people have found themselves since the arrival of European settlers. The Community lacks resources and knowledge to provide an informed critique of the scientific issues involved in the NSP proposal. The Community . . . demand[s] access to the information crucial to a complete understanding of this story." The Mdewakanton connected the local situation to other examples of environmental racism, from the siting of waste dumps to the health risks associated with radiation.[28]

Claims about the environmental injustice of the Prairie Island plant took on added meaning when NSP announced that it had reached a preliminary agreement with the Mescalero Apache tribe in New Mexico for the construction of a facility that could store the waste generated by NSP and other utilities until the federal government completed the Yucca Mountain repository. To the Prairie Island Coalition (PIC)—a group that grew to include over thirty Indigenous, environmental, consumer, and student groups—the proposal seemed to be yet another example of "radioactive colonialism" and a society that forced its most vulnerable members to face greater risks from radiation. Not surprisingly, the Mescalero initiative proved controversial, and the plan eventually failed. Back in Minnesota, opponents of the ISFSI at Prairie Island alleged that the proposal had never been more than a ruse by NSP to convince the public that Minnesota's waste would eventually be moved.[29]

Arguments about environmental justice came not just from the Mdewakanton community, but from the environmental groups involved in the PIC as well. The heart of the argument against the proposal for dry cask storage came down to the belief that once put in dry casks, the waste would remain at Prairie Island indefinitely, causing an already underprivileged group to bear a disproportionate risk. "The Mdewakanton Sioux Community is being asked to bear increased risks because government has failed to provide for adequate permanent storage of spent nuclear fuel," stated the Joint Religious Legislative Coalition, a group of religious organizations that had come together in opposition to the Prairie Island ISFSI. All members of society—including the most marginalized—had a right to participate in decisions about such topics as energy and waste disposal. "Participation means that energy decisions are made not by a small body of experts, planners or entrepreneurs but by the public at large armed with essential information about the costs and benefits of various energy choices. To the fullest extent possible, the public must participate in decisions about energy use patterns and supply choices, especially the allocation of costs and harms." Concerns about the permanence of the facility and its impact on public and environmental health remained, but they were filtered through the targeted impact of the waste on the Indian community at Prairie Island.[30]

It is certainly possible that non-Indian antinuclear activists viewed the discourse of environmental justice in pragmatic terms and utilized it in their arguments against the ISFSI because they believed this strategy offered the best hope for success. But the discourse of environmental justice also connected to traditions of antinuclear rhetoric that had been in place for decades. In its focus on participatory democracy, its anticorporate tenor, its emphasis on shared burdens of risk, and its concerns with environmental and public health, the PIC drew on these earlier arguments, fusing them with the more modern framework of environmental justice.

Amid the controversy legislators continued to work on the issue. For much of the legislative session, it appeared that no common ground could be found. If the legislature failed to act in the 1994 session, most experts agreed that NSP would be forced to close the plant, or at least significantly curtail power generation. Finally, just hours before the close of the session, the legislature passed a bill that seemed to offer something to both sides: NSP would get authority to build seventeen dry casks, enough to run the reactor through 2000, but would then shut down the plant if it had no alternative storage for spent fuel. NSP would also develop renewable energy capacity so that it could decrease its reliance on nuclear power. The compromise emerged after continued pressure from NSP. In the days before the vote, NSP chairman James Howard met privately with ten senators who had opposed the bill, and his pledge to close the plant if needed convinced them to change their votes. Opponents of ISFSI claimed that these "backroom dealings" once again demonstrated the power imbalances at work. A spokesman for the Prairie Island community commented, "It proves that back-room deals work. . . . We've always been willing to talk, but unfortunately the Indian community didn't count." Still, the PIC claimed that the promised phase-out of nuclear power—if the promise was kept—represented a significant victory.[31]

The PIC continued to organize and protest the decision after the passage of the compromise bill. The group staged rallies at the capitol and at the power plant for months, and again when NSP actually loaded its first dry cask in May 1995. The environmental justice implications of the issue continued to serve as the central point of these protests. In 1996 the PIC hosted a symposium titled "Confronting Nuclear Racism" that brought together representatives of seven communities facing risks to their health and environment from every step of the nuclear fuel cycle. Speakers discussed the impact of uranium mining on Native American communities in the US Southwest and Canada; the proposed storage facilities at the Mescalero reservation and at Yucca Mountain, which lies within the territory of the Western Shoshone; and a uranium enrichment facility planned for an African American community in Louisiana. The symposium underscored the contention that "at every link in the nuclear chain, communities of color bear a disproportionate share of the destruction and risks associated with radiation exposures from nuclear waste and failing nuclear technology. . . . Nuclear racism is a fundamental link in the global nuclear chain. Without nuclear racism, the whole nuclear industry could not continue to exist." Activists at Prairie Island connected their struggle to national and international struggles for environmental justice.[32]

The coalition had good reason to maintain its active presence. As the opponents of the ISFSI had feared, the 1994 legislation did not end the controversy. By the late 1990s the federal government appeared no closer to a waste storage solution. The seventeen casks at Prairie Island began to fill, and NSP returned to the legislature to again ask for expanded capacity at the ISFSI. After another contentious round of legislative debates and the prospect of the forced shutdown of Prairie Island, the legislature in 2003 approved up to forty-eight casks and stipulated that any future requests could be handled by the Public Utilities Commission and would not need to come before the legislature. The 2003 law also required Xcel Energy (NSP merged with another company and changed its name in 2000) to pay the Prairie Island Indian community $2.5 million per year so that the tribe could purchase land away from the reactor for tribal members who wished to relocate. In 2008 Xcel requested and received permission to construct an additional thirty-five dry casks. In 2011 the company secured license renewals to continue running the Prairie Island plant through 2033. A 2012 decision by the federal government to cancel plans for the Yucca Mountain repository ensures that the radioactive waste will remain at the Prairie Island ISFSI for the foreseeable future—just as opponents of the storage facility had long predicted.[33]

Nearly fifty casks have been filled at the Prairie Island ISFSI, where the Minnesota Department of Health measures their radiation levels daily. The casks sit on an oblong concrete pad just to the southwest of the two reactors, surrounded by an earthen berm and two security fences. The pad lies approximately one mile from the tribal casino. The presence of radioactive waste—which will remain toxic for generations and on-site indefinitely—both underscores and explains Prairie Island's geography, its clash of the technological and the bucolic.[34]

The long-term presence of radioactive waste is only one of the legacies of this story. For almost forty years, Prairie Island served as a bellwether for the antinuclear movement and the nuclear power industry. National questions about the storage of spent fuel, the preparation of environmental impact statements, and the burdens of risk from radiation played out at Prairie Island. The resolutions to these questions reverberated around the country; dry cask storage is now the industry standard for storing fuel at reactors until the development of other options. If the storage of spent fuel once seemed the "Achilles' heel" of commercial nuclear power, that tender spot has been protected; in February 2012 the NRC approved the first operating and construction license for a new nuclear power plant in over three decades.

After nearly forty years of protest and public debate, the generation of nuclear power at Prairie Island will continue. This has been a frequent result of antinuclear protests in the United States—power generation continues at Three Mile Island, Seabrook Station, and Diablo Canyon, as well. Indeed, one of the legacies of the Prairie Island controversy is the questions that it raises about the effectiveness of environmental justice as a mainstream protest strategy. But the fusion of social and environmental concern that charged the Prairie Island debate has its own significance, even if it did not end reliance on nuclear power. The emergence of sustainability as the focus of environmental discourse over the past two decades has demanded far greater attention to questions of social equity and environmental justice, and it has merged these concerns with the more traditional emphases of environmentalism. The debates over nuclear power production and radioactive waste at Prairie Island represent an early and important example of this fusion, and another chapter in the ongoing struggle to define and achieve the public good.

AFTERWORD

Minnesota's Many
Intersecting Crossroads

Kathleen A. Brosnan

How do we communicate in written language a sense of a particular place? A sense of place, scholars have argued, is a "tangible phenomenon," a compilation of the "interacting influences" of human societies and their built environments, plant and animal communities, water and terrestrial features, and energy sources.[1] How do we convey this sense of place through carefully mediated histories of a given region such as the greater Twin Cities? As historians we contextualize the changes over time that give meaning to our stories. We explain the contingencies and complexities that shaped the multifarious causes of past events that we recognize as significant.[2] As environmental historians we also work to convey and balance the dynamism of diverse human societies and the dynamism of the varied ecosystems of any region. The editors here face another challenge—the same one encountered by their predecessors in this esteemed History of the Urban Environment series from the University of Pittsburgh Press. Anthologies such as this one present myriad voices that create the potential for a fragmented narrative. In the end, do the editors unite those voices in a meaningful ensemble that give us a sense of place?

With *Nature's Crossroads*, I am happy to report, the whole is indeed greater than the sum of its parts, and the parts—the individual chapters—are excellent.

Using Minneapolis and Saint Paul as the linchpins, George Vrtis and Christopher Wells establish the chronological bounds of this regional study with the cities' mid-nineteenth-century origins, although in their own essay, they also point to the *longue durée* of the Dakota land on which the cities stand.[3] However, the editors purposely leave the geographic margins decidedly less clear, and in the end this choice adds a distinctive nuance to the book. Rather than confine themselves to narrowly defined political boundaries that, from an environmental perspective, are most often arbitrary, these contributors contemplate historical phenomena that occur in disparate, ever-shifting spaces. Each contributor has adopted, for the most part, a local, regional, or global scale for his or her essay. Taking the essays together as a whole, however, the editors have woven them into a more comprehensive spatial analysis—what we might think of as the greater Twin Cities' many intersecting crossroads.

The geographer Anne Kelly Knowles observes that, as practiced, "history is the study of when, geography the study of where."[4] The historian Richard White contends that historians have too often treated space as nothing more than "a simple container for the political, social, or cultural."[5] Nonetheless, the chosen spatial scale shapes the narratives that historians craft. Each frame offers strengths and weaknesses. Historical phenomena look different when interpreted at different scales. Yet, like the human societies and natural environments that jointly produce it, space is dynamic and integrated rather than static and segregated. Thus, while historical phenomena occur on a series of scales, these scales overlap and permeate each other as space shifts with environmental change and new cultural mores.[6] Wells and Vrtis explore these interpenetrations in their opening essay as they contrast the commercial steamboat landings of Saint Paul and the industrial waterfront of Minneapolis, both of which sought to exploit the larger region's natural resources and link them to national and international markets. From the interconnected landscapes captured in their essay and the other essays, *Nature's Crossroads* advances a more sophisticated spatial-historical analysis. While the limited confines of this Afterword prevent me from addressing each of the excellent essays, I will highlight a few that illustrate this analysis and what I perceive as the transformative "interacting influences" that emerged to produce the overlapping spaces which constitute the greater Twin Cities. I will address three of these influences: the centrality of waters, the illusion of human control over nature, and the perseverance of Native American communities despite ongoing settler colonialism.

John Anfinson, for example, primarily employs a local scale in his essay to explain both the centrality of waters and the illusion of human control over nature, focusing on the disease implications of the choices Minneapolis and Saint Paul officials made in building public water systems. Responding to typhoid epidemics and increasing population density, the cities replaced dangerous hodgepodge systems of wells and privies. Over four decades Saint Paul transformed a 137-square-mile watershed and some twenty lakes to quench its residents' thirst. Creating this system entailed dams, reservoirs, and dredging that transformed the watershed, provided relatively safe water, and cemented the city's dominance over its hinterlands. Emphasizing quantity, Minneapolis officials alternatively turned to Saint Anthony Falls and the seemingly abundant flow of the Mississippi River. Mounting evidence of unchecked river pollution and recurrent disease episodes forced the city to repeatedly relocate its pumping stations, but the chosen water source remained the issue. In the end, Anfinson concludes, Minneapolis experienced a higher contagion rate because "without adequate knowledge of disease transmission, Minneapolis officials inadvertently replaced older water and waste disposal methods with a modern system of pumps and sewers that themselves became a vector for spreading disease." Anfinson's focus on municipal infrastructure utilizes a spatial scale that shifts from the local to the regional by examining the origins of the water in the pipes. Recognizing this shift, as White explains, does not necessarily suggest greater complexity.[7] Saint Paul's system was not inherently more complex than the one its neighbor employed, but its use of more distant waters does reveal different spatial relations with distinctive environmental consequences. Finally, local officials also operated in theoretical national and global spaces in which ideas about and technologies for urban sanitation were shared, debated, and implemented.[8] With each effort, sanitarians believed they had contained the underlying problems, only to discover that control was illusory. The systems required constant rejiggering.

In her essay Karen Wellner tells the story of Loring Park, "Minneapolis's original grand jewel park." Her analytical scale is a decidedly and appropriately local one, focused on thirty acres on the once western edge of the city. The French philosopher and sociologist Henri Lefebvre argues that economics, politics, and the social markers of race, class, and gender, among other factors, shape social spaces, such as Loring Park.[9] On these thirty acres the park's builders created their vision of a nature in which children could be safe and grow into better citizens. Over time the meaning and function of this social space changed, and officials' ability to shape it proved elusive. Local residents repeatedly asserted new visions over time, ensuring the dynamism of Loring Park and its "nature" persisted. These visions bespoke the ways in which this physical location intersected with a dynamic national cultural space where the perceived value of urban parks shifted from recreation to open space to sustainability.[10] As Wellner also highlights, the national also traversed this local space through concrete infrastructure that embodied an increasingly suburban vision of the United States. Completed in 1971 and stretching from the eastern Great Lakes to the western Great Plains, the I-94 highway sidelined the park as it cut through Minneapolis, altering the visual and auditory environment of the park as well as local spatial relations by cutting off access from some surrounding neighborhoods.

William Barnett takes the readers to different Minneapolis neighborhoods to explain the origins of the American Indian Movement (AIM), underscoring the perseverance of Native American communities despite ongoing settler colonialism. He analyzes "the local problems that led Indians to organize, including environmental justice issues, and reveals the web of connections that bound far-flung Indian communities together." Although the term *environmental justice* had not yet been coined, AIM's demands reflected the reality that the United States, a settler colonial nation, consigned Native Americans, whether in urban or rural places, to "environmentally impoverished lands." Led by Ojibwe men with experience resisting white authority, AIM formed in the late 1960s to address the untenable conditions of a Minneapolis ghetto. Developing a "warrior image" in response to police abuses was essential to AIM's early efforts, but as Barnett explains, community organization efforts also emphasized improving the built environment, particularly the substandard housing that posed the greatest threat to the community's health. "From this urban base," Barnett concludes, "AIM built bridges to the reservations, developing a national movement that asserted a shared Indian identity across tribal and geographic boundaries and challenged centuries of mistreatment." The activists had reorganized local space and created a nexus that transcended it and worked across broader spatial scales.[11]

Broader spatial scales also help illuminate the transformative "interacting influences" that have created the greater Twin Cities. Well before AIM reclaimed Minneapolis as a pan-Indian space in the twentieth century and before Congress granted Minnesota statehood in 1858, different bands of the Dakota called home a region that stretched from the Dakotas to Wisconsin and from Iowa into the Canadian provinces of Manitoba, Saskatchewan, and Alberta. Through activities and annual movements, Wells and Vrtis explain, the Dakota linked the diverse ecosystems of this vast region before European competition over North America and the emergence of a settler colonial American state brought dramatic changes, such as the introduction of devastating diseases, the migration of and competition from the Ojibwe (pushed from their own land by white settlers), and the assertion of US political, military, and economic authority. "The primary object of settler colonization is the land itself rather than the surplus value to be derived from mixing native labour with it."[12] Land cession treaties stripped the Dakota and Ojibwe of most of their land, paving the way for an economic and ecological transformation. Separated by only fourteen miles, the cities of Saint Paul and Minneapolis adopted different strategies in their effort to anchor the new economy. The former emerged as a transportation hub, linking steamboats and express routes, while the latter became an industrial center to process the region's wheat production. Each depended on the Mississippi, the riverine artery that gave life to the region as they both repurposed the Dakota's rich ecosystems and advanced new spatial relations.

Thomas Finger expands this regional narrative. While wheat—"a product of specific soil types, weather patterns, latitudes, and genetic composition"— belongs to a particular place, it provided, via the river-powered mills of Minneapolis, a connection from the farms of the Upper Midwest to international commodity markets. Decisions made on a global scale, such as responding to European crop failures with greater reliance on America's bonanza farms, had local impacts. Recognizing their inability to control ecological forces such as floods in the mill district and stem rust and pests in the fields, regional entrepreneurs attempted to minimize their financial risk by utilizing British capital to expand agricultural hinterlands, retool factories, and construct transportation networks that linked it all. And the eventual intervention of the US Army Corps of Engineers to construct new dams for flood control made the Upper Mississippi part of national riverine space.[13] Finger's complex nineteenth-century story succeeds by operating on a series of scales that permeate each other. As he notes, "historians would have an incomplete story of wheat agriculture in Minnesota if we focused on either grasshoppers or the export of British capital at the expense of the other."

An expanded regional narrative also illuminates why some activists contend that, in the twenty-first century, the "fight for environmental justice must be framed, first and foremost, *as* a struggle for Indigenous sovereignty."[14] This struggle emerges from a past in which settler colonial societies took possession of Indigenous land and resources and exiled to reservations those Native Americans who survived violent conquest. Persistent colonial practices that supported the economic infrastructure also led the US government to assert dominion over reservation lands, often contributing to their environmental degradation.[15] Michael McNally explores the construction of Mississippi River dams by the Army Corps of Engineers. These dams and their impacts occurred primarily on a regional scale, but as noted above, they intersected with a national riverine space. The settler colonial state constructed this space to promote commercial development and navigability that further entrenched its control. "Crucially," McNally highlights, "the flooded lands were all on reservations, cherished homelands not ceded in treaties by Anishinaabe . . . communities who continued to live in profound relationship with the ecosystems on and around the lakes." The construction of downstream locks, greater acknowledgment of tribal governments, and a financial settlement with the Leach Lake Tribal Council in the 1980s seemed to lessen the impacts of the federal dams, but settler colonialism remains ongoing.[16] When a 1988 drought threatened the

Twin Cities' water and sanitation systems, and thus their commercial viability, the governor demanded access to the headwaters over which the Ojibwe held sovereignty. He did so at the expense of the integrative Ojibwe economy that relied on the ecological integrity of the lakes. The Ojibwe had constructed a tribal space shaped by *"bimaadiziwin . . .* the Ojibwe term that describes the integration of the economic and the ecological, the moral, and the spiritual." As political and economic actors, people intentionally produce and reproduce spaces, albeit often with unintended results.[17] In this case, we can see the ways that managing riverine space to serve national and regional priorities came into direct conflict with local autonomy and control over local space. And, of course, the crisis demonstrated again the limits to humans' mastery of nature.

We see the intersection of local, national, and Indigenous space perhaps most vividly in Chantal Norrgard's chapter on the reconstruction of the Grand Portage National Monument. The conservation and historical preservation movements at the core of this project, she observes, "became instruments for their proponents to exercise political influence and to control land use and the actions of local Indigenous people." These activists promoted what Lefebvre might have called a "representation of space," a conceptualized space offered by experts to identify what is lived and what is perceived.[18] In Norrgard's words, this conception was "a particular association between fur trade post and wilderness in need of protection" and inconsistent with Indian control. The creation of the monument in 1958 ended Ojibwe ownership, and although the Ojibwe retained economic benefits from and access to the land, the National Park Service's dominion constituted a continuation of settler colonialism. Subsequently, with assertions of Ojibwe self-determination and shifting cultural mores, the NPS heritage center, which opened in 2007, also included exhibits on the Ojibwe, but the lived space that is the monument continues to be contested ground reflecting the larger tensions that define the Greater Twin Cities.

In *Nature's Crossroads* the contributors give us a notion of the larger Twin Cities' physicality by exploring the different, but often intersecting scales (local, regional, global) and spaces (urban, national, Indigenous) in which its history occurred. I have highlighted a handful of essays to illustrate three of the influences that, from my perspective, interacted to shape these spaces and transform the diverse environments of this region, but these themes are present in other essays as well. The centrality of water involves not only the Mississippi River, but the region's many lakes, the Grand Portage, a campus "watershed," an island reservoir of radioactive waste, and acid rain. Other essays confirm the folly of humans assuming that they have conquered nature, but many are equally adept at erasing other illusions, from notions of entrepreneurial control of financial systems to planners' power to dictate the pace and manner of suburban development to officials' and activists' ability to regulate destructive mining practices. America's settler colonial practices are essential to under-standing the emergence of the Twin Cities region, from the dispossession of Native American lands to the reconceptualization of their natural resources as economic means to enhancing state power to create a built environment that facilitated the commercial, residential, and recreational needs of settlers and their progeny. Yet the authors also magnify the persistent historical and contemporary Indigenous presence, particularly in more recent decades as the Ojibwe pursued self-government and expanded sovereignty over their lands.

Some twenty-five years ago I biked from the Twin Cities to Chicago. We followed the bike paths along the Mississippi and drifted into the suburbs. Lat-er, along the country highways we saw fields cleared of timber and under culti-vation. We witnessed passing trains filled with iron ore. And one of those July days shifted from the nineties and sunshine to the seventies and rain in less than an hour, giving the riders a quick and intimate idea of the region's variable weather. I experienced the region through my five senses on this bicycle jour-ney, but this book has completed and complicated my sense of place. Deeply researched and forcefully argued chapters extrapolate the historical forces that created the Twin Cities and shaped human relations with its nature over time. They recognize the dynamic societies that contested natural resources in ever-changing environments. In doing so, they also evince the power of good history to inform, enlighten, and elevate our knowledge of the world around us.

NOTES

INTRODUCTION: UNEARTHING NATURE'S CROSSROADS

1. The essential source for the MPR raccoon story on June 12–15, 2018, is MPR reporter Tim Nelson's Twitter feed: see Tim Nelson, Twitter posts, June 12, 2018, 8:04–10:24 a.m., https://twitter.com/hashtag/MPRraccoon. (Note: the Twitter time stamps are all PST and thus need to be adjusted by adding two hours to reflect Minnesota time. The original time stamps are maintained here and in the notes that follow, but the narrative reflects the actual local Minnesota time.)

2. Tim Nelson, Twitter posts, June 12, 2018, 1:26–6:42 p.m., quote at 6:42 p.m., https://twitter.com/hashtag/MPRraccoon.

3. Mary McGuire, "St. Paul's Skyscraper-Scaling Raccoon Captivates Internet," CBSN Minnesota (WCCO), June 12, 2018, 9:02 p.m., https://minnesota.cbslocal.com /2018/06/12/raccoon-st-paul/.

4. James Gunn, Twitter post, June 12, 2018, 4:08 p.m., https://twitter.com/hash tag/MPRraccoon.

5. Tim Nelson, Twitter posts, June 12, 2018, 5:35 p.m., June 13, 2018, 12:45 a.m.–12:36 p.m., https://twitter.com/hashtag/MPRraccoon.

6. Lester Holt and Ron Mott, *NBC Nightly News with Lester Holt*, NBC Universal, June 13, 2018.

7. Ros Atkins, *Outside Source with Ros Atkins*, BBC World News, June 13, 2018.

8. Trevor Noah, *The Daily Show with Trevor Noah*, Comedy Central, June 13, 2018.

9. Tim Harlow, "Skyscraper-Climbing Raccoon Rescued, Set Free in the Wild," *Star Tribune* (Minneapolis), June 13, 2018, 10:55 p.m., https://www.startribune.com /skyscraper-climbing-raccoon-rescued-will-be-set-free-in-the-wild/485398071/.

10. The classic work on this is Raymond Williams, *The Country and the City* (New York: Oxford University Press, 1973).

11. Bill McKibben, *The End of Nature* (New York: Random House, 1989).

12. There is a rich and growing historiography on urban environmental history that seeks to collapse the often rigid nature–city (or nature–culture) divide to better under-

stand how cities and the natural world interact and shape one another. The works not-
ed here offer valuable recent contributions to these conversations: Catherine McNeur, *Taming Manhattan: Environmental Battle in the Antebellum City* (Cambridge, MA: Harvard University Press, 2014); Michael Rawson, *Eden on the Charles: The Making of Boston* (Cambridge, MA: Harvard University Press, 2010); Anthony N. Penna and Conrad Edick Wright, eds., *Remaking Boston: An Environmental History of the City and Its Surroundings* (Pittsburgh: University of Pittsburgh Press, 2009); Brian C. Black and Mi-
chael J. Chiarappa, eds., *Nature's Entrepôt: Philadelphia's Urban Sphere and Its Environ-
mental Thresholds* (Pittsburgh: University of Pittsburgh Press, 2012); William Cronon, *Nature's Metropolis: Chicago and the Great West* (New York: W. W. Norton, 1991), 19; Kathleen A. Brosnan, William C. Barnett, and Ann Durkin Keating, eds., *City of Lake and Prairie: Chicago's Environmental History* (Pittsburgh: University of Pittsburgh Press, 2020); Craig E. Colton, *An Unnatural Metropolis: Wresting New Orleans from Nature* (Baton Rouge: Louisiana State University Press, 2005); Martin V. Melosi and Joseph A. Pratt, eds., *Energy Metropolis: An Environmental History of Houston and the Gulf Coast* (Pittsburgh: University of Pittsburgh Press, 2007); Char Miller, ed., *Cities and Nature in the American West* (Reno: University of Nevada Press, 2010); William Deverell and Greg Hise, eds., *Land of Sunshine: An Environmental History of Metropolitan Los Angeles* (Pittsburgh: University of Pittsburgh Press, 2005), esp. Jenny Price's "Thirteen Ways of Seeing Nature in LA"; Richard A. Walker, *The Country in the City: The Greening of the San Francisco Bay Area* (Seattle: University of Washington Press, 2007); Christopher J. Castenada and Lee M. A. Simpson, *River City and Valley Life: An Environmental History of the Sacramento Region* (Pittsburgh: University of Pittsburgh Press, 2013); Matthew Klingle, *Emerald City: An Environmental History of Seattle* (New Haven, CT: Yale University Press, 2007); and Frederick L. Brown, *The City Is More than Human: An Animal History of Seattle* (Seattle: University of Washington Press, 2016). More broadly still, see the innovative and insightful essays in William Cronon, ed., *Uncommon Ground: Rethinking the Human Place in Nature* (New York: W. W. Norton, 1995); and John P. Herron and Andrew G. Kirk, eds., *Biology, Culture, and Environmental History* (Albuquerque: University of New Mexico Press, 1999). On the vast literature on the connected coexistence between wildlife and our urban human lives, see two insightful recent sourc-
es, one nonfiction and the other fiction: Dan Flores, *Coyote America: A Natural and Supernatural History* (New York: Basic Books, 2016); and Mark Kurlansky, *City Beasts: Fourteen Stories of Uninvited Wildlife* (New York: Riverhead Books, 2015).

   13. For a thorough compilation of the secondary sources on the environmental his-
tory of the Twin Cities, see George Vrtis and Christopher W. Wells, eds., *Twin Cities Environmental History: A Bibliography of Published and Unpublished Sources* (St. Paul: Minnesota Historical Society, 2012), available at https://www.mnhs.org/library/tips
/history_topics/TwinCitiesEnvironHistory.php.

1: A TALE OF TWO WATERFRONTS

1. For an influential treatment of the underlying logic of the "settler-colonial state," see Patrick Wolfe, *Settler Colonialism and the Transformation of Anthropology: The Politics and Poetics of an Ethnographic Event* (London: Cassell, 1999), esp. 163.

2. Owamniyomni translates variously as turbulent water, whirlpool, eddy. By 1680, when Father Louis Hennepin renamed the waterfall Saint Anthony Falls, it had moved roughly eight miles upstream from its original site. It continued to retreat a few feet per year thereafter, reaching a point within a thousand feet of its modern location by 1850. See John O. Anfinson, "Spiritual Power to Industrial Might: 12,000 Years at St. Anthony Falls," *Minnesota History* 58, no. 5/6 (2003): 252–69; Scott F. Anfinson, "Archaeology of the Central Minneapolis Riverfront," *Minnesota Archaeologist* 48, no. 1–2 (1989): 14.

3. Gwen Westerman and Bruce M. White, *Mni Sota Makoce: The Land of the Dakota* (Saint Paul: Minnesota Historical Society Press, 2012), 15; "Mnisota Makoce: A Dakota Place," Bdote Memory Map, accessed November 13, 2021, http://bdotememorymap .org/mnisota/; Waziyatawin, *What Does Justice Look Like? The Struggle for Liberation in Dakota Homeland* (Saint Paul: Living Justice Press, 2008).

4. Westerman and White, *Mni Sota Makoce*, 14, 85, 89–90, 94–103, 108–11; Michael J. Lansing, "From Sustenance to Leisure on Minnesota Land," MNopedia, accessed November 13, 2021, https://www.mnopedia.org/sustenance-leisure-minnesota-land; Eileen M. McMahon and Theodore J. Karamanski, *North Woods River: The St. Croix River in Upper Midwest History* (Madison: University of Wisconsin Press, 2009), 2–3, 33; Mary Lethert Wingerd, *North Country: The Making of Minnesota* (Minneapolis: University of Minnesota Press, 2010), 4–5; Gary Clayton Anderson, *Kinsmen of Another Kind: Dakota-White Relations in the Upper Mississippi Valley, 1650–1862* (Lincoln: University of Nebraska Press, 1984), 2–8.

5. Wingerd, *North Country*, 61–64; McMahon and Karamanski, *North Woods River*, 24–28; Anderson, *Kinsmen of Another Kind*, 14–28.

6. Wingerd, *North Country*, 33–37; William E. Lass, *Minnesota: A History*, 2nd ed. (New York: W. W. Norton, 1998), 43–45; Peter DeCarlo, *Fort Snelling at Bdote: A Brief History* (Saint Paul: Minnesota Historical Society Press, 2016), 18.

7. On the fur trade in the Upper Mississippi Valley, see Rhoda R. Gilman, "The Fur Trade in the Upper Mississippi Valley, 1630–1850," *Wisconsin Magazine of History* 58, no. 1 (1974): 2–18. On the fur trade in the broader Great Lakes region, see esp. Richard White, *The Middle Ground: Indians, Empires, and Republics in the Great Lakes Region, 1650–1815* (New York: Cambridge University Press, 1991).

8. On the jockeying for control among competing empires, see esp. Wingerd, *North Country*, 45–74. For a fuller explanation of Pike's Treaty, see DeCarlo, *Fort Snelling at Bdote*, 20–21; Lass, *Minnesota*, 82–83. On the distinction between purchasing western lands from France in the Louisiana Purchase versus acquiring lands through treaties with

Native Americans, see Robert Lee, "Accounting for Conquest: The Price of the Louisiana Purchase of Indian Country," *Journal of American History* 103, no. 4 (2017): 921–42.

9. DeCarlo, *Fort Snelling at Bdote*, 20–29.

10. Matthew Cassady and Peter J. DeCarlo, "Fort Snelling in the Expansionist Era, 1819–1858," MNopedia, accessed November 13, 2021, https://www.mnopedia.org/place /fort-snelling-expansionist-era-1819-1858; Lass, *Minnesota*, 77–87; DeCarlo, *Fort Snelling at Bdote*, 25; Wingerd, *North Country*, 82–83; Westerman and White, *Mni Sota Makoce*, 144–45.

11. Gilman, "Fur Trade in the Upper Mississippi Valley," 16–17; DeCarlo, *Fort Snelling at Bdote*, 35–37; Wingerd, *North Country*, 83–85; Jocelyn Wills, *Boosters, Hustlers, and Speculators: Entrepreneurial Culture and the Rise of Minneapolis and St. Paul, 1849–1883* (Saint Paul: Minnesota Historical Society Press, 2005), 21–24; David Lanegran, *Minnesota on the Map: A Historical Atlas* (Saint Paul: Minnesota Historical Society Press, 2008), 111.

12. Gilman, "Fur Trade in the Upper Mississippi Valley," 18.

13. On changes in the fur trade, see esp. Rhoda R. Gilman, "Last Days of the Upper Mississippi Fur Trade," *Minnesota History* 42, no. 4 (1970): 122–40; Wingerd, *North Country*, 96–100. On treaties and treaty making generally, see esp. Martin Case, *The Relentless Business of Treaties: How Indigenous Land Became U.S. Property* (Saint Paul: Minnesota Historical Society Press, 2018). On the 1837 treaties specifically, see Case, *Relentless Business of Treaties*, 37–38; Wingerd, 128–34; DeCarlo, *Fort Snelling at Bdote*, 37–40; "1837 Land Cession Treaties with the Ojibwe & Dakota," Treatiesmatter.org, accessed November 13, 2021, http://treatiesmatter.org/treaties/land/1837-ojibwe-dakota; Westerman and White, *Mni Sota Makoce*, 155–63; Wills, *Boosters, Hustlers, and Speculators*, 24–25. Notably, the Ojibwe did not agree to removal and retained traditional use rights to the land.

14. According to the Northwest Ordinance of 1787, settlers could claim any part of the public domain once it was surveyed, but formal land surveys took time. This did not stop settlers from rushing to stake claims to unsurveyed lands as soon as ratified treaties made them part of the public domain. Land offices tended to honor such claims, allowing squatters to purchase them after they were formally surveyed, although it took until 1841, when Congress passed the Preemption Act, for squatters' claims to be recognized as legally binding. See Wills, *Boosters, Hustlers, and Speculators*, 13–14; D. W. Meinig, *The Shaping of America: A Geographical Perspective on 500 Years of History*, vol. 2, *Continental America, 1800–1867* (New Haven, CT: Yale University Press, 1993), 240–45. On the earliest land claims in what became downtown Saint Paul, see Virginia Brainard Kunz, *St. Paul: Saga of an American City* (Woodland Hills, CA: Windsor Publications, 1977), 6–10. The story of the race to claim the prime location on the east bank of Saint Anthony Falls is frequently recounted. See, for example, Wingerd, *North Country*, 159–60; Shannon M. Pennefeather, ed., *Mill City: A Visual History of the Minneapolis Mill District* (Saint Paul: Minnesota Historical Society Press, 2003), 26–27; Lass, *Minne-*

sota, 100–101; Carol Ryrie Brink, *The Twin Cities* (New York: Macmillan, 1961), 27–28; Ronald Abler, John S. Adams, and John R. Borchert, *The Twin Cities of St. Paul and Minneapolis* (Cambridge, MA: Ballinger, 1976), 13; Lucile M. Kane, *The Falls of St. Anthony: The Waterfall That Built Minneapolis* (Saint Paul: Minnesota Historical Society Press, 1987), 14–16.

15. The terminology of "east bank" and "west bank" here is potentially confusing, but is important for understanding local geography. Because the Mississippi generally flows from north to south, settlers looking upstream referred to all land on the river's right bank as the "east bank" and all land on the left bank as the "west bank," regardless of the river's actual geographic orientation in any particular place. Because the river twists and turns—and does not follow a north–south line at the central business district of either city—this makes for some interesting place-names. In Saint Paul, for example, the river flows downstream to the east-northeast. This means that West Saint Paul is located directly south of downtown Saint Paul, while South Saint Paul is located east of West Saint Paul. In Minneapolis, where the river flows east-southeast over Saint Anthony Falls, the "east bank" at the falls is oriented more north than east and the "west bank" is more south than west.

16. On the importance of early steamboat traffic in Saint Paul, see John O. Anfinson, *The River We Have Wrought: A History of the Upper Mississippi* (Minneapolis: University of Minnesota Press, 2003), 4–5. In addition to steamboat traffic, the fur trade helped sustain Saint Paul's economy during this period, especially after traders from Pembina, near the Canadian border in the Red River Valley, began to stop in Saint Paul rather than Mendota in 1847. In the early 1850s annuity payments from new treaties transformed the old fur trade into what came to be known as the "Indian trade," in which cash payments for goods became more common than trading furs, much to the benefit of Saint Paul merchants. See Bruce M. White, "The Power of Whiteness: Or, the Life and Times of Joseph Rolette Jr.," *Minnesota History* 56, no. 4 (1998): 178–97, esp. 186. On Steele, see Kane, *Falls of St. Anthony*, 14–16.

17. Kunz, *St. Paul: Saga of an American City*, 13.

18. See especially Wingerd, *North Country*, 176–84; Wills, *Boosters, Hustlers, and Speculators*, 34–39.

19. Wingerd, *North Country*, 203 (quote), 185–204. See also Westerman and White, *Mni Sota Makoce*, 163–95; DeCarlo, *Fort Snelling at Bdote*, 44–46; Lass, *Minnesota*, 108–12.

20. Wingerd, *North Country*, 212–19; Lass, *Minnesota*, 120–26.

21. Anfinson, *River We Have Wrought*, 5; Wingerd, *North Country*, 247.

22. On the complicated politics leading to statehood, see esp. Wingerd, *North Country*, 225–54. As the state's western border, Congress designated a line following the Red and Bois de Sioux Rivers to Lake Traverse, across the portage to Big Stone Lake, and then due south from its outlet to Iowa.

23. On the US–Dakota war, see especially Wingerd, *North Country*, 258–345; Ken-

neth Carley, *The Dakota War of 1862*, 2nd ed. (Saint Paul: Minnesota Historical Society Press, 2001); Duane P. Schultz, *Over the Earth I Come: The Great Sioux Uprising of 1862* (New York: St. Martin's Press, 1992); Gary Clayton Anderson and Alan R. Woolworth, *Through Dakota Eyes: Narrative Accounts of the Minnesota Indian War of 1862* (Saint Paul: Minnesota Historical Society Press, 1988). On the broader context of federally sanctioned violence as a means to dispossess Indians as a key component of US expansion, see Jeffrey Ostler, *Surviving Genocide: Native Nations and the United States from the American Revolution to Bleeding Kansas* (New Haven, CT: Yale University Press, 2019).

24. On the ways that similar processes shaped Chicago, which played a role similar to that of the Twin Cities across a much larger region, see esp. William Cronon, *Nature's Metropolis: Chicago and the Great West* (New York: W. W. Norton, 1991).

25. Wills, *Boosters, Hustlers, and Speculators*, 75 (statistics); Paul Donald Hesterman, "Interests, Values, and Public Policy for an Urban River: A History of Development along the Mississippi River in Saint Paul, Minnesota" (PhD diss., University of Minnesota, 1985), 88–90; Kunz, *St. Paul: Saga of an American City*, 10–11; Paul Hesterman, "The Mississippi and St. Paul: Change Is a Constant for River and the City That Shaped It," *Ramsey County History* 21, no. 1 (1986): 9.

26. Wills, *Boosters, Hustlers, and Speculators*, 60–63, 65–67, 75–82, 111; Anfinson, *River We Have Wrought*, 6–7. See also Merrill E. Jarchow, *Amherst H. Wilder and His Enduring Legacy to St. Paul* (Saint Paul: Amherst H. Wilder Foundation, 1981).

27. Virginia Brainard Kunz, *Saint Paul: The First 150 Years* (Saint Paul: Saint Paul Foundation, 1991), 27, 34; Kunz, *St. Paul: Saga of an American City*, 28, 38, 41.

28. Hesterman, "Mississippi and St. Paul," 4; John O. Anfinson, *River of History: A Historic Resources Study of the Mississippi National River and Recreation Area* (Saint Paul: US Army Corps of Engineers, 2003), 1–2, 26–27, 40–44, 77–79; Hesterman, "Interests, Values, and Public Policy," 88–90; Jarchow, *Amherst H. Wilder*, 8.

29. Lass, *Minnesota*, 139; Wills, *Boosters, Hustlers, and Speculators*, 109–19, 133, 138–39; Don L. Hofsommer, *Minneapolis and the Age of Railways* (Minneapolis: University of Minnesota Press, 2005), xi–xii; Anfinson, *River of History*, 76; John L. Work, *Cargill Beginnings: An Account of Early Years* (Minneapolis: Cargill, 1965), 77–78; Hesterman, "Interests, Values, and Public Policy," 92–93.

30. Wills, *Boosters, Hustlers, and Speculators*, 113; Lass, *Minnesota*, 136–37.

31. Wills, *Boosters, Hustlers, and Speculators*, 86–87, 109–13; Abler, Adams, and Borchert, *Twin Cities of St. Paul and Minneapolis*, 16; Richard White, *Railroaded: The Transcontinentals and the Making of Modern America* (New York: W. W. Norton, 2012), 23–26.

32. *Executive Documents of the State of Minnesota for the Fiscal Year Ending July 31, 1894*, vol. 2 (Saint Paul: Pioneer Press, 1895), 857; Lass, *Minnesota*, 136.

33. Kunz, *St. Paul: Saga of an American City*, 41–42, 49–52; Hesterman, "Interests,

Values, and Public Policy," 95–96; Kunz, *Saint Paul: The First 150 Years*, 45; Anfinson, *River of History*, 157.

34. Anfinson, *River We Have Wrought*, 53–88, 81 (quote).

35. On the life and career of James J. Hill, see esp. Albro Martin, *James J. Hill and the Opening of the Northwest* (New York: Oxford University Press, 1976); Michael P. Malone, *James J. Hill: Empire Builder of the Northwest* (Norman: University of Oklahoma Press, 1996).

36. On the evolution of Saint Paul's built environment, see the chapters in Part II of this volume. See also Kunz, *Saint Paul: Saga of an American City*, 54, 65.

37. Kane, *Falls of St. Anthony*, 16–19, 26, 57.

38. Kane, *Falls of St. Anthony*, 30–39, 42.

39. Kane, *Falls of St. Anthony*, 48–61. For a broader environmental history of the mills in the Merrimack River Valley, including Lowell, see Theodore Steinberg, *Nature Incorporated: Industrialization and the Waters of New England* (New York: Cambridge University Press, 1991).

40. On the seasonal nature of the logging industry and the environmental context in which loggers operated, see Kevin Brown, "Making Stumps and Fields," in this volume; and Pennefeather, *Mill City*, 48–59.

41. Kane, *Falls of St. Anthony*, 63–64; Karel D. Bicha, *C.C. Washburn and the Upper Mississippi Valley* (New York: Garland, 1995), 133–34.

42. On the tunnel break, see esp. The Historyapolis Project, "'Minneapolis Is Ruined': The Tunnel Disaster of 1869," accessed November 13, 2021, https://umn.maps.arcgis .com/apps/MapJournal/index.html?appid=77b5d6b348834ad2ad8bbc91218bc4be#map.

43. Kane, *Falls of St. Anthony*, 62–80; Historyapolis Project, "Minneapolis Is Ruined."

44. Anfinson, "Spiritual Power to Industrial Might," 9.

45. Kane, *Falls of Saint Anthony*, 58; Anfinson, *River of History*, 132.

46. Wills, *Boosters, Hustlers, and Speculators*, 162.

47. For more on the transformation of Minneapolis's flour mills, their interconnec-tions to surging wheat production across the region, and the expansion of regional railroad systems, see Thomas Finger, "Down to the Farm," in this volume. See also Wills, *Boosters, Hustlers, and Speculators*, 162; Kane, *Falls of St. Anthony*, 101; National Park Service, "Wheat Farms, Flour Mills, and Railroads: A Web of Interdependence," accessed June 3, 2022, https://www.nps.gov/articles/wheat-farms-flour-mills-and-railroads-a-web-of -interdependence-teaching-with-historic-places.htm.

48. Kane, *Falls of St. Anthony*, 57, uses the phrase "sawdust cities" to describe Saint Anthony and Minneapolis.

49. Kane, *Falls of St. Anthony*, 106–7.

50. Kane, *Falls of St. Anthony*, 102–3, Pennefeather, *Mill City*, 100.

51. Kane, *Falls of St. Anthony*, 104–5; Pennefeather, *Mill City*, 106.

52. Kane, *Falls of St. Anthony*, 99; John S. Adams and Barbara J. VanDrasek,

Minneapolis-St. Paul: People, Place, and Public Life (Minneapolis: University of Minnesota Press, 1993), 36.

53. See Agnes M. Larson, *History of the White Pine Industry in Minnesota* (Minneapolis: University of Minnesota Press, 1949), 19–20, 148, 159, 355; Also Scott F. Anfinson, "Archaeology of the Central Minneapolis Riverfront," *Minnesota Archaeologist* 49, no. 2 (1990): 23.

54. Anfinson, *River of History*, 140–41.

55. Anfinson, "Archaeology of the Central Minneapolis Riverfront," 21; Abler, Adams, and Borchert, *Twin Cities of St. Paul and Minneapolis*, 17–18.

56. Kane, *Falls of St. Anthony*, 86, 147; Pennefeather, *Mill City*, 80, 82; Ray Lowry, "'Hill's Folly': The Building of the Stone Arch Bridge," *Hennepin County History* 47, no. 1 (1988): 18–26.

57. For an in-depth analysis of the headwaters dams and their negative effects on the Ojibwe residents of the headwaters region, see esp. Michael D. McNally, "Upstream, Downstream," in this volume. See also Jane Lamm Carroll, "Dams and Damages: The Ojibway, the United States, and the Mississippi Headwaters Reservoirs," *Minnesota History* 52 (1990): 2–15; Kane, *Falls of St. Anthony*, 128–33; Merritt, *Corps, the Environment, and the Upper Mississippi River Basin*, 3–11, 8 (quote); David R. Treuer, "The Rhetoric of Reservoirs," *Minnesota History* 53, no. 4 (1992): 156–62.

58. Kane, *Falls of St. Anthony*, 146–57; Anfinson, "Spiritual Power to Industrial Might," 264–66.

59. See esp. Finger, "Down to the Farm," in this volume; and Kane, *Falls of St. Anthony*, 146.

60. Ralph W. Hidy, *The Great Northern Railway: A History* (Boston, MA: Harvard Business School Press, 1988), 92.

61. Malone, *James J. Hill*, 243; Ralph W. Hidy, Frank Ernest Hill, and Allan Nevins, *Timber and Men: The Weyerhaeuser Story* (New York: Macmillan, 1963), 212–14, 224; Larson, *History of the White Pine Industry in Minnesota*, 398; Jeff Forester, *The Forest for the Trees: How Humans Shaped the North Woods* (Saint Paul: Minnesota Historical Society Press, 2004), 60; Martin, *James J. Hill and the Opening of the Northwest*, 465; Judith Healey, *Frederick Weyerhaeuser and the American West* (Saint Paul: Minnesota Historical Society Press, 2013), 140–44.

2: DOWN TO THE FARM

1. For an environmental history that underscores the place of Chicago in the American wheat trade, see William Cronon, *Nature's Metropolis: Chicago and the Great West* (New York: W. W. Norton, 1991), 97–147.

2. William Edgar, *The Story of a Grain of Wheat* (New York: D. Appleton, 1925), 24–26.

3. See A. L. Olmstead and P. W. Rhode, *Creating Abundance: Biological Innovation and American Agricultural Development* (Cambridge: Cambridge University Press,

2008), 17–63; William Edgar, *The Medal of Gold: A Story of Industrial Achievement* (Minneapolis: Bellman, 1925), 31–40.

4. Rudolph Peterson, *Wheat: Botany, Cultivation and Utilization* (London: L. Hill Books, 1965), 48–55.

5. Peterson, *Wheat*, 128.

6. Olmstead and Rhode, *Creating Abundance*, 24–25; Peterson, *Wheat*, 53–55; Edgar, *Story of a Grain of Wheat*, 26.

7. Olmstead and Rhode, *Creating Abundance*, 17–65.

8. While the region was popularly called "the Northwest" during the late nineteenth century, I have decided to label it the "Upper Midwest" to avoid confusion with other regions that have been called the Northwest throughout American history.

9. Benjamin Horace Hibbard, "The History of Agriculture in Dane County, Wisconsin" (PhD diss., University of Wisconsin-Madison, 1902), 125. Farmers and breeders also began developing harder breeds of winter wheat, whose hard casing could better withstand prolonged freezing.

10. *History of Steele and Waseca Counties, Minnesota* (Chicago: Union, 1887), 515.

11. Merrill E. Jarchow, *The Earth Brought Forth: A History of Minnesota Agriculture to 1885* (Saint Paul: Minnesota Historical Society, 1949), 180; B. F. Johnson, "Hard or Soft Wheat?" *Northwestern Miller*, February 14, 1879.

12. On the massive efforts to reengineer the waterfall, see Wells and Vrtis, "A Tale of Two Waterfronts," in this volume.

13. Olmstead and Rhode, *Creating Abundance*, 27.

14. Edgar, *Medal of Gold*; Herman Steen, *Flour Milling in America* (Westport, CT: Greenwood Press, 1973); Charles Byron Kuhlmann, *The Development of the Flour-Milling Industry in the United States: With Special Reference to the Industry in Minneapolis* (Clifton, NJ: A. M. Kelley, 1973).

15. Edgar, *Medal of Gold*, 42–45; Steen, *Flour Milling in America*, 43–45.

16. Peterson, *Wheat*, 139–40.

17. Edgar, *Medal of Gold*, 95.

18. Kate Roberts and Barbara Caron, "'To the Markets of the World': Advertising in the Mill City, 1880-1930," *Minnesota History* 58, no. 5 (2003): 310.

19. Eugene Smalley, *History of the Northern Pacific Railroad* (New York: G. P. Putnam's Sons, 1883); Hiram M. Drache, *The Day of the Bonanza: A History of Bonanza Farming in the Red River Valley of the North* (Fargo: North Dakota Institute for Regional Studies, 1964), 34–67.

20. Harold E. Briggs, "Grasshopper Plagues and Early Dakota Agriculture, 1864–1876," *Agricultural History* 8 (April 1934): 51–63.

21. James B. Power to Fredrick Billings, September 2, 1877, box 1, bolder 5, James B. Power Papers, North Dakota State Institute for Regional Studies, Fargo, ND.

22. C. G. Moody, in the *Yankton Weekly Dakotan*, July 30, 1864, qtd. in Briggs, "Grasshopper Plagues," 52.

23. Briggs, "Grasshopper Plagues," 56.

24. For a detailed history of bonanza farming and farmers, see Drache, *Day of the Bonanza*; and Hiram Drache, *The Challenge of the Prairie: Life and Times of Red River Pioneers* (Fargo: North Dakota Institute for Regional Studies, 1970).

25. Briggs, "Grasshopper Plagues," 60–61.

26. William F. Dalrymple, January 1, 1880, box 29, 1880 diary, William F. Dalrymple Papers, 1836–1916, Wisconsin Historical Society Archives, Madison.

27. Laura M. Hamilton, "Stem Rust in the Spring Wheat Area in 1878," *Minnesota History* 20 (June 1939): 158.

28. Peterson, *Wheat*, 160–65.

29. Hamilton, "Stem Rust in the Spring Wheat," 162–64.

30. Drache, *Day of the Bonanza*, 132.

31. Drache, *Day of the Bonanza*, 132.

32. Annual Reports of Northern Pacific to the State of Minnesota, 1876, 135.c.5.313, Minnesota Historical Society, Saint Paul (hereafter MHS).

33. "Reminisces," A/.R981f, Folder 2, 22, Fred S. Rutledge and Family Papers, MHS.

34. Drache, *Day of the Bonanza*, 210–14.

35. Drache, *Day of the Bonanza*, 204–17.

36. Albro Martin, *James J. Hill and the Opening of the Northwest* (New York: Oxford University Press, 1976), 414, 462–63, 509–10.

37. James J. Hill Papers" 1823–1985, Farms (Series F), M458, MHS (hereafter Hill Papers); James Hill, *Highways of Progress* (New York: Doubleday, Page, 1912); Ralph Hidy et al., *The Great Northern Railway: A History* (Boston: Harvard Business School Press, 1988), 67, 78.

38. "Red River Rolling Mill," 20.G.6.1–3, 1823–1985; "St. Anthony Falls Water Power Company," 20.H.4.3, both in Hill Papers.

39. Christabel Susan Lowry Orwin, *History of British Agriculture, 1846–1914* (London: Longmans, 1964), 242–43.

40. Brian Mitchell and Phyllis Deane, *Abstract of British Historical Statistics* (Cambridge: Cambridge University Press, 1962), 6.

41. E. P. Thompson, *The Making of the English Working Class* (New York: Pantheon Books, 1964).

42. While Great Britain was experiencing a demographic boom, England received much of the highest-quality wheat imports, particularly from the United States, while Scotland, Ireland, and Wales were generally the destination for lower-quality wheat. Thus I can make a distinction between the population growth, industrialization, and trade policy of Great Britain, while isolating England as the destination of the majority of wheat imports from the United States.

43. H. M. Larson, *The Wheat Market and the Farmer in Minnesota* (New York: Longmans, Green, 1926), 25.

44. Sir William Ashley, *The Bread of Our Forefathers: An Inquiry in Economic History*

(Oxford: Clarendon Press, 1928), 4–10; Christian Petersen, *Bread and the British Economy, c. 1770–1870* (Hants, UK: Scolar Press, 1995), 150–81.

45. R. Perren, "Structural Change and Market Growth in the Food Industry: Flour Milling in Britain, Europe, and America, 1850–1914," *Economic History Review* 43 (August 1990): 422–23.

46. Larson, *Wheat Market and the Farmer in Minnesota*, 141.

47. Perren, "Structural Change and Market Growth in the Food Industry," 423–24.

48. Larson, *Wheat Market and the Farmer in Minnesota*, n. 140.

49. William H. Dunwoody to Washburn Crosby & Co., February 22, 1878, box 2, folder 1, William H. Dunwoody and Family Papers, 1837–1915, MHS.

50. William H. Dunwoody to J. J. & R. Walker, May 23, 1881, box 2, folder 1, William H. Dunwoody and Family Papers, 1915–1837, MHS.

51. Edgar, *Medal of Gold*, 70. The description of Dunwoody's trip was taken from 67–70.

52. Stolterfoht, Sons & Co. to Rd River Rolling Mills, April 1, 1884, 20G.6.2, Hill Papers.

53. Orwin, *History of British Agriculture, 1846–1914*, 242–43.

54. James J. Hill, qtd. in Martin, *James J. Hill and the Opening of the Northwest*, 277.

55. US Bureau of the Census, *Historical Statistics of the United States: Colonial Times to 1970* (Washington, DC: Government Printing Office, 1975), 510–12, 898–99; Mitchell and Deane, *Abstract of British Historical Statistics*, 97–102.

56. New York State Legislature, *Report of the Special Committee on Railroads* (Albany, NY: Weed, Parsons, 1880), 1139.

57. Mira Wilkins, *The History of Foreign Investment in the United States to 1914* (Cambridge, MA: Harvard University Press, 1989), 319.

58. Wilkins, *History of Foreign Investment*, 319.

59. *The Times* (London), November 1, 1889; "A Big Deal Closed: English Capital Controlling the Mills of the Northwest," *New York Times*, December 24, 1889.

60. Wilkins, *History of Foreign Investment*, 319–21.

61. *The Times* (London), November 1, 1889.

62. US Bureau of Statistics, "The Grain Trade of the United States, and the World's Wheat Supply and Trade," in *Monthly Summary of Commerce and Finance of the United States, January 1900* (Washington, DC: Government Printing Office, 1900), 1995.

63. US Bureau of Statistics, "Grain Trade of the United States," 1998.

64. US Bureau of Statistics, "Grain Trade of the United States," 1997.

65. Steen, *Flour Milling in America*, 50.

66. US Bureau of Statistics, "Grain Trade of the United States," 2012.

67. US Bureau of Statistics, "Grain Trade of the United States," 2009.

68. Allan Bogue, *From Prairie to Corn Belt: Farming on the Illinois and Iowa Prairies in the Nineteenth Century* (Chicago: University of Chicago Press, 1963); Cronon, *Nature's Metropolis*; E. West, *The Contested Plains: Indians, Goldseekers, and the Rush to*

Colorado (Lawrence: University of Kansas Press, 1998); Sterling Evans, *Bound in Twine: The History and Ecology of the Henequen-Wheat Complex for Mexico and the American and Canadian Plains, 1880–1950* (College Station: Texas A&M University Press, 2007); Julie Courtwright, *Prairie Fire: A Great Plains History* (Lawrence: University Press of Kansas, 2011).

69. Washburn, who also built considerable wealth from lumbering, is a possible exception.

70. Katherine G. Morrissey, *Mental Territories: Mapping the Inland Empire* (Ithaca, NY: Cornell University Press, 1997).

71. Anna Lowenhaupt Tsing, *Friction: An Ethnography of Global Connection* (Princeton, NJ: Princeton University Press, 2005).

3: COMPETING HINTERLANDS

1. William Cronon, *Nature's Metropolis: Chicago and the Great West* (New York: W. W. Norton, 1991).

2. John R. Borchert, *America's Northern Heartland: An Economic and Historical Geography of the Upper Midwest* (Minneapolis: University of Minnesota Press, 1987).

3. David A. Lanegran, *Minnesota on the Map: A Historical Atlas* (Saint Paul: Minnesota Historical Society Press, 2008), 39–47.

4. Lawrence Martin, *The Physical Geography of Wisconsin* (Madison: University of Wisconsin Press, 1965).

5. Perry John Hickerson Jr. and Jay Allen Hickerson, eds., *Hickerson Family History* (Marshall, MN: Henle's Speedy Print, 1988), 160.

6. Hickerson and Hickerson, *Hickerson Family History*, 159.

7. William Henry, *Northern Wisconsin: A Hand-Book for the Homeseeker* (Madison, WI: Democrat Printing, State Printer, 1896), 6.

8. Henry, *Northern Wisconsin*, 62.

9. Edward L. Peet, *Official Report of Burnett County Immigration Board* (Grantsburg, WI: Journal Printing, 1902), 14.

10. Peet, *Official Report of Burnett County Immigration Board*, 23.

11. Several copies of newspaper reports are reproduced in Hickerson and Hickerson, *Hickerson Family History*, 150–75.

12. Hickerson and Hickerson, *Hickerson Family History*, 160.

13. Peet, *Official Report of Burnett County Immigration Board*, 27.

14. Hickerson and Hickerson, *Hickerson Family History*, 193–94.

15. On the larger environmental history of this subject, see Sterling D. Evans, *Bound in Twine: The History and Ecology of the Henequen-Wheat Complex for Mexico and the American and Canadian Plains, 1880–1950* (College Station: Texas A&M University Press, 2007).

16. Paul D. Nelson, "The Greatest Single Industry? Crex, Created Out of Nothing," *Ramsey County History* 40 (Winter 2006): 4–15.

17. Nelson, "Greatest Single Industry?" 10–11.

18. Nelson, "Greatest Single Industry?" 10–11.

19. St. Paul Pioneer Press, *The Book of Minnesota: Development, Resources, Enterprise and Beauty of the North Star State* (Saint Paul: St. Paul Pioneer Press, 1903), n.p.

20. Hickerson and Hickerson, *Hickerson Family History*, 170–71.

21. Timothy Bawden, "The Northwoods Back to Nature?" in *Wisconsin Land and Life*, ed. Robert Ostergren and Thomas Vale (Madison: University of Wisconsin Press, 1997), 460.

22. Robert Gough, *Farming the Cutover: A Social History of Northern Wisconsin 1900–1940* (Lawrence: University of Kansas Press, 1997), 106–67.

23. "About Crex Meadows," Friends of Crex, November 13, 2021, https://www.crex meadows.org/about-crex-meadows; and "Other Wildlife Areas," Friends of Crex, November 13, 2021, https://www.crexmeadows.org/other-wildlife-areas.

24. "Glacial Lake Grantsburg Wildlife Management Complex," Friends of Crex, December 27, 2011, http://www.crexmeadows.org/wildlifeareas.htm.

4: UPSTREAM, DOWNSTREAM

1. Initial lake level estimate by Carswell McClellan, Army Corps of Engineers, 48th Cong., 1st sess., 1884, House Executive Document, no. 76, "Damages to Chippewa Indians," 29 (serial 2200), 20–24.

2. Jane Lamm Carroll, "Dams and Damages: The Ojibway, the United states and the Mississippi Headwaters Reservoirs," *Minnesota History* 52 (Spring 1990): 2–15.

3. David Treuer, "The Rhetoric of Reservoirs," *Minnesota History* 53 (Winter 1992): 158.

4. Carroll, "Dams and Damages," 3. See also John O. Anfinson, *The River We Have Wrought: A History of the Upper Mississippi* (Minneapolis: University of Minnesota Press, 2005).

5. Carroll, "Dams and Damages," 4.

6. Jane Lamm Carroll, "Lake Winnibigoshish Reservoir Dam," *Mississippi River Headwaters Reservoirs*, Historic American Engineering Record MN-65 (1993), 2, National Park Service.

7. *Statutes at Large*, 21:193; 47th Congr., 1st sess., 1881, House Executive Document no. 1, as cited in Carroll, "Dams and Damages," 6.

8. H. B. Whipple to Hiram Price, Commissioner of Indian Affairs, August 17, 1883, "Damages to Chippewa Indians," 18.

9. The cap of $22,500 was raised when the appropriation increased to include Leech Lake.

10. "Damages to Chippewa Indians," 3.

11. Sturgeon Man, Memorandum of a Council Held Nov. 8, 1883, "Damages to Chippewa Indians," 27.

12. Namewinini (Sturgeon Man) made explicit that Ojibwe leaders had never given their consent to the Northwest Indian Commission's proposed settlement. See "Eighth Council at Leech Lake, August 16, 1889," 51st Cong., 1st sess., 1890, House Executive Document, no. 247, "Chippewa Indians in Minnesota," 4 (serial 2747), 142.

13. Carroll, "Dams and Damages," 13.

14. See, for example, Harold Hickerson, *The Chippewa and Their Neighbors: A Study in Ethnohistory* (New York: Holt, Rinehart, and Winston, 1970).

15. An oft-cited written account of this oral tradition is found in Eddie Benton-Banai, *The Mishomis Book: The Voice of the Ojibway* (Minneapolis: University of Minnesota Press, [1988] 2010), 94-101.

16. Marvin Manypenny, address to Camp Justice Assembly, White Earth, Minn., August 1991, as cited in Michael D. McNally, *Ojibwe Singers: Hymns, Grief, and a Native Culture in Motion* (New York: Oxford University Press, 2000; repr. Saint Paul: Minnesota Historical Society, 2009), 185n37.

17. Frederic Baraga, *Dictionary of the Ojibway Language* (Saint Paul: Minnesota Historical Society, [1878] 1992), 220.

18. A. I. Hallowell, *Culture and Experience* (Philadelphia: University of Pennsylvania Press, 1955), 171; see also Michael McNally, *Honoring Elders: Aging, Authority, and Ojibwe Religion* (New York: Columbia University Press, 2009), esp. 48-62.

19. Densmore was characterizing what she had been told by Noodinens, a White Earth consultant. Frances Densmore, *Chippewa Customs* (Saint Paul: Minnesota Historical Society Press, [1929] 1979).

20. Basil Johnson, *Ojibway Heritage* (Lincoln: University of Nebraska Press, 1990), 33.

21. Thomas Vennum Jr., *Wild Rice and the Ojibway People* (Saint Paul: Minnesota Historical Society Press, 1988), 62.

22. White Earth Nation Tribal Council, "Manoomin and the Anishinaabeg," 64, white paper submitted to International Indian Treaty Council for Shadow Report to the United Nations Committee on the Elimination of all Forms of Racial Discrimination, 2008.

23. Andy Favorite, qtd. in Jill Doerfler, *Where the Food Grows on Water: The Continuance of Scientific Racism and Colonization*, CAP Report no. 48 (Minneapolis: University of Minnesota Center for Urban and Regional Affairs, 2006).

24. This interpretation runs against the grain of that argued by Calvin Martin, *Keepers of the Game: Indian-Animal Relationships and the Fur Trade* (Berkeley: University of California Press, 1982). See also Shepard Krech, *The Ecological Indian* (New York: Norton, 2000).

25. Treaty with the Chippewa, February 22, 1855, ratified March 3, 1855, 10 Stat., 1165. The Leech Lake Reservation was subsequently expanded Treaty with the Chippewa of the Mississippi and the Pillager and Lake Winnibigoshish Bands, March 11, 1863, ratified, March 13, 1863, 12 Stat., 1249.

26. See Melissa Meyer, *The White Earth Tragedy: Ethnicity and Dispossession at a Minnesota Anishinaabe Reservation* (Lincoln: University of Nebraska Press, 1999).

27. Second Council at Lake Winnibigoshish, Sept. 2, 1889, US Department of the Interior, "Chippewa Indians in Minnesota," 154.

28. Second Council at Lake Winnibigoshish, Sept. 2, 1889, 154.

29. Fourth Council at Leech Lake, Aug. 2, 1889, "Chippewa Indians in Minnesota," 122–23.

30. Second Council at Lake Winnibigoshish, Sept. 2, 1889, 154.

31. Vennum, *Wild Rice and the Ojibway People*, 20.

32. Vennum, *Wild Rice and the Ojibway People*, 20–27.

33. Eighth Council at Leech Lake, August 16, 1889, "Chippewa Indians in Minnesota," 142.

34. Report of the D. B. Herriman, Indian Agent at Chippewa Agency, *Annual Report of the Commissioner of Indian* Affairs, 34th Cong, 1st sess., 1855, H. Doc. 1, 373, as cited in Vennum, *Wild Rice and the Ojibway People*, 27.

35. Memorandum of a Council Held Nov. 8, 1883, 29.

36. Memorandum of a Council Held Nov. 8, 1883, 27.

37. See, for example, the exchange between Commissioner Blakeley and Flatmouth the Younger, Memorandum of a Council Held Nov. 8, 1883, 26–30; and Henry B. Whipple, cited in Hiram Price to Henry Teller, November 18, 1882, "Damages to Chippewa Indians," 11.

38. William Marshall to H[enry]. B. Whipple, August 15, 1883, "Damages to Chippewa Indians," 18–19.

39. Marshall to Whipple, August 15, 1883, 18–19.

40. Enmegabowh to H[enry]. B. Whipple, December 18, 1881, box 15, H. B. Whipple Papers, Minnesota Historical Society (hereafter Whipple Papers).

41. Enmegabowh to Whipple, December 18, 1881.

42. Sturgeon Man, qtd. in Memorandum of a Council Held Nov 8, 1883, 26.

43. Third Council at Lake Winnibigoshish, September 2, 1889, "Chippewa Indians in Minnesota," 156–57.

44. Marshall to Whipple, August 15, 1883, 19.

45. The Rice Commission in particular was keen to seize on the despair of Ojibwe whose homelands were flooded in order to persuade them of the urgency of moving to White Earth, where the fuller plan of assimilation in Minnesota was to be carried out. See, for example, Second Council at Lake Winnibigoshish, Sept. 2, 1889, 154.

46. See Frederick Hoxie, *A Final Promise: The Campaign to Assimilate the Indians, 1880–1920* (Lincoln: University of Nebraska Press, 1984).

47. White Cloud to H[enry]. B. Whipple, n.d., "Damages to Chippewa Indians," 8.

48. Pillager and Mississippi Band Chiefs and Headmen to Hiram Price, August 2, 1883, "Damages to Chippewa Indians," 17. See also, for example, Enmegabowh to Whipple, December 18, 1881.

49. Hiram Price to H[enry]. B. Whipple, August 28, 1883, box 16, folder 8, Whipple Papers.

50. Memorandum of a Council Held Nov. 8, 1883, 31. Blakeley ignored the possibility of legal action, for which Ojibwe in the 1880s would not have had standing.

51. Second Council at Lake Winnibigoshish, September 2, 1889, 154.

52. Treuer, "Rhetoric of Reservoirs," 156–57.

53. Bruce White, "Criminalizing the Seasonal Round," paper presented at the Annual Meeting of the Ethnohistory Society, Springfield, Missouri, 1998. See also Karl Jacoby, *Crimes against Nature: Squatters, Poachers, Thieves, and the Hidden History of American Conservation* (Berkeley: University of California Press, 2003).

54. See Melissa Meyer, *The White Earth Tragedy: Ethnicity and Dispossession at a Minnesota Anishinaabe Reservation* (Lincoln: University of Nebraska Press, 1999).

55. Enmegabowh to H[enry]. B. Whipple, December 6, 1881, box 15, Whipple Papers.

56. Brenda Child, *Holding Our World Together: Ojibwe Women and the Survival of Community* (New York: Viking Penguin, 2012).

57. See Tsegaye Nega, "Saving Wild Rice: The Rise and Fall of the Nett Lake Dam," *Environment and History* 14 (2008): 5–39, for how comanagement of dams can promote wild rice flourishing.

58. See also Minnesota Chippewa Tribe et al. v. the United States of America, 29 Indian Claims Commission 211 (1972), esp. 230–32. Federal consultation with recognized tribes was further mandated in 2000 by President Clinton's Executive Order 13175, Consultation and Coordination with Indian Tribal Governments.

59. Kevin Pollard, "Summer Dates of '88 Drought Marked by Sprinkling Bans, River's Rise and Fall," *Star Tribune* (Minneapolis), August 17, 1988, 7a; Ron Nargang, interviewed by Fred Sam Lazaro in "Scorched," transcript, *MacNeil Lehrer NewsHour*, PBS, August 3, 1988.

60. As cited in "Scorched."

61. Jim Parsons, "Hot and Dry: What Do You Do?" *Star Tribune*, July 29, 1988, b1. Under pressure from such activists, the *Star Tribune* reported that the Leech Lake tribal council threatened lawsuits if Winnie and Leech were lowered, "citing possible damage to their wild rice crop and to fish and wildlife. The Corps apparently didn't relish getting involved in such a suit, although it took the position that it had the legal authority to release the water"; Jim Parsons, "Corps Denies Request to Open Reservoirs," *Star Tribune*, August 4, 1988, 1a.

62. In the tense week when the Army Corps of Engineers was weighing Governor Perpich's request, the *Star Tribune* reported that the Corps was hoping that as part of the plan, the state of Minnesota would compensate the tribe for any losses from the requested drawdown. The paper also reported that "state officials have said that they probably would agree to restore an area on the reservation that once contained natural beds of wild rice and was breeding habitat for waterfowl"; Parsons, "Hot and Dry," b1.

5: MAKING STUMPS AND FIELDS

1. Samuel B. Green, *Forestry in Minnesota* (Saint Paul: Pioneer Press, 1902), 164.

2. Committee on Land Utilization, *Land Utilization in Minnesota: A State Program for the Cutover Lands* (Minneapolis: University of Minnesota Press, 1934), 117.

3. Committee on Land Utilization, *Land Utilization in Minnesota*, 6.

4. Richard White, *The Organic Machine: The Remaking of the Columbia River* (New York: Hill and Wang, 1995), ix.

5. Neil Smith, *Uneven Development: Nature, Capital, and the Production of Space*, 3rd ed. (1984; Athens: University of Georgia Press, 2008), 116. This paper is grounded, in part, in efforts since the mid-1990s to bridge the concerns of environmental and labor history through the examination of the simultaneous construction of the material and social-cultural conditions of work. This project has been advanced best by White, *Organic Machine*; John Soluri, *Banana Cultures: Agriculture, Consumption, and Environmental Change in Honduras and the United States* (Austin: University of Texas Press, 2005), esp. 139–52; Myrna Santiago, *The Ecology of Oil: Environment, Labor, and the Mexican Revolution* (Cambridge: Cambridge University Press, 2006); Thomas G. Andrews, *Killing for Coal: America's Deadliest Labor War* (Cambridge, MA: Harvard University Press, 2008), esp. 124–25; and Arthur F. McEvoy, "Working Environments: An Ecological Approach to Health and Safety," *Technology and Culture* 36, no. 2, Supplement (1995): S145–73.

6. Henry B. Steer, "Lumber Production in the United States, 1799–1946," US Department of Agriculture, Miscellaneous Publication no. 669 (Washington, DC: Government Printing Office, 1948), 11.

7. Per capita annual lumber consumption in the United States exploded during industrialization, peaking in 1906 at 82 cubic feet. By the late 1970s it had fallen to about 30 cubic feet. See Michael Williams, *Americans and Their Forests: A Historical Geography* (Cambridge: Cambridge University Press, 1989), 487–88.

8. Agnes M. Larson, *The White Pine Industry in Minnesota: A History* (1949; repr., Minneapolis: University of Minnesota Press, 2007), 124.

9. Larson, *White Pine Industry in Minnesota*, 232–33.

10. On capital mobility and the American lumber industry, see Thomas R. Cox, *The Lumberman's Frontier: Three Centuries of Land Use, Society, and Change in America's Forests* (Corvallis: Oregon State University, 2010); William Cronon, *Nature's Metropolis: Chicago and the Great West* (New York: W. W. Norton, 1991); Larson, *White Pine Industry in Minnesota*; William G. Robbins, *Lumberjacks and Legislators: Political Economy of the U.S. Lumber Industry, 1890–1940* (College Station: Texas A&M University Press, 1982); and Williams, *Americans and Their Forests*. More than a "wasteland," the cutover should be examined with attention not only to the absence of old growth forests, but to the new and different working environments forged there.

11. William B. Greeley, *Forests and Men* (1951; repr., New York: Arno Press, 1972), 40.

12. Williams, *Americans and Their Forests*, 310.

13. Williams, *Americans and Their Forests*, 437.

14. Richard White, "Poor Men on Poor Lands: The Back-to-the-Land Movement of the Early Twentieth Century," in *Experiences in a Promised Land: Essays in Pacific Northwest History*, eds. G. Thomas Edwards and Carlos A. Schwantes (Seattle: University of Washington Press, 1986), 287–303; and Robert Gough, *Farming the Cutover: A Social History of Northern Wisconsin, 1900–1940* (Lawrence: University of Kansas Press, 1997).

15. Though in this essay I present a relatively static view of the working environment, the particular methods of logging varied by place, and did change some over time. Firms curtailed the use of rivers to drive logs during the 1890s, instead deploying more logging railroads. After 1900 the use of steam haulers and caterpillars began replacing logging with horses.

16. Sawyers would not cut completely through the tree. A strip of "holding wood" was left to ensure a predictable direction for the tree's fall. Herman Haupt Chapman, *Report on the Methods of Lumbering in Minnesota, with special reference to the tract operated by the Duluth Logging and Contracting Company, at Island Lake, north of Duluth, Minnesota during the winter of 1903–04*, 1 [numbering of this document restarted approximately halfway through, beginning with his discussion of felling trees], Herman Haupt Chapman Papers (MS 134), Box 52, Sterling Memorial Library, Yale University, New Haven, CT.

17. Chapman, *Report on the Methods of Lumbering in Minnesota*, 1.

18. Don D. Lescohies, "Industrial Accidents and Employers' Liability," in *The Twelfth Biennial Report of the Bureau of Labor, Industries, and Commerce of the State of Minnesota, 1909–1910* (Saint Paul, 1910), 267.

19. Lescohies, "Industrial Accidents and Employers' Liability," 270.

20. Leonard Costley, interview by Bruce C. Harding, August 3, 1957, transcript, 8, Minnesota Historical Society, Saint Paul (hereafter MHS).

21. Louis Heinzer, interview by John Esse, February 26, 1976, transcript, 57–58, Interviews with Pioneer Lumbermen (P2385), MHS.

22. "Dangerous Results of Incorrect Logging," Occupational Safety and Health Administration, March 14, 2012, http://www.osha.gov/SLTC/etools/logging/manual/felling/cuts/dangers.html.

23. Chapman, *Report on the Methods of Lumbering in Minnesota*, 3.

24. Chapman, *Report on the Methods of Lumbering in Minnesota*, 3.

25. Costley interview, 8.

26. Frank Werthner, interview by John Esse with William Rajala, May 10, 1977, transcript, 22, MHS.

27. Werthner interview, 21–22. In the Bureau of Labor, Industry and Commerce's 1910 report on accidents, author Don Lescohies noted that horses themselves presented

a danger in the working environment, as the panic of horses in the woods or mills resulted in several injuries to workers. See his "Industrial Accidents and Employers' Liability,"
273.

28. Lescohies, "Industrial Accidents and Employers' Liability," 280.

29. Chapman, *Report on the Methods of Lumbering in Minnesota*, 27.

30. Jacob Pete, interview by John Esse, May 27, 1976, transcript, 19, MHS.

31. Pete interview, 19.

32. Costley interview, 2.

33. James Reid, interview, July 25, 1973, transcript, 9, Beltrami County History Center, Bemidji, MN.

34. "Northern Minnesota Is Soon Coming into Its Own," *Bemidji Daily Pioneer*, July
12, 1907.

35. The fourteen cutover counties, as defined by Jesness and Nowell, were: Aitkin,
Beltrami, Carlton, Cass, Clearwater, Cook, Crow Wing, Hubbard, Itasca, Koochiching,
Lake, Lake of the Woods, Pine, and Saint Louis. Oscar B. Jesness and Reynolds I. Nowell,
A Program for Land Use in Northern Minnesota: A Type Study in Land Utilization (Min-
neapolis: University of Minnesota Press, 1935), 10.

36. Jesness and Nowell, *Program for Land Use in Northern Minnesota*, 49–50.

37. Jesness and Nowell, *Program for Land Use in Northern Minnesota*, 47.

38. US Bureau of the Census, *Fourteenth Census of the United States Taken in the
Year 1920*, vol. 6, pt. 1 (Washington, DC: Government Printing Office, 1922), 487, 496.

39. Roy G. Blakey, *Taxation in Minnesota* (Minneapolis: University of Minnesota
Press, 1932), 130–31.

40. Kevin C. Brown, "'All That Country Will Be Taken Up by the Thrifty Settler':
Migration, Environment, and the Cutover Lands of Minnesota, USA from the 1890s to
the 1930s," *Global Environment: A Journal of History and Natural and Social Sciences* 9
(2012): 160–80. The Weeks Act of 1911 and the Clarke-McNary Act of 1924 both gave
significant and growing aid to states for forest fire protection. See Robbins, *Lumberjacks
and Legislators*; and Stephen J. Pyne, *Fire in America: A Cultural History of Wildland
and Rural Fire* (Princeton, NJ: Princeton University Press, 1982), esp. 346–57. The proper history of forestry in Minnesota began the year after the Hinkley fire in 1895, when
the state made the state auditor also the forest commissioner and created the position
of chief fire warden. For this and other legislation in the state related to forests, see Elizabeth Bachmann, *A History of Forestry in Minnesota* (Saint Paul: Minnesota State Department of Conservation, 1965). On taxes, see Blakey, *Taxation in Minnesota*, 147–69.

41. Werthner interview, 6.

42. John Ollila, interview by Robert Wheeler, May 5, 1976, transcript, 2, MHS.

43. "State Should Aid in Giving Cheap Dynamite," *Bemidji Daily Pioneer*, December 10, 1908. See also "Charles S. Carter Talks of Cheap Dynamite Bill," *Bemidji Daily
Pioneer*, February 23, 1909; and "'Dynamite Bill' Killed," *Bemidji Daily Pioneer*, April 17,
1909.

44. "Why Not?," editorial, *Bemidji Daily Pioneer*, May 3, 1910. In addition to their efforts to improve the cutover through state aid in the form of dynamite, cutover counties succeeded in securing funding for drainage ditches to improve swampy land. By the 1930s the main legacy of these projects was their contribution to the indebtedness of cutover counties.

45. Clifford E. Ahlgren, interview by Newell Searle, May 11, 1976, transcript, 13–14, MHS.

46. "Sells His Mechanical Mule," *Princeton Union*, November 11, 1909.

47. Alexander Carno, interview by John Esse, August 4, 1976, transcript, 10, MHS.

48. George Frederick Eitel, interview by Elwood R. Maunder, n.d. (ca. 1950s), transcript, 1, Forest History Society, Durham, NC.

49. Eitel interview, 1.

50. William Kaukola, interview by Robert Wheeler, February 17, 1977, transcript, 38, MHS.

51. Ollila interview, 6.

52. J. C. McDowell and W. B. Walker, "Farming on the Cut-Over Lands of Michigan, Wisconsin, and Minnesota," US Department of Agriculture, Bulletin no. 425 (Washington, DC: Government Printing Office, 1916), 13.

53. "Logged Off Land Sales, 1917–1919," Box 6, Immigration Land Company papers (P940), MHS.

54. W. A. Hartman and J. D. Black, "Economic Aspects of Land Settlement in the Cut-Over Region of the Great Lakes States," US Department of Agriculture, Circular no. 160 (Washington, DC: Government Printing Office, 1931), 45.

55. "No Help for Him," *Minneapolis Journal*, September 6, 1902. After the First World War the federal government appropriated money for disabled veterans to establish agricultural colonies. Many of these veterans' experiences working cutover land in northern Minnesota seem not to have differed much from that of the writer to the *Minneapolis Journal* a decade and a half earlier. See Bill G. Reid, "Colonies for Disabled Veterans in Minnesota," *Minnesota History* 39 (1965): 241–51.

6: "FOLLOW THE ARROWS TO THE ARROWHEAD"

1. Jack Ruttger, interview by Margaret Robertson, June 11, 1991, Interview 79.8, Minnesota Resort Industry Oral History Project (hereafter ROHP), Minnesota Historical Society, Saint Paul (hereafter MHS).

2. Collection Description and Biography, Roy E. Dunn Papers, MHS; Roger Pinckney, "No Dull Days at Dunn's," December 5, 2016, http://www.pelicanrapidschamber.com/historyhappenedhere/no_dull_days_at_dunn.htm; *Detroit Lakes, Minnesota: 412 Lakes within 25 Miles: Make This Your Summer Playground; The Outing Paradise of the Northwest* (Detroit Lakes, MN: Civic & Commerce Association, 1937), MHS.

3. Tourism's influence on the region remains less understood than other industries. On tourism, see Larry Lankton, *Hollowed Ground: Copper Mining and Community*

Building on Lake Superior, 1840s–1990s (Detroit: Wayne State Press, 2010); Benjamin Johnson, "Conservation, Subsistence, and Class at the Birth of the Superior National Forest," *Environmental History* 4 (1999): 80–99; Greg Summers, *Consuming Nature: Environmentalism in the Fox River Valley, 1850–1950* (Lawrence: University Press of Kansas, 2006).

4. On the relationship between Chicago and its hinterland and the idea that city and country share a common history, see William Cronon, *Nature's Metropolis: Chicago and the Great West* (New York: W. W. Norton, 1991).

5. On post–World War II deindustrialization, see Jefferson Cowie and Joseph Heathcott, eds., *Beyond the Ruins: The Meanings of Deindustrialization* (Ithaca, NY: ILR Press, 2003).

6. On interwar environmentalism, see Paul Sutter, *Driven Wild: How the Fight against Automobiles Launched the Modern Wilderness Movement* (Seattle: University of Washington Press, 2002); Neil Maher, *Nature's New Deal: The Civilian Conservation Corps and the Roots of the American Environmental Movement* (New York: Oxford, 2008). On connections among work, leisure, and environmental consciousness, see Richard White, "'Are You an Environmentalist or Do You Work for a Living?' Work and Nature," in *Uncommon Ground: Toward Reinventing Nature*, ed. William Cronon (New York: W. W. Norton, 1995), 171–85; Connie Chiang, *Shaping the Shoreline: Fisheries and Tourism on the Monterey Coast* (Seattle: University of Washington Press, 2008). On summer camps, see Abigail A. Van Slyck, *A Manufactured Wilderness: Summer Camps and the Shaping of American Youth, 1890–1960* (Minneapolis: University of Minnesota Press, 2010).

7. Hal Rothman, *Devil's Bargains: Tourism in the Twentieth-Century American West* (Lawrence: University Press of Kansas, 1998). On tourism history, see Warren Belasco, *Americans on the Road: From Autocamp to Motel, 1910–1945* (Cambridge, MA: MIT Press, 1979); Dona Brown, *Inventing New England: Regional Tourism in the Nineteenth Century* (Washington, DC: Smithsonian Institution Press, 1995); Cindy Aron, *Working at Play: A History of Vacations in the United States* (New York: Oxford University Press, 1999); Marguerite Shaffer, *See America First: Tourism and National Identity, 1880–1940* (Washington, DC: Smithsonian Institution Press, 2001); Blake Harrison, *The View from Vermont: Tourism and the Making of an American Rural Landscape* (Burlington, VT: University Press of New England, 2006); Susan Sessions Rugh, *Are We There Yet? The Golden Age of Family Vacations* (Lawrence: University Press of Kansas, 2008); Karen L. Cox, ed., *Destination Dixie: Tourism and Southern History* (Gainesville: University Press of Florida, 2012).

8. On New Deal conservation, see Sara Gregg, *Managing the Mountains: Land Use Planning, the New Deal, and the Creation of a Federal Landscape in Appalachia* (New Haven, CT: Yale University Press, 2010); Maher, *Nature's New Deal*; Sarah Phillips, *This Land, This Nation: Conservation, Rural America and the New Deal* (Cambridge: Cambridge University Press, 2007). On local engagement with conservation regulations, see Karl Jacoby, *Crimes against Nature: Squatters, Poachers, Thieves, and the Hidden History*

of American Conservation (Berkeley: University of California Press, 2001); Louis War-ren, *The Hunter's Game: Poachers and Conservationists in 20th Century America* (New Haven, CT: Yale University Press, 1997); Mark Spence, *Dispossessing the Wilderness: Indian Removal and the Making of the National Parks* (New York: Oxford University Press, 1999).

9. Hal Barron, *Mixed Harvest: The Second Great Transformation in the Rural North, 1870–1930* (Chapel Hill: University of North Carolina Press, 1997); Belasco, *Americans on the Road*; Sutter, *Driven Wild*; Christopher W. Wells, *Car Country: An Environmental History* (Seattle: University of Washington Press, 2013); David Louter, *Windshield Wilderness: Cars, Roads, and Nature in Washington's National Parks* (Seattle: University of Washington Press, 2009); Richard West Sellars, *Preserving Nature in the National Parks: A History* (New Haven, CT: Yale University Press, 1997). On 1920s vacations, see Daniel Horowitz, *The Morality of Spending: Attitudes toward the Consumer Society in America, 1875–1940* (Baltimore: Johns Hopkins University Press, 1985), 140.

10. The *Minneapolis Daily Star* cartoon appeared on the cover of the Ten Thousand Lakes-Greater Minnesota Association, *Annual Report* (1929), MHS.

11. George Wehrwein and Hugh Johnson, "A Recreation Livelihood Area," *Journal of Land and Public Utility Economics* 19 (May 1943): 193–206, reported 1938 tourist expenditures of $129 million in Minnesota. In 1928 Roy Dunn reported that tourists spent over $90 million in Minnesota; Dunn to Matthias Koll, March 29, 1929, "Koll, Ten Thousand Lakes of Minnesota Association, 1928–1933," box 10, Matthias Koll Papers, MHS (hereafter Koll Papers). Other indicators include out-of-state automobile traffic, ferry crossings, and fishing and hunting licenses issued.

12. "20,000 More Tourists Attracted to Minnesota's Vacation Playground by 10,000 Lakes Boosters," *St. Paul Pioneer Press*, December 28, 1919; Merrill Cragun, "Reminiscences," Merrill Cragun Papers, 1934–1986, MHS.

13. Ten Thousand Lakes of Minnesota Association, "A Review of the Campaign of Advertising Conducted by the Ten Thousand Lakes of Minnesota Association, 1919–1920"; Ten Thousand Lakes of Minnesota Association, *Annual Report*, 1919; *A Report to the Members of the 10,000 Lakes of Minnesota Association* (Saint Paul: Corning, 1919), all in MHS.

14. Ten Thousand Lakes of Minnesota Association, *Annual Report*, 1923, MHS.

15. Northern Minnesota Development Association Convention, Declaration of Principles, 20–21, June 1919, "NMDA, 1919–22-Minutes, Resolutions, Financial Statements," box 9, Koll Papers.

16. Responses to M[atthias] N. Koll's Minnesota Scenic Highway Questionnaire, March 27, 1917, "Minnesota Scenic Highway Association, March 1917," and "Minnesota Scenic Highway Association, April-July 1917," box 8, ; Responses to M[atthias] N. Koll's Minnesota Scenic Highway Questionnaire, February 10, 1920, "Minnesota Scenic Highway Association, Questionnaires (Towns) A-Z, 1920," box 8, all in Koll Papers.

17. Minnesota Board of Immigration, *Minnesota: The Land of Opportunity* (Saint Paul: State Bureau of Immigration, 1919). Also see Minnesota Board of Immigration, *Northeastern Minnesota: Land of Certainties* (Minneapolis: Great West Printing, 1919); Hiram Drache, *Taming the Wilderness: The Northern Border Country, 1910–1939* (Danville, IL: Interstate Publishers, 1992), 267–73.

18. US Forest Service, *A Vacation Land of Lakes and Woods: The Superior National Forest* (Washington, DC: Government Printing Office, 1919), 10; Johnson, "Conservation, Subsistence, and Class at the Birth of the Superior National Forest." On the Term Permit Act, see Sutter, *Driven Wild*, 60–62.

19. Minnesota Arrowhead Association, *1925 Annual Report*, "Minnesota Arrowhead Association Annual Reports," n.p., box 7, Koll Papers; Timothy Brady, "The Road to the Lake," *Minnesota Monthly* (May 1987), box 3, Ruttger Family uncataloged material, MHS. Citations reflect the material as I encountered it before cataloging. The collection is now cataloged as Ruttger Resorts Records.

20. "50 Years of Service to the Vacation Travel Industry, Minnesota Arrowhead Association 1924–74," "Minnesota Arrowhead Association, Miscellaneous 1938–76," box 2, Conservation Printed Matter from Minnesota Organizations, 1921–1981, MHS; David Backes, *Canoe Country: An Embattled Wilderness* (Minocqua, WI: Northword Press, 1991), 55; Minnesota Arrowhead Association, *President's Report* (1936–1938), MHS; Writers' Program of the Work Projects Administration in the State of Minnesota, *The WPA Guide to the Minnesota Arrowhead Country* (Chicago: Albert Whitman, 1941).

21. Minnesota Commission of Conservation, *Biennial Report of the Commission of Conservation*, 1932, 55–57 (hereafter MCC, *Biennial Report*); MCC, *Biennial Report* (1934), 194–200; MCC, *Biennial Report*, 1936, 269–80; Minnesota Department of Highways, *1934 Condition Map of Minnesota Trunk Highways*; Minnesota State Planning Board, *Report of the Minnesota State Planning Board*, 1938, all in MHS.

22. Christine Bold, *The WPA Guides: Mapping America* (Jackson: University Press of Mississippi, 1999); Maher, *Nature's New Deal*; Barbara Sommer, *Hard Work and a Good Deal: The Civilian Conservation Corps in Minnesota* (Saint Paul: Minnesota Historical Society Press, 2008).

23. Writers' Program of the Works Projects Administration in the State of Minnesota, *The WPA Guide to the Minnesota Arrowhead Country*; Work Progress Administration in the State of Minnesota, *Minnesota: A State Guide* (New York: Viking, 1938).

24. Minnesota Department of Transportation, *Official Road Map of Minnesota*, 1940, MHS.

25. Sigurd Olson, "Reflections of a Guide," *Field and Stream*, June 1928, repr. in David Backes and Sigurd Olson, eds., *The Meaning of Wilderness Essential Articles and Speeches* (Minneapolis: University of Minnesota Press, 2001), 13; Sigurd Olson, "Why Wilderness?" *American Forests*, 1938, repr. in Backes and Olson, *Meaning of Wilderness*, 47.

26. Sigurd Olson, "The Superior National Forest Must Be Saved," 1929, Sigurd F. Olson Speeches, accessed June 4, 2022, https://www.northland.edu/centers/soei/sigurd -legacy/sigurd-speeches/#1929-the-superior-national-forest-must-be-saved.

27. Frank Waugh, *Recreation Uses on the National Forests* (Washington, DC: US Department of Agriculture, Government Printing Office, 1918), 26; William Tweed, *Recreation Site Planning and Improvement in the National Forests, 1891–1942* (Washington, DC: US Department of Agriculture, Forest Service, 1981). On Forest Service recreation policy, see Henry Graves, *A Policy of Forestry for the Nation* (Washington, DC: US Department of Agriculture, Government Printing Office, 1918); Henry Graves, "A Crisis in National Recreation," *American Forestry* 26 (1920): 391–400. On interagency battles, see Hal Rothman, "'A Regular Ding-Dong Fight': Agency Culture and Evolution in the NPS-USFS Dispute, 1916–1937," *Western Historical Quarterly* 20 (May 1989): 141–61.

28. Arthur Carhart, "An Outline Plan for the Recreational Development of the Superior National Forest," 1921, "Carhart-Miscellaneous" box 15, US Forest Service Collection (hereafter USFS), Iron Range Research Center, Chisholm, MN (hereafter IRRC); Aldo Leopold, "The Wilderness and Its Place in Forest Recreational Policy," *Journal of Forestry* 19 (1921): 718–21.

29. Ernest Oberholtzer to USDA Secretary William Jardine, November 7, 1927, and Jardine to Oberholtzer, November 26, 1927, both in "Correspondence and Miscellaneous Papers, 1927–1939 (1)," box 2, President's Quetico-Superior Committee Collection (hereafter QSC), MHS. On Oberholtzer, see Joe Paddock, *Keeper of the Wild: The Life of Ernest Oberholtzer* (Saint Paul: Minnesota Historical Society Press, 2001).

30. Oberholtzer to Joseph Klobucher [Secretary, Arrowhead Sportsmen's Association], February 23, 1933, "Arrowhead Sportsmen's Association, 1933," box 4; Joe Pluth to Oberholtzer, December 31, 1928, folder 2, box 61, both in Quetico-Superior Council Papers, MHS. Pluth reported he had four thousand people take canoe trips and that ten thousand canoeists passed through Ely the previous year.

31. Oberholtzer to Charles Kelly, July 6, 1932, and Oberholtzer to Jardine, November 7, 1927, both in "Correspondence and Miscellaneous Papers, 1927–1939 (1)," box 2, QSC. Also see Edward M. "Ted" Hall Oral History Interviews with Margaret Robertson, February 5, 6, and 8, 1987, Interview 58.8, Oral History Interviews of the Minnesota Environmental Issues, Oral History Project, 1986–1990, MHS; Oberholtzer to Seth Gordon (Izaak Walton League Conservation Director), "Wilderness Areas-General Correspondence," box 15, Chippewa National Forest Collection (hereafter CNF), IRRC.

32. Will Dilg, quoted in Michael Furtman, "Izaak Walton League of America: 75 Years of Canoe Country Advocacy," accessed June 4, 2022, https://sites.google.com/view/minnesotaikes/who-we-are/about-us/canoe-country-advocacy.

33. Arrowhead Division of the Izaak Walton League of America, *The Issue of Conservation*, brochure, "Quetico-Superior Politics, 1927–34," n.p., box 4, QSC, MHS. The W. J. McCabe chapter of the Izaak Walton League of America website, accessed Decem-

ber 5, 2016, http://www.duluthikes.org/, contains information on the League's Superior National Forest efforts.

34. Biographical Information on Arthur Carhart, "Carhart-Miscellaneous," box 15, USFS, IRRC. On Carhart, see Donald Baldwin, *The Quiet Revolution: Grass Roots of Today's Wilderness Preservation Movement* (Boulder, CO: Pruett Publishing, 1972).

35. Arthur Carhart quoted in George A. Selke, Report of December 15, 1964, 3–9, "Subject Files, BWCA (2)," box 3, QSC, MHS; Arthur Carhart, "An Outline Plan for the Recreational Development of the Superior National Forest, Preliminary Prospectus," "Carhart-Miscellaneous," box 15, USFS, IRRC.

36. Arthur Carhart, "Recreation in the Forests," *American Forestry* 26 (May 1920): 268–72.

37. [Arthur] Carhart to [W. F.] Ramsdell, February 25, 1928, "Carhart-Miscellaneous," box 15, USFS, IRRC. Also see Carhart to James Gould, January 28, 1928, "Carhart-Miscellaneous," box 15, USFS, IRRC; David Backes, "Wilderness Visions: Arthur Carhart's 1922 Proposal for the Quetico Superior Wilderness," *Forest & Conservation History* 35 (1991): 128–37.

38. Arthur Carhart, "Ely's Opportunity," *Ely Miner*, July 15, 1921, in "Carhart-Miscellaneous," box 15, USFS, IRRC.

39. Carhart to Gould, January 28, 1928; Burntside Lake and Lodge, Clippings File, IRRC. Many northern Saint Louis County resorts were built in the 1930s. Before World War II, the Ely area contained approximately thirty resorts.

40. Hunting and fishing licenses provided funds to purchase public access sites, develop stocking programs, and employ game wardens. On licensing, see state conservation department biennial reports: Commission of Conservation, State of Minnesota, *Biennial Report*, 1932; Minnesota Department of Conservation, *Biennial Report*, 1934–1960. On stocking requests and fishing, see Backes, *Canoe Country*, 91–92; and Margaret Beattie Bogue, *Fishing the Great Lakes: An Environmental History, 1783–1933* (Madison: University of Wisconsin Press, 2000).

41. Helen Beebe to Roy Dunn, June 29, 1938, and July 12, 1938, Resort Files, "1938 Dunn's Lodge, A-C," box 8; Mr. Kline to Dunn, August 6, 1941, Resort Files, "1940–41 Dunn's Lodge, Jon-Le," box 9; Philip Haser to Dunn, September 26, 1941, Resort Files, "1940–41 Dunn's Lodge, Hos-Joh," box 9; letters from Sydney Anderson and Hugo J. Bredehoeft to Dunn, n.d., Resort Files, "1937 Dunn's Lodge, A-D" box 8; Leslie Setzer to Dunn, Resort Files, "1937 Dunn's Lodge, N-S," n.d., box 8, all in Roy Dunn Papers, MHS (hereafter Dunn Papers).

42. Duluth & Iron Range Rail Road Co., *The Playground of a Nation in the Arrowhead Country: Via the Vermilion Route* (Duluth: Duluth & Iron Range Railroad Company, 1926); Saint Louis County Club, *Sportland of the North: Cool Minnesota Arrowhead Country* (1936), n.p. (quote), MHS.

43. Roy Dunn to Curt Teich Company, March 3, 1937; Curt Teich Company to

Dunn, March 31, 1937; and Dunn to George Bradley, July 2, 1937, Resort Files, "1937 Dunn's Lodge, A-D," box 8, all in Resort Files, "1937 Dunn's Lodge, A-D," box 8; Dunn to George Etzell, June 4, 1941, Resort Files, "1940–41 Dunn's Lodge, E-Har," box 9, all in Dunn Papers.

44. "Visit Va-Ka-Shun Land in 1929," folder 1, Pamphlets Relating to Description and Travel in Minnesota, MHS.

45. "History of Gunflint and the Kerfoots," December 5, 2016, http://www.gunflint .com/about_us/about_us.htm; *Justine Kerfoot* (Minnetonka, MN: Hennepin County Library, 1988), Videotape 274, MHS; Justine Kerfoot, interviews by Margaret Robertson, November 1986 and March 1987, Interview 58.15, Oral History Interviews, Minnesota Environmental Issues Oral History Project, 1986–1990, MHS; Mark Lowry, *Lady of the Gunflint Trail* (Saint Paul: Twin Cities Public Television, 1980), videotape 131, MHS.

46. *Gunflint Gossips*, 8 October 1938, February 5, 1939, and February 17, 1940, MHS; *Justine Kerfoot*; Justine Kerfoot interviews; Gunflint Trail Association Cash Book, 1937–1969, "Gunflint Trail Association, 1937–1986," box 2, Kratoska Papers, MHS.

47. Bert Pfeifer, interview by James Fogerty, November 17, 1999, Interview 79.6, ROHP.

48. Jack Stedman, interview by James Fogerty, October 17, 1995, Interview 79.14, ROHP.

49. Stedman interview.

50. Ed and Kay Gilman, interview by James Fogerty, June 24, 1993, Interview 79.3, ROHP.

51. Carol Crawford Ryan, interview by James Fogerty, June 14, 1993, Interview 79.12, ROHP.

52. Jack Ruttger interview.

53. Jack Ruttger interview.

54. Max "Buzz" Ruttger Jr., interview by Theresa Ruttger, August 19, 1997, Interview 79.9, ROHP. Buzz took over in 1947 and sold the resort in 1969 to another pioneering resort family, the Maddens.

55. Max "Buzz" Ruttger Jr. interview.

56. Max "Buzz" Ruttger Jr. interview; Rollis Bishop, interview by James Fogerty, November 17, 1999, Interview 79.1, ROHP; Summary and Description, Warmington Papers, MHS; Photo of musicians at Grand View Lodge, Resorts and Recreational Facilities in Saint Louis County, ca. 1925, Album 162, MHS.

57. On Star Island, see Carol Crawford Ryan, *Star Island: A Minnesota Summer Community* (Saint Paul: Pogo Press, 2000). Commercial Club of Cass Lake and Committee representing the Government Tenants on Star Island to the National Bureau of Forestry, August 19, 1918, box 11, Koll Papers; Star Island Protective League, *The Loon*, MHS.

58. Lydick Mercantile Co. to A. J. Starr, Esq., "Koll, Correspondence, March 1914," box 3, Koll Papers; Commercial Club of Cass Lake and Committee representing the

Government Tenants on Star Island to the National Bureau of Forestry, August 19, 1918; Recreational Use Policy (1926), "Recreation, Minnesota National Forest," box 7, CNF, IRRC.

59. "Cass Lake, the Permanent Home of the Pine," in "January-May, 1932," box 5, Koll Papers; "Forest Recreational Plan-A Foundation," in "Recreation Plans-Forest," box 7, CNF, IRRC. "Cass Lake, Minnesota in the Chippewa National Forest (1941)," box 6, Travel Literature Collection, Research Center, Henry Ford Museum, Dearborn, MI; US Department of Agriculture, Forest Service, *Chippewa National Forest* (Washington, DC: Government Printing Office, 1942), 12, box 3, CNF, IRRC.

60. List of employment inquiries, Resort Files, 1937, "Letters of Application," box 8; list of potential employees and Roy E. Dunn's remarks, Resort Files, 1941, "Letters of Application," box 9; Dunn to Margo Cairns, May 2, 1941, Resort Files, 1941, "Letters of Application," box 9; letter from Dunn [form letter], July 14, 1938, Resort Files, 1938, "Dunn's Lodge, H-L," box 8; Dunn to Alice Gerold, March 19, 1945, Resort Files, "1942–49 Dunn's Lodge," box 9, all in Dunn Papers.

61. Ruby Treloar, interview by Theresa Ruttger, August 19, 1997, Interview 79.15, ROHP.

62. Commissioner J. W. Clark's quote opening the 1949 guide appears on the project description sheet in Ted Leagjeld, interview by Margaret Robinson, June 11, 1991, Interview 79.4, ROHP; Robert Johnson, *Tour Guide of Minnesota* (Saint Paul: Department of Business Research and Development, 1949).

63. On capitalism and nature, see Jacoby, *Crimes against Nature*; William G. Robbins, *Landscapes of Promise: The Oregon Story, 1800–1940* (Seattle: University of Washington Press, 1997); Donald Worster, *Dust Bowl: The Southern Plains in the 1930s* (New York: Oxford, 1979); Cronon, *Nature's Metropolis*.

64. Cartoon reprinted on back cover of Ten Thousand Lakes-Greater Minnesota Association, *Annual Report* (1929).

65. US Forest Service, *Vacation Land of Lakes and Woods*.

66. "Facts about Your Minnesota Arrowhead Association, What Is the Minnesota Arrowhead Association," and "50 Years of Service to the Vacation Travel Industry, Minnesota Arrowhead Association, 1924–74," all in "Minnesota Arrowhead Association, Miscellaneous, 1938–1976," box 2, Conservation Printed Matter from Minnesota Organizations, 1921–1981, MHS.

67. Ely Commercial Club, *Ely, Minnesota: The Playground of a Nation* (1932), MHS; Arthur Carhart, "Ely's Opportunity," *Ely Miner*, July 15, 1921, "Carhart-Miscellaneous," box 15, USFS, IRRC. On the Iron Range, see Paul H. Landis, *Three Iron Mining Towns: A Study in Cultural Change* (Ann Arbor, MI: Edwards Brothers, 1938).

68. Minnesota Arrowhead Association, *Main Highways and Some By-ways Duluth* (1928), MHS.

69. Minnesota Arrowhead Association, *Minnesota Arrowhead Country: Hotel and Resort Directory* (1938), MHS.

70. *Tips on Minnesota Motor Trips*, in "Pamphlets Relating to Tour Guides," MHS.

71. *The Land of Ten Thousand Lakes over Jefferson Highway Minnesota* (1920 and 1922); "Making Money for the State: Ten Thousand Lakes of Minnesota Association Has Justified Its Creation and Existence," *Western Magazine* 17 (January 1921): 11–13; Ivan Coppe, "Minnesota Opens Wide Her Gates: Ten Thousand Lakes Association Organized to Make Tourists Welcome," *Western Magazine* 11 (May 1918): 170–73; Ten Thousand Lakes of Minnesota Association, *Minnesota Canoe Trails* (1918), MHS.

72. Ed Shave, "Touring the Long Bow Country," in *The Long Bow Country of Minnesota* (Hackensack: Northern Minnesota Publishing, 1938).

73. Minnesota Arrowhead Association, *Tourist Guide and Map of Minnesota Arrowhead Country*; Ten Thousand Lakes-Greater Minnesota Association, *Minnesota: Land of Ten Thousand Lakes* (1928 and 1931); Minnesota Arrowhead Association, *The Minnesota Arrowhead Country: Follow the Arrows to the Arrowhead* (Duluth: Minnesota Arrowhead Association, 1926); *Vacationland in Northern Minnesota: In the Land O'Pines* (Pequot, MN: Benn A. Wagner Service, 1933); W. A. Fisher Company, *Ely: Map of Resort and Canoe Area; Vacationland in Northern Minnesota: In the Land O'Pines* (Virginia, MN: W. A. Fisher, 1944); Adolph A. Toftey, *Cook County, Minnesota: Nature's Gift to the Arrowhead Country* (1920 and 1930); Auto Road Map Company, *Sportsman's and Automobile Map of the Upper Peninsula of Michigan*; Rand McNally and Company, *Rand McNally Official 1924 Auto Trails Map* (Chicago: Rand McNally, 1924), all in MHS.

74. Grand Marais Chamber of Commerce, *Come See: The Gunflint Trail* (1938); "Call of the Open," *St. Paul Pioneer Press*, Spring 1934.

7: FOUNTAINS OF LIFE AND DEATH

1. "About on a Par," *Minneapolis Tribune*, January 7, 1902.

2. John O. Anfinson, with Hemispheres Inc. and Pat Nunnally, *River of History: Historic Resources Study of the Mississippi National River and Recreation Area* (Saint Paul: National Park Service and Corps of Engineers, 2003), 76, 79, 89; Frank Haigh Dixon, *A Traffic History of the Mississippi River System* (Washington, DC: Government Printing Office, 1909), 20; Mildred Hartsough, *From Canoe to Steel Barge* (Minneapolis: University of Minnesota Press, 1934), 100.

3. Martin Melosi, *The Sanitary City: Environmental Services in Urban America from Colonial Times to the Present*, abridged ed. (Pittsburgh: University of Pittsburgh Press, 2008), 19; Stuart Galishoff, "Triumph and Failure: The American Water Supply Problem, 1860–1923," in *Pollution and Reform in American Cities, 1870–1930*, ed. Melosi (Austin: University of Texas, 1980), 33; Joel Tarr, James McCurley, and Terry F. Yosie, "The Development and Impact of Urban Wastewater Technology: Changing Concepts of Water Quality Control, 1850–1930," in Melosi, *Pollution and Reform*, 60–62.

4. *Annual Reports of the Various City Officers of the City of Minneapolis, Minnesota, for the Year Ending April 1, 1879* (Minneapolis: Johnson, Smith & Harrison, 1879), 102; Melosi, *Sanitary City*, 31; Tarr et al., "Development and Impact," 61.

5. *Annual Reports*, Minneapolis, 1879, 109; Kate Foss-Mollan, *Hard Water: Politics and Water Supply in Milwaukee, 1870–1995* (West Lafayette, IN: Purdue University Press, 2001), 10.

6. *Eighth Annual Report of the State Board of Health of Minnesota, for the Years 1879–1880* (Saint Peter: J. K. Moore, 1881), 173.

7. *Eighth Annual Report of the State Board of Health of Minnesota*, 174.

8. World Health Organization, "Typhoid Fever," November 13, 2021, http://www.who.int/topics/typhoid_fever/en. On Typhoid Mary, see Judith Walzer Leavitt, *Typhoid Mary: Captive to the Public's Health* (Boston: Beacon Press, 1996); Minnesota Department of Health, *Report of Investigation of the Typhoid Fever Epidemic, 1935* (Minneapolis: Minneapolis Department of Health, 1938), 95.

9. *Annual Reports of the Various City Officers of the City of Minneapolis, Minnesota, for the Year 1893* (Minneapolis: Harrison & Smith, 1894), 360.

10. *Nineteenth Report of the State Board of Health and Vital Statistics of Minnesota, 1901–1902* (Saint Paul: Pioneer Press, 1902), 10–11.

11. *Annual Reports of the Various City Officers of the City of Minneapolis, Minnesota, for the Year 1903* (Minneapolis: Heywood Manufacturing, 1904), 558.

12. *Third Biennial Report (New Series) of the State Board of Health and Vital Statistics of Minnesota, Sanitary Engineering Division, 1909–1910* (Minneapolis: Syndicate Printing, 1911), 38.

13. "The Water Question Solved," *Minneapolis Tribune*, June 18, 1867.

14. "The City Water Works," *Minneapolis Tribune*, June 18, 1872.

15. "Sewerage and Water Works," *Minneapolis Tribune*, August 25, 1876.

16. "Minneapolis Water," *Minneapolis Tribune*, December 21, 1876.

17. "Peckham on River Water," *Minneapolis Tribune*, March 19, 1877.

18. Saint Anthony Falls Water Power Co. v. St. Paul, 168 U.S. 349 (1897); "East Side Water Works," *Minneapolis Tribune*, November 29, 1878.

19. "The Water Works: Talks with Superintendent Johnson, Chief Engineer Brackett and Others on the Subject; The Three Phases of the Discussion Extension, Enlargement, and Removal," *Minneapolis Tribune*, December 9, 1879.

20. *Ninth Report of the State Board of Health of Minnesota, for the Years 1881 and 1882* (Minneapolis: Johnson, Smith & Harrison, 1883), 172–73.

21. *Ninth Report of the State Board of Health of Minnesota*, 173, 174.

22. "The Water Works: The Investigating Committee Still at Work," *Daily Minnesota Tribune*, October 16, 1883; *The Water Works of the City of Minneapolis, Minnesota: A Brief Historical Sketch and a Description of the Present Water Works* (1919), 3, call #: TD225.M6 M76 1919, Minnesota Historical Society, Saint Paul (hereafter MHS); James A. Dodge, C. L. Herrick, and C. W. Hall, "Paper C Water Supply of Minneapolis Committee," March 6, 1883, *Bulletin of the Minnesota Academy of Natural Sciences* 3, no.1 (1889): 39, Proceedings and Accompanying Papers, 1883–1886, ed. C. W. Hall.

23. Dodge et al., "Paper C," 41–42, 41 (quote), emphasis in original.

24. *"The Water Works of the City of Minneapolis,"* 5.

25. *Annual Reports of the Various City Officers of the City of Minneapolis, Minnesota, for the Year 1889* (Minneapolis: Harrison & Smith, 1890), 360.

26. *Annual Reports*, Minneapolis, 1889, 114–15; *Annual Reports of the Various City Officers of the City of Minneapolis, Minnesota, for the Year 1900* (Minneapolis: Harrison & Smith, 1901), 254; *Annual Reports of the Various City Officers of the City of Minneapolis, Minnesota, for the Year 1902* (Minneapolis: Harrison & Smith, 1903), 238.

27. *Annual Reports*, Minneapolis, 1902, 115; "It Is Expected that Minneapolis Will Have Pure Water When the North Side Pump Station Is Completed," *Minneapolis Tribune*, November 22, 1888.

28. Merrill E. Jarchow, "Charles Gilfillan: Builder behind the Scenes," Minnesota History (Spring 1967): 224; Saint Paul Regional Water Services, "Beyond the Faucet: The Story of the Saint Paul Regional Water Services," (n.d.), 4, accessed June 4, 2022, https://www.stpaul.gov/sites/default/files/Media%20Root/Water%20Services/BeyondTheFaucet.pdf.

29. Jarchow, "Charles Gilfillan," 224–25.

30. Jarchow, "Charles Gilfillan," 226.

31. Jarchow, "Charles Gilfillan," 229–30; Philip D. Jordan, *The People's Health: A History of Public Health in Minnesota to 1948* (Saint Paul: Minnesota Historical Society Press, 1953), 103.

32. *Fourth Annual Report of the Board of Water Commissioners of the City of St. Paul, December 1, 1885* (Saint Paul: Globe Job Office, D. Ramaley & Son, 1886), 32.

33. *Attorney and Engineer's Report*, in *Fifth Annual Report of the Board of Water Commissioners of the City of St. Paul, December 1, 1886* (Saint Paul: Globe Job Office, D. Ramaley & Son, 1886), 77.

34. *Annual Reports of the City Officers and City Boards of the City of Saint Paul, from for November 1, 1886, to December 31, 1887* (Saint Paul: Globe Job Office, D. Ramaley & Son, 1889), 4; *Annual Reports of the City Officers and City Boards of the City of St. Paul for the Fiscal Year Ending December 31, 1888* (Saint Paul: Globe Job Office, D. Ramaley & Son, 1889), 341.

35. *Annual Reports of the City Officers and City Boards of the City of St. Paul for the Fiscal Year Ending December 31, 1889* (Saint Paul: Pioneer Press, 1890), 369.

36. *Annual Reports*, Saint Paul, 1889, 276.

37. *Annual Reports*, Saint Paul, 1886–1887, 443, 453.

38. *Annual Reports*, Saint Paul, 1888, 405.

39. *Annual Reports*, Saint Paul, 1888, 404.

40. *Annual Reports*, Saint Paul, 1886–1887, 453, 457–58; Secretary of the State Board of Health and Vital Statistics, arranged and ed., *Third Biennial Report: The Vital Statistics of the State of Minnesota, for the Years 1890–91* (Minneapolis: Harrison & Smith, 1893), 132; *Annual Reports*, Saint Paul, 1888, 404–5; *Annual Reports*, Saint Paul, 1889, 492.

41. *Annual Reports of the City Officers and City Boards of the City of St. Paul for the Fiscal Year Ending December 31, 1891* (Saint Paul: Herald Print, 1892), 342, 347. *Merriam-Webster* defines *cholera infantum* as "an acute noncontagious intestinal disturbance in infants." Many of the cases identified as cholera infantum may have been typhoid fever; thus the number of cases and deaths due to typhoid may have been higher than stated.

42. *Annual Reports*, Saint Paul, 1891, 327; *Annual Reports of the City Officers and City Boards of the City of St. Paul for the Fiscal Year Ending December 31, 1893* (Saint Paul: Saint Paul Herald Print, 1894), 409, 520–21, 711.

43. Caroline Bartlett Crane, *Report on a Campaign to Awaken Public Interest in Sanitary and Sociological Problems in the State of Minnesota* (Saint Paul: Volkszeitung, 1911), 30.

44. Crane, *Report on a Campaign*, 29–30.

45. *Annual Reports*, Minneapolis, 1891, 371, 385 (quote).

46. *Annual Reports of the Various City Officers of the City of Minneapolis, Minnesota, for the Year 1892* (Minneapolis: Harrison & Smith, 1893), 399, 402.

47. *Annual Reports*, Minneapolis, 1893, 360, 365–72.

48. *Annual Reports*, Minneapolis, 1893, 361.

49. *Annual Reports of the Various City Officers of the City of Minneapolis, Minnesota, for the Year 1894* (Minneapolis: Harrison & Smith, 1895), 419; *Annual Reports of the Various City Officers of the City of Minneapolis, Minnesota, for the Year 1896* (Minneapolis: Harrison & Smith, 1897), 424.

50. *Annual Reports*, Minneapolis, 1896, 140.

51. Galishoff, "Triumph and Failure," 44–45; Melosi, *Sanitary City*, 58–59, 93.

52. *Annual Reports of the Various City Officers of the City of Minneapolis, Minnesota, for the Year 1897* (Minneapolis: Harrison & Smith, 1898), 514–15.

53. *Annual Reports*, Minneapolis, 1897, 77–78; *Annual Reports of the Various City Officers of the City of Minneapolis, Minnesota, for the Year 1898* (Minneapolis: Harrison & Smith, 1899), 188; *Water Works of the City of Minneapolis*, 5.

54. *Annual Reports of the Various City Officers of the City of Minneapolis, Minnesota, for the Year 1899* (Minneapolis: Tribune Printing, 1900), 316; *Annual Reports*, Minneapolis, 1900, 154–55, 255; *Annual Reports*, Minneapolis, 1902, 240.

55. *Annual Reports*, Minneapolis, 1903, 550.

56. "Talked It Over: Council Committee Discusses Proposed Enlargement of Waterworks," *Minneapolis Tribune*, October 13, 1900; February 6, 1901; "City Water Betterments," December 11, 1901; "Water Not Wholly Bad," February 21, 1904; "It's Strong Drink," *Minneapolis Journal*, September 28, 1901; *Annual Reports*, Minneapolis, 1902, 237.

57. *Annual Reports of the Various City Officers of the City of Minneapolis, Minnesota, for the Year 1904* (Minneapolis: Heywood Manufacturing, 1905), 238, 506–7, 506 (quote).

58. Frank H. Castner and James H. Duryea, "Pure Water for Minneapolis," (1905), 1–7, 1 (quote), Call #: TD225.M6 C27 1905, MHS Collections.

59. *Annual Reports of the Various City Officers of the City of Minneapolis, Minnesota, for the Year 1905* (Minneapolis: Heywood Manufacturing, 1906), 540.

60. *Annual Reports of the Various City Officers of the City of Minneapolis, Minnesota, for the Year 1906* (Minneapolis: Heywood Manufacturing, 1907), 383.

61. *Annual Reports of the Various City Officers of the City of Minneapolis, Minnesota, for the Year 1909* (Minneapolis: Syndicate Printing, 1910), 8k.

62. Galishoff, "Triumph and Failure," 39–44; "Typhoid Fever Prevalent: Unusual Amount of Disease throughout the Country," *New York Times*, October 15, 1901.

63. *Report of Pure Water Commission*, Minneapolis, July 16, 1909, Call # TD225.M6 P87 1909, MHS Collections; *Annual Reports*, Minneapolis, 1909, 26g.

64. Rudolph D. Herring, *Report on an Improved Water Supply for the City of Minneapolis*, March 17, 1910, 22 (quote), Call #: TD225.M6 H47 1910, MHS Collections.

65. *Annual Reports of the Various City Officers of the City of Minneapolis, Minnesota, for the Year 1910* (Minneapolis: Syndicate Printing, 1911), 10k.

66. *Annual Reports*, Minneapolis, 1910, 10k; *Water Works of the City of Minneapolis*, 15.

67. Saint Paul Regional Water Services, "Beyond the Faucet," 5.

68. Robert D. Morris, *The Blue Death: The Intriguing Past and Present Danger of the Water You Drink* (New York: Harper, 2008), 278.

69. Saint Paul Regional Water Services, *Saint Paul Regional Water Services Water Supply Plan*, 2016, https://www.stpaul.gov/sites/default/files/2021-06/WSP_Saint%20Paul_1975-6227_05-02-2019%20Public%20Version.pdf.

8: URBAN ENVIRONMENTAL HISTORY AND LORING PARK

1. Theodore Wirth, *Minneapolis Park System 1883–1944* (Minneapolis: Board of Park Commissioners, 1945).

2. Galen Cranz, "Four Models of Municipal Park Design in the United States," in *Denatured Visions. Landscape and Culture in the Twentieth Century*, ed. Stuart Wrede and William H. Adams (New York: Museum of Modern Art, 1991), 118–23. For Cranz's more recent fifth model, see Galen Cranz and Michael Boland, "Defining the Sustainable Park: A Fifth Model for Urban Parks," *Landscape Journal* 23 (2004): 102–20. For further examination of how economics and technologies influence urban public space, see Paige S. Warren et al., "Urban Ecology and Social Organization," *Urban Ecology*, ed. Kevin J. Gaston (Cambridge: Cambridge University Press, 2010), 172–201.

3. For further insight into Olmsted's thoughts about parks and cities, see Frederick Law Olmsted, *Civilizing American Cities: Writings on City Landscapes*, ed. S. B. Sutton (New York: De Capo Press, 1997). For Horace Cleveland's Minneapolis parkway plans, see Horace W. S. Cleveland, "Suggestions for a System of Parks and Parkways for the City of Minneapolis," Minneapolis Park Commission Meeting, June 2, 1883, http://books

.google.com/books/about/Suggestions_for_a_system_of_parks_and_pa.html?id=mLQ1 AAAAMAAJ.

4. Horace W. S. Cleveland, "Influence of Parks on the Character of Children," in *Second Report of the American Park and Outdoor Art Association* (Boston: Rockwell and Churchill, 1898).

5. Lance M. Neckar, "Fast-Tracking Culture and the Landscape: Horace William Shaler Cleveland and the Garden in the Midwest," in *Regional Garden Design in the United States*, ed. Therese O'Malley and Marc Treib (Washington, DC: Dumbarton Oaks Research Library and Collection, 1995), 90.

6. Theodore Wirth, qtd. in Minneapolis Park Board, *Superintendent's Report: Annual Report on Minneapolis Parks*, Minneapolis (1906), 43 (hereafter *Annual Report* and the date).

7. Minneapolis Park Board, *Annual Report*, 1910, 58.

8. Minneapolis Park Board, *Annual Report*, 1902, 48.

9. Theodore Wirth, "Floral Decorations in Parks and City or Village Squares," *Bulletin of the American Institute of Park Superintendents* 1 (1906): 2.

10. Peter J. Schmitt, *Back to Nature: The Arcadian Myth in Urban America* (New York: Oxford University Press, 1990).

11. Horace W. S. Cleveland to Frederick Law Olmsted, Minneapolis, June 6, 1893, in Olmsted, *The Papers of Frederick Law Olmsted: The Last Great Projects, 1890-1895*, ed. David Schuyler, Gregory J. Kaliss, and Jeffrey Schlossberg, vol. 9 of *The Papers of Frederick Law Olmsted* (Baltimore: Johns Hopkins University Press, 2015), 649–50.

12. Minneapolis Park Board, *Annual Report*, 1902, 19.

13. For accounts of the rational city planning movement and urban parks, see Galen Cranz, *The Politics of Park Design: A History of Urban Parks in America* (Cambridge, MA: MIT Press, 1982); and Roderick Frazier Nash, *Wilderness and the American Mind*, 4th ed. (New Haven, CT: Yale University Press, 2001).

14. Cranz, *Politics of Park Design*, 80–81.

15. Minneapolis Park Board, *Annual Report*, 1910, 20.

16. Galen Cranz, "Urban Parks as a Mechanism of Social Control," 6, paper presented at the American Sociological Association Meeting, San Francisco, 1975.

17. G. Stanley Hall, "Boy Life in a Massachusetts Country Town Thirty Years Ago," *Proceedings of the American Antiquarian Society* 7 (1892): 107–28.

18. Wirth, *Minneapolis Park System*, 220.

19. Minneapolis Spirit, "The Story of a City of Lakes and Gardens—A Half Century of Progress in Making Good Citizenship," *American City* 6 (1912): 398.

20. Minneapolis Park Board, *Annual Report*, 1911, 83.

21. Robin F. Bachin, "Cultivating Unity: The Changing Role of Parks in Urban America," *Places* 15, no. 3 (2003): 14.

22. Wirth, *Minneapolis Park System*, 67.

23. Wirth, *Minneapolis Park System*, 68.

24. Calvin F. Schmid, *Social Saga of Two Cities: An Ecological and Statistical Study of Social Trends in Minneapolis and St. Paul* (Minneapolis: Minneapolis Council of Social Agencies, 1937), 69.

25. Cranz, *Politics of Park Design*, 101.

26. Schmid, *Social Saga of Two Cities*, 69.

27. Cranz, *Politics of Park Design*, 110.

28. David C. Smith, *City of Parks: The Story of Minneapolis Parks* (Minneapolis: Foundation for Minneapolis Parks, 2008).

29. V. Musselman, "Teen Trouble: What Recreation Can Do about It," 5, National Recreation Association, 1943.

30. Cranz, *Politics of Park Design*, 137.

31. Cranz, *Politics of Park Design*, 137.

32. Minneapolis Planning Commission, *Comprehensive Planning for the Loring Park Neighborhood*, Neighborhood Series no. 6 (Minneapolis: City Planning Department, 1959).

33. Minneapolis Planning Commission, *Loring Park Development Report*, Neighborhood Series no. 6 (Minneapolis: City Planning Department, 1979).

34. Albert J. Rutledge, *A Visual Approach to Park Design* (New York: Garland STPM Press, 1981).

35. Anna Chiesura, "The Role of Urban Parks for the Sustainable City," *Landscape and Urban Planning* 68 (2004): 129–38.

36. Minneapolis Park and Recreation Board, "Sustainability 2013," July 15, 2014, http://www.minneapolisparks.org.

37. Cranz and Boland, "Defining the Sustainable Park," 104.

38. Joan I. Nassauer, "Cultural Sustainability: Aligning Aesthetics with Ecology," in *Placing Nature: Culture and Landscape Ecology*, ed. Joan I. Nassauer (Washington, DC: Island Press, 1997).

9: "AWHEEL FROM CHICAGO TO THE TWIN CITIES"

1. "Cycle Paths of City," *St. Paul Globe*, June 8, 1902 (hereafter *SPG*); Horace B. Hudson, *Hudson's Dictionary of Minneapolis and Vicinity* (Minneapolis: Hudson, 1906), 12.

2. Horace B. Hudson, "Awheel from Chicago to the Twin Cities," *Minneapolis Journal*, January 5, 1901 (hereafter *MJ*).

3. Steve Friedman, "Minneapolis," *Bicycling*, May 26, 2010, 56; Copenhagenize Index, accessed November 13, 2021, https://copenhagenizeindex.eu/; Stephanie Pearson, "Minneapolis Is a Big Wheel in the Urban Cycling Movement," *National Geographic*, September 16, 2021.

4. For example, the word *sidepath* does not appear in Bob Mionske's exhaustive *Bicycling & The Law: Your Rights as a Cyclist* (Boulder, CO: Velo Press, 2007). Only a few academic or popular histories mention them: Ross D. Petty, "Bicycling in Minneapolis in the Early 20th Century," *Minnesota History* 62, no. 3 (2010): 90–91; Karen McCal-

ley, "Bloomers & Bicycles: Health and Fitness in Victorian Rochester," *Rochester History* 69, no. 2 (2008): 14; Ron Spreng, "The 1890s Bicycling Craze in the Red River Valley," *Minnesota History* 54, no. 6 (1995): 280; Evan Friss, "The Cycling City: Bicycles and the Transformation of Urban America in the 1890s" (PhD diss., City University of New York, 2011). Sidepath tags are mislabeled in Minneapolis Public Works Department, "Minneapolis Bicycle Master Plan," draft, August 2010, 11, 13; and "Bicycle License Plate," both in the ephemera collection of the Minnesota Historical Society.

5. David V. Herlihy, *Bicycle: The History* (New Haven, CT: Yale University Press, 2004), chap. 11.

6. Clay McShane, *Down the Asphalt Path: The Automobile and the American City* (New York: Columbia University Press, 1994), 6–7, 19, 63–73; Michael R. Fein, *Paving the Way: New York Road Building and the American State, 1880–1956* (Lawrence: University Press of Kansas, 2008), 24, 26; Christopher W. Wells, "The Changing Nature of Country Roads: Farmers, Reformers, and the Shifting Uses of Rural Space, 1880–1905," *Agricultural History* 80 (Spring 2006): 148–51.

7. *Annual Report of the City Engineer of Minneapolis* (Minneapolis: Harrison and Smith, 1897), 36 (hereafter *Annual Report . . . Minneapolis* and the year); "The Wheelmen's Joy," *LAW Bulletin*, October 16, 1896, 526 (quote); "The Cycle-Paths of St. Paul," *LAW Bulletin*, September 30, 1898, 262.

8. *Annual Report . . . Minneapolis*, 1897, 61; "Construction of Bicycle Paths," May 1896, *Minneapolis City Charter and Ordinances* (1905), 859; see also "Cycle Paths of City," *SPG*, June 8, 1902.

9. "Charles T. Raymond: A Brief Sketch . . .," *Sidepaths*, February 1901, 72 (quote); Act of March 4, 1896, ch. 68, 1896 N.Y. Laws 90. For more on New York, see James Longhurst, "The Sidepath Not Taken: Bicycles, Taxes and the Rhetoric of the Public Good in the 1890s," *Journal of Policy History* 25, no. 4 (2013): 557–86.

10. Act of March 27, 1899, ch. 152, 1899 NY Laws 301; updated in Act of April 24, 1900, ch. 640, 1900 N.Y. Laws 1393; Harry Noyes Greene et al., *The Highway Law of the State of New York . . .* (Albany, NY: Matthew Bender, 1902), 235–38 (quote).

11. Act of March 27, 1899; Greene, *Highway Law*, 246; Ryan v. Preston, 10 N.Y. Ann. Cas. 5 (N.Y.S. 1901).

12. "Monroe County Sidepath Guide," 1900, Pamphlet Folder, Rochester Public Library, Rochester, NY; Longhurst, "Sidepath Not Taken," 570–71; "No Let Up," *Sidepaths*, December 1900, 436; *LAW Magazine*, June 1901, 16.

13. *Annual Report . . . Minneapolis*, 1897, 151 (quote); "News Nuggets," *SPG*, July 4, 1897; "Cycle-Paths of St. Paul," 262.

14. "New Laws Needed," *SPG*, March 11, 1900, 10; "Cycle Path Movement," *SPG*, March 25, 1900, 10.

15. "Sold Lucky Numbers," *SPG*, March 13, 1900, 5 (quotes); "Wrong in Principle," *SPG*, April 7, 1900, 4. See also Horace B. Hudson, ed., *A Half Century of Minneapolis* (Minneapolis: Hudson, 1908), 501.

16. "May Open Circuit Here," *SPG*, May 13, 1900, 7; and "After 500-Mile Record," *SPG*, June 10, 1900, 9 (quotes). See also "Money for Cycle Paths," *SPG*, June 3, 1900, 11; "Wheelmen Ask for $5,000," *SPG*, June 5, 1900, 2.

17. "Keeping Up the Paths," *SPG*, February 30, 1901, 4; "The Cycle-Path System," *MJ*, April 15, 1899; "Wrong in Principle," *SPG*, April 7, 1900, 4; "Record for 1,000 Miles," *SPG*, June 24, 1900, 9.

18. "Work of City Engineer," *SPG*, March 9, 1901, 2; Isaac Houlgate, "Guide to Minneapolis Bicycle Paths," 1902, Maps Collection, Minnesota State Archives, St. Paul.

19. "Cycle Path Movement," *SPG*, March 25, 1900, 10; and "Bicycle Sidepaths," *Genesee (NY) Daily News*, April 18, 1900 (quotes). See also "Cycle Paths," *LAW Bulletin*, July 1, 1898, 42; "Cycle Path Legislation," *LAW Bulletin*, December 30, 1898, 476; "Director Dodge Interested," *LAW Magazine*, October 1900, 2; "Bicycling," *Outing*, October 1900, 115.

20. "Wheels Again in Sight," *SPG*, April 1, 1900; "Cycle Route to Omaha," *SPG*, April 15, 1900; "To Build to Chicago," *SPG*, February 8, 1900, 5 (quotes). See also Hudson, "Awheel."

21. Hudson, "Awheel."

22. "Wheels Again in Sight," *SPG*, April 1, 1900 ("so nearly perfect"); "State Cycle License," *MJ*, January 28, 1901, 12 ("to provide for"); for the text of the legislation, see *Journal of the House of the Thirty-Second Session of the Legislature of the State of Minnesota* (St. Paul: McGill-Warner, 1901), 330.

23. "Plan for Next Season," *SPG*, March 28, 1901, 5; "Side-Path Commissioners Go," *MJ*, March 27, 1901, 2. For more on Hurd, see Folder "General Correspondence, 1900–1901," Box 2, Rukard Hurd Papers, 1895–1911, Minnesota State Archives, St. Paul; "Rukard Hurd," *SPG*, September 14, 1902, 18; *Journal of the Senate of the Thirty-Second Session of the Legislature of the State of Minnesota* (St. Paul: McGill-Warner, 1901), 658; chap. 126 of the *General Laws of Minnesota* (1901), 153.

24. "Want Commissioners Named," *SPG*, April 14, 1901; "Cycle Paths of City," *SPG*, June 8, 1902, 12.

25. "Cycle Path Plans," *MJ*, April 15, 1901, 9.

26. "Cycle Paths of City," *SPG*, June 8, 1902, 12; "Paths Need Repairing," *SPG*, April 29, 1900, 9 (quote).

27. "Bicycle Notes," *SPG*, April 21, 1901, 11 (quote); "Paths for Workingmen," *MJ*, April 25, 1901, 9 (quote); "Why East Siders Are Tagless," *MJ*, April 29, 1901, 8; "East Side Cycle Paths," *MJ*, May 4, 1901, 6; "Difference of Opinion," *MJ*, May 11, 1901, 9.

28. "Cycle Path Fight," *MJ*, November 13, 1901, 6; "The City May Be Liable," *MJ*, November 21, 1901; "No Sidepath Commission," *MJ*, December 25, 1901, 6.

29. "St. Paul Has Many Miles," *SPG*, June 14, 1903, 24; "Cycle Paths Passing," *MJ*, May 18, 1904, 7.

30. "Raise Price of Tags," *SPG*, February 21, 1904, 15; "Cyclists Protest," *SPG*, March

1, 1904, 2; "Wheelmen Protest," *SPG*, April 4, 1904, 5; "Must Pay for Tags," *SPG*, April 7, 1904, 2; "Cycle Dealers Sore," *SPG*, April 10, 1904; "Taylor Explains," *SPG*, April 13, 1904, 2; "Want Them Ousted," *SPG*, April 14, 1904, 10; "Dealers in Wheels Scorn Commissions," *SPG*, April 18, 1904, 2; "County Solons Hear Wheelmen," *SPG*, April 19, 1904, 10; "Wheelmen Are Hot after Three Scalps," *SPG*, April 21, 1904, 2; "Auto Owners Are after Cycle Paths," *SPG*, April 6, 1904, 2 (quote).

31. "Wheelmen Will Repair," *SPG*, April 23, 1904, 3; "Commissioners Cut Price," *SPG*, April 30, 1904; "Path Fund Runs Out," *SPG*, June 23, 1904 (quote); "Cycle Tags Cause No Great Furore [*sic*]," *SPG*, April 24, 1905, 8; "Bicycle Paths," approved February 13, 1905, *Compiled Ordinances of the City of St. Paul* (1908), 29.

32. Andrew Rinker, qtd. in "Cycle Paths Are a Bone of Contention," *MJ*, April 15, 1906, 1; "New Bike Paths," *MJ*, May 7, 1901, 12; "Minneapolis Sidepath Work," *Sidepaths*, February 1901, 54; "Are the Bike Paths of Too Cheap Construction?" *MJ*, April 22, 1901, 13; "Contract by Guess," *MJ*, April 30, 1902, 8; "Cycle Paths Need Fixing," *MJ*, April 11, 1906, 10 (quote).

33. "Save the Cycle Paths," *MJ*, April 26, 1905, 4; "Hears Path Protest," *MJ*, May 12, 1905, 6; Hudson, *Hudson's Dictionary of Minneapolis*, 12.

34. G. W. Sublette, qtd. in "Brisk Boom," *MJ*, March 14, 1902, 15.

35. "Left to Pay Bill," *MJ*, May 29, 1902, 9; "Other Sporting News," *MJ*, June 9, 1902, 12; *Proceedings of the City Council of the City of Minneapolis*, (Minneapolis: City of Minneapolis, 1906), 22a; *Proceedings of the Common Council of the City of St. Paul* (St. Paul: Pioneer Press, 1906), 25, 173, 204.

36. Andrew Rinker, *Annual Report of the City Engineer . . . 1908* (Minneapolis: Heywood Manufacturing, 1909), table 52, 94; Rinker, *Annual Report of the City Engineer . . . 1910* (Minneapolis: Heywood Manufacturing, 1911), 8e.

37. Rinker, *Annual Report . . . 1910*, table 6, 12e.

38. David C. Smith, *City of Parks: The Story of Minneapolis Parks* (Minneapolis: Foundation for Minneapolis Parks, 2008), ix–xi; "The Wheelmen's Joy," *LAW Bulletin*, October 16, 1896, 526; compare Smith, *City of Parks*, 66–67.

39. *Eighteenth Annual Report of the Board of Park Commissioners of the City of Minneapolis* (Minneapolis: Harrison & Smith, 1901), 15 (quote); "Cycle Path Is Praised," *MJ*, July 25, 1902, 8; "Minneapolis Bicycle Master Plan," draft, August 2010, 19, 24, accessed March 14, 2012, https://www2.minneapolismn.gov/government/projects/public-works /completed-projects/bike-peds-projects/bike-master-plan/; compare against Theodore G. Wirth, "Retrospective Glimpses into the History of the Board of Park Commissioners . . . ," Board of Park Commissioners, July 16, 1945, 130–32; Board of Park Commissioners, "The Story of W.P.A. and Other Federal Aid Programs in the Minneapolis Parks, Parkways and Playgrounds," January 1, 1937, both in Minnesota State Archives.

40. David Paul Nord, "Minneapolis and the Pragmatic Socialism of Thomas Van Lear," *Minnesota History* 45, no. 1 (1976); compare Nord, "The Paradox of Municipal

Reform in the Late Nineteenth Century," *Wisconsin Magazine of History* 66, no. 2 (1982): 11; Carl H. Chrislock, *The Progressive Era in Minnesota, 1899–1912* (St. Paul: Minnesota Historical Society, 1971), 4, 26.

41. "City of Saint Paul Bicycle Feasibility Study," February 1974, 22, City of Saint Paul Planning Department, Minnesota State Department of Transportation Library, St. Paul.

10: THE SUBURB OF MINNEAPOLIS

1. Christopher W. Wells, *Car Country: An Environmental History* (Seattle: University of Washington Press, 2012), 132; Robert M. Fogelson, *Downtown: Its Rise and Fall, 1880–1950* (New Haven, CT: Yale University Press, 2001), 249–80.

2. John A. Jakle and Keith A. Sculle, *Lots of Parking: Land Use in Car Culture* (Charlottesville: University of Virginia Press, 2004), 243.

3. Nathan Cherry and Kurt Nagle, *Grid/Street/Place: Essential Elements of Sustainable Urban Districts* (Chicago: American Planning Association Press, 2009), 6; Douglas Farr, *Sustainable Urbanism: Urban Design with Nature* (Hoboken, NJ: John Wiley & Sons, 2007), 23–29; Ellen Dunham-Jones and June Williamson, *Retrofitting Suburbia: Urban Design Solutions for Redesigning Suburbs* (Hoboken, NJ: John Wiley & Sons, 2009), 10.

4. H. G. Benton, "The Establishment of Industrial and Residential Districts by Ordinance in Minneapolis," in *American City*, ed. Arthur Hastings Grant and Harold Sinley Buttenheim, vol. 9 (New York: Civic Press, 1913), 519–21; Edward H. Bennett and Andrew Wright Crawford, *Plan of Minneapolis* (Minneapolis: Civic Commission, 1917), 103.

5. Bennett and Crawford, *Plan of Minneapolis*, 108.

6. Milton Coleman, "Nicollet-Lake Project Lures Two Major Stores," *Minneapolis Star*, March 2, 1976; M. Howard Gelfand, "K mart May Build at Nicollet Project," *Minneapolis Tribune*, March 3, 1976; Norman Draper, "Near North Side: Time to Clean Up an Image—Residents Celebrate to Boost Area Pride," *Star Tribune* (Minneapolis), September 28, 1986.

7. Robert Fishman, *Bourgeois Utopias: The Rise and Fall of Suburbia* (New York: Basic Books, 1987), x.

8. Maya Rao, "Minneapolis Looks to Move Kmart, Restore Nicollet Avenue," *Star Tribune*, April 14, 2012; Jake Weyer, "City Developing Work Plan to Reopen Nicollet," *Southwest Journal*, December 12, 2012; "Planning Blunder #7: Redevelopment of Lake and Nicollet," Getting around Minneapolis, April 13, 2011, http://gettingaroundmpls. wordpress.com/2011/04/13/planning-blunder-7-redevelopment-of-lake-and-nicollet/.

9. This essay is not in any way a discourse on the goods or evils of suburbia. For an essay compiling works on suburbia and its critics, see John Archer, "The Place We Love to Hate: The Critics Confront Suburbia, 1920–1960," in *Constructions of Home: Interdisciplinary Studies in Architecture, Law, and Literature*, ed. Klaus Stierstorfer (Brooklyn, NY: AMS Press, 2010).

10. For a brief discussion of superblocks, see Sarah Whiting, "Chicago—Superblockism: Chicago's Elastic Grid," in *Shaping the City: Studies in History, Theory, and Urban Design*, ed. Edward Robbins and Rodolphe El-Khoury (New York: Routledge, 2004).

11. E. T. Abbott, *Davison's Pocket Map of Minneapolis* (Minneapolis: C. Wright Davison, 1884); George B. Wright, *George B. Wright's Map. Minneapolis. 1873* (Saint Paul: Rice, 1873); Theodore C. Blegen, *Minnesota: A History of the State* (Minneapolis: University of Minnesota Press, 1975), 178; John H. Stevens, *Personal Recollections of Minnesota and Its People, and Early History of Minneapolis* (Minneapolis: Marshall Robinson, 1890), 233–34.

12. A. W. Warnock, *The Twin Cities and Surroundings* (New York: Kendrick-Odell Press, 1915); Twin City Rapid Transit, *Twin City Trolley Trips* (Buffalo, NY: Matthews Northrup Works, 1906); H. W. Benneche, *Atlas of Minneapolis, Hennepin County, Minnesota. Including Parts of St. Louis Park and Golden Valley Township in Hennepin County. Also Part of Ramsey County Known as the Midway District* (Minneapolis: Minneapolis Real Estate Board, 1914); Vincent Oredson, "Planning a City: Minneapolis, 1907–17," *Minnesota History* 33, no. 8 (Winter 1953): 331–32; Kristin M. Anderson and Christopher W. Kimball, "Designing the National Pastime: Twin Cities Baseball Parks," *Minnesota History* 58, no. 7 (Fall 2003): 340; Steve Berg, *Target Field: The New Home of the Minnesota Twins* (Minneapolis: MVP Books, 2010), 84, 125; Edward H. Bennett and Andrew Wright Crawford, *Plan of Minneapolis* (Minneapolis: Civic Commission, 1917), 15–16.

13. Walter Curt Behrendt, "Off-Street Parking: A City Planning Problem," *Journal of Land & Public Utility Economics* 16, no. 4 (November 1940): 464–66.

14. Frederick L. Johnson with Thomas U. Tuttle, *Suburban Dawn: The Emergence of Richfield, Edina and Bloomington* (Richfield, MN: Richfield Historical Society, 2009), 126–29; Behrendt, "Off-Street Parking," 464–66; Charles S. LeCraw, "An Economic Study of Interior Block Parking Facilities" (graduate course paper, Yale University, 1946), 2–3; State of Minnesota, Session Laws of Minnesota for 1937: Chapter 464—S.F. No. 142, 756–57; F. K. Dhainin, *A Statement on the Parking Problems in the Central Business District, Minneapolis, Minnesota* (Minneapolis: City Planning Commission, 1945); Fogelson, *Downtown*, 218–19.

15. Fogelson, *Downtown*, 218–19, 297–98; John W. Diers and Aaron Isaacs, *Twin Cities by Trolley: The Streetcar Era in Minneapolis and St. Paul* (Minneapolis: University of Minnesota Press, 2007), 7.

16. Sears, Roebuck & Co., *Modern Homes* (Chicago: Sears, Roebuck & Co., 1936), 40; Aladdin Homes, *Individual Homes or Complete Cities* (Toronto: Canadian Aladdin Co., 1920), 92–93.

17. Edward H. Bennett and Andrew Wright Crawford, *Plan of Minneapolis* (Minneapolis: Civic Commission, 1917), xii, 107; Wells, *Car Country*, 157.

18. Wells, *Car Country*, 124, 251–53.

19. US Bureau of the Census, "No. HS-7. Population of the Largest 75 Cities: 1900

to 2000," *Statistical Abstract of the United States, 2003*, 13–14, https://www.census .gov/history/pdf/los_angeles_pop.pdf; Robert E. Lang and Meghan Zimmerman Gough, "Growth Counties: Home to America's New Suburban Metropolis," in *Redefining Urban and Suburban America: Evidence from Census 2000*, ed. Alan Berube, Bruce Katz, and Robert E. Lang (Washington, DC: Brookings Institution Press, 2006), 3:65, 78; Lisa Plank and Thomas Saylor, "Constructing Suburbia: Richfield in the Postwar Era," *Minnesota History* 61, no. 2 (Summer 2008): 50–51.

20. Johnson and Tuttle, *Suburban Dawn*, 179–85; "Rapid Conversion," *Mass Transit Magazine*, July 1954, 32–40.

21. Judith A. Martin and Antony Goddard, *Past Choices/Present Landscapes: The Impact of Urban Renewal on the Twin Cities* (Minneapolis: Center for Urban and Regional Affairs, 1989), 60–69; Maya Rao, "Downtown Minneapolis Is Seen as Drowning in Sea of Parking Lots," *Star Tribune*, July 25, 2012.

22. M. Jeffrey Hardwick, *Mall Maker: Victor Gruen, Architect of an American Dream* (Philadelphia: University of Pennsylvania Press, 2004), 142–55; Johnson and Tuttle, *Suburban Dawn*, 168–73; Jakle and Sculle, *Lots of Parking*, 207.

23. Hardwick, *Mall Maker*, 142–55; Johnson and Tuttle, *Suburban Dawn*, 168–73; Gregory T. Peerbolte, *Randhurst: Suburban Chicago's Grandest Shopping Center* (Charleston, SC: History Press, 2011), 51.

24. Alan Ehrenhalt, "Minneapolis Revises Zoning Codes," in *The Inner City: A Handbook for Renewal*, ed. Roger L. Kemp (Jefferson, NC: McFarland, 2001), 290; Martin and Goddard, *Past Choices/Present Landscapes*, 64–65.

25. Pearson et al., "Evolution of the Whittier Neighborhood," 38.

26. Draper, "Near North Side."

27. City of Minneapolis Planning Commission, *Decision '67 Action for Better Living—System for Social Improvements*, 1967; Bennett and Crawford, *Plan of Minneapolis*, 25–26; Irwin Unger, *The Best of Intentions: The Triumph and Failure of the Great Society under Kennedy, Johnson, and Nixon* (Saint James, NY: Brandywine Press, 1996), 223–33.

28. "Ministers Aid Unusual Fight for Tavern License in 'Model City,'" *Jet*, May 7, 1964, 48–51.

29. Minneapolis Planning Commission, *Decision '67*.

30. Pearson et al., "Evolution of the Whittier Neighborhood," 40–41; Gelfand, "K mart"; "Planning Blunder #7."

31. Coleman, "Nicollet-Lake Project"; M. Howard Gelfand, "K mart May Build at Nicollet Project," *Minneapolis Tribune*, March 3, 1976; "Nicollet-Lake K mart Pledged," *Minneapolis Star*, May 10, 1976; Peg Meier, "K mart Plan Opposed in Nicollet Area," *Minneapolis Tribune*, March 30, 1976.

32. M. Howard Gelfand, "City-Council Approves K mart Construction at Nicollet-Lake Area," *Minneapolis Tribune*, May 15, 1976.

33. Joseph S. Wood, "Suburbanization of Center City," *Geographical Review* 78, no. 3 (July 1988): 326–27; Ruth Hammond, "Lake St. K mart, Famous for Its Backside, Puts on a Warm Heart for Its Neighborhood," *Minneapolis Tribune*, July 8, 1978; Gelfand, "City-Council Approves K mart."

34. Hammond, "Lake St. K mart."

35. Office of Community Planning and Economic Development (CPED), *The Minneapolis Plan for Sustainable Growth*, 2009, P 2–5 Section 2.3.4.

36. Stevens, *Personal Recollections*, 233–34.

37. Kate Brown, "Gridded Lives: Why Kazakhstan and Montana Are Nearly the Same Place," *American Historical Review* (February 2001): 22; James C. Scott, *Seeing Like a State: How Certain Schemes to Improve the Human Condition Have Failed* (New Haven, CT: Yale University Press, 1998), 59–63.

38. Kenneth T. Jackson, *Crabgrass Frontier: The Suburbanization of the United States* (Oxford: Oxford University Press, 1987), 150.

39. Kira Obolensky, *Garage: Reinventing the Place We Park* (Newton, CT: Taunton Press, 2003), 13–16, 23; Burton Ashford Bugbee, "The Garage's Place Is in the Home," *House Beautiful* February 1932), 136.

40. Nicollet Avenue Task Force Report, *Nicollet Avenue: The Revitalization of Minneapolis' Main Street*, May 2000, 39–40.

41. CPED, *Minneapolis Plan*, i-1.

42. David Frank, qtd. in Rao, "Minneapolis Looks to Move."

43. William Morris, "Minneapolis Pursues Bid for Lake Street Kmart," *Finance and Commerce*, December 5, 2018, https://finance-commerce.com/2018/12/minneapolis -pursues-bid-for-lake-street-kmart/.

44. Solomon Gustavo, "Are They Ever Going to Tear Down the Old Kmart Blocking Nicollet Avenue?" *MinnPost*, October 1, 2021, https://www.minnpost.com/metro/2021 /10/are-they-ever-going-to-tear-down-the-old-kmart-blocking-nicollet-avenue/.

45. Cliff Ellis, "The New Urbanism: Critiques and Rebuttals," *Journal of Urban Design* 7, no. 3 (2002): 261–62.

46. Kim Dovey, *Becoming Places: Urbanism, Architecture, Identity, Power* (New York: Routledge, 2010), 4–5.

11: THE CAMPUS AS WATERSHED

1. World Commission on the Environment and Development, *Our Common Future* (New York: Oxford University Press, 1987); James Speth, *A Bridge at the End of the World: Capitalism, the Environment, and Crossing from Crisis to Sustainability* (New Haven, CT: Yale University Press, 2005). In its planning document for water supply management for the greater Twin Cities metropolitan area, the Metropolitan Council's *Metropolitan Area Master Water Plan*, following Minnesota State law, states that their goal is a "sustainable water supply." Metropolitan Council, *Metropolitan Area Master Water*

Plan, 2010. For helpful discussions of resilience and adaptability, see Andrew Zolli and Ann Marie Healey, *Resilience: Why Things Bounce Back* (New York: Free Press, 2012); Peter Newman, Timothy Beatley, and Heather Boyer, *Resilient Cities: Responding to Peak Oil and Climate Change* (Washington, DC: Island Press, 2009).

2. William Cronon, *Nature's Metropolis: Chicago and Great West* (New York: Norton, 1992); Richard White, *The Organic Machine: The Remaking of the Columbia River* (New York: Hill and Wang, 1995); Timothy Mitchell, *Rule of Experts: Egypt, Techno-Politics, Modernity* (Berkeley: University of California Press, 2002); Bill McKibben, *The End of Nature* (New York: Random House, 2006); Speth, *Bridge at the End of the World*.

3. In 2017 Augsburg College changed its name to Augsburg University. For the sake of consistency, I use Augsburg College throughout this chapter.

4. Aldo Leopold, *A Sand County Almanac and Sketches Here and There* (New York: Oxford University Press, 1949); Wendell Berry, *The Art of the Commonplace: The Agrarian Essays of Wendell Berry* (Berkeley, CA: Counterpoint Press, 2003). This essay, by focusing on the author's own local ecological context, reflects this approach.

5. Madhu Prakash and Gustavo Esteva, *Escaping Education: Living as Learning with Grassroots Cultures*, 2nd ed. (New York: Peter Lang, 2008).

6. Paulo Freire, *Pedagogy of the Oppressed*, 30th anniversary ed. (New York: Continuum, 2000); H. Giroux, *Teachers as Intellectuals* (South Hadley, MA: Bergin Garvey, 1988); D. A. Gruenewald, "The Best of Both Worlds: A Critical Pedagogy of Place," *Environmental Education Research* 14 (2008): 308–24.

7. David Orr, *Earth in Mind: On Education, Environment, and the Human Prospect* (Washington, DC: Island Press, 2004); James J. Farrell, *The Nature of College: How a New Understanding of Campus Life Can Change the World* (Minneapolis: Milkweed Press, 2010); American Association for Sustainability in Higher Education, *Sustainability Curriculum in Higher Education: A Call to Action* (Denver: AASHE, 2010), https://www.google.com/url?sa=t&rct=j&q=&esrc=s&source=web&cd=&ved=2ahUKEwizrKn MuJvzAhULV8oKHVEAC_QQFnoECAYQAQ&url=https%3A%2F%2Finterdisciplinary studies.org%2Fdocs%2FConferences%2F2010_Documents_A_Call_to_Action.pdf&usg= AOvVaw2hckeTF9VWVhkN_LM-6TkB.

8. The Seminary began to split off from the college in the 1920s and would eventually merge with several other local Lutheran seminaries to form Luther Seminary in Saint Paul by 1963.

9. Odd S. Lovoll, *The Promise of America: A History of the Norwegian-American People* (Minneapolis: University of Minnesota Press, 1984), 13–15; Jon Gjerde, *From Peasants to Farmers: The Migration from Balestrand, Norway, to the Upper Middle West* (Cambridge: Cambridge University Press, 1985), 8.

10. Augsburg College catalog, 1902–1903, 6, Augsburg College Archives, Minneapolis (hereafter College [or Seminarium] Catalog and year).

11. Seminarium catalog, 1897–1898, 3.

12. Carl Chrislock, *From Fjord to Freeway: 100 Years, Augsburg College* (Minneapolis: Augsburg College, 1969), 9.

13. Theodore Steinberg, *Nature Incorporated: Industrialization and the Waters of New England* (Cambridge: Cambridge University Press, 1991).

14. Chrislock, *From Fjord to Freeway,* 9.

15. Calvin Fremling, *Immortal River: The Upper Mississippi in Ancient and Modern Times* (Madison: University of Wisconsin Press, 2005), chap. 3.

16. These patterns have now been cast into doubt by climate change, but to date rainfall amounts have not shown any clear trends in the last decade.

17. Chrislock, *From Fjord to Freeway,* 99.

18. City of Minneapolis, *Annual Reports of the Various City Officers of the City of Minneapolis for the Year 1890* (Minneapolis: Harrison & Smith, 1891), 369, City of Minneapolis Archives, Minneapolis.

19. See John Anfinson's chapter in this volume for more on the problems of waterborne disease in the Twin Cities; and Martin Melosi, *Precious Commodity: Providing Water for America's Cities* (Pittsburgh: University of Pittsburgh Press, 2011), 40–51. In September 1881, 450–600 cases of typhoid were reported in Minneapolis, with a 10% mortality rate.

20. "What We Drink: That Is, What Those of Us Drink Who Drink Water; Some More Analyses from the Board of Health—Rivers and Wells—Chemistry and Comparison," *Minneapolis Tribune,* March 22, 1880; "River Water," *Minneapolis Tribune,* September 21, 1893, 8.

21. Chrislock, *From Fjord to Freeway,* 97.

22. College Catalog, 1913–1914, 15.

23. College Catalog, 1892–1893, 5; College Catalog, 1897–1898, 3.

24. For background on the development of ecology as a discipline, see Donald Worster, *Nature's Economy: A History of Ecological Ideas,* 2nd ed. (New York: Cambridge University Press, 1994).

25. City of Minneapolis, *Annual Reports of the Various City Officers of the City of Minneapolis for the Year 1890* (Minneapolis: Harrison & Smith, 1891), 231, 259; Andrew Connor et al., "From Rural to Urban: An Environmental History of Augsburg College, 1872–2005" (undergraduate course paper, Augsburg College, 2006), 8, https://www.augsburg.edu/wp-content/uploads/2012/08/Auggie_environment2-1.pdf.

26. Minnesota Pollution Control Agency, *Report on Upper Mississippi River Basin,* accessed November 13, 2021, http://www.pca.state.mn.us/index.php/view-document.html?gid=6035; J. R. Stark et al., *Water Quality in the Upper Mississippi River Basin, Minnesota, Wisconsin, South Dakota, Iowa, and North Dakota, 1995–98,* U.S. Geological Survey Circular 1211, 35, http://pubs.water.usgs.gov/circ1211/.

27. US Geological Service, National Water Information System, "USGS Water Data for the Nation," accessed November 27, 2021, http://waterdata.usgs.gov/nwis/.

28. College catalog, 1901–1902, 7. Water mains reached the campus in 1890, but originally connected to fire hydrants only. In 1900 the new Main Building was the first to have its plumbing attached to the municipal water supply.

29. College Catalog, 1904–1905; Chrislock, *From Fjord to Freeway*, 90–95.

30. John O. Anfinson, *The River We Have Wrought: A History of the Upper Mississippi* (Minneapolis: University of Minnesota Press, 2003), chap. 11; Fremling, *Immortal River*.

31. College Catalog, 1930–1931, 84.

32. See Andrew Karvonen, *Politics of Urban Runoff: Nature, Technology, and the Sustainable City* (Cambridge, MA: MIT Press, 2011).

33. James Wiener and Mark Sandheinrich, "Contaminants in the Upper Mississippi River: Historic Trends, Responses to Regulatory Controls, and Emerging Concerns," *Hydrobiologia* 640 (2010): 49–70.

34. City of Minneapolis, *Annual Report of 2012 Combined Sewer Overflow Program and 2011 Activities* (Minneapolis, April 16, 2012), 7.

35. Minneapolis Water Department, *Annual Report*, 1988, 11. The volume and complexity of the supply network were such, however, that an impressive 3.6 billion gallons of that total were unaccounted for due to leaks and waste in the system.

36. Metropolitan Council, Water Supply Planning in the Twin Cities Metropolitan Area Technical Report, 2007, https://metrocouncil.org/Wastewater-Water/Planning /Water-Supply-Planning.aspx; Minneapolis Water Department, *Annual Report*, 1988, 11–14.

37. To water the campus grounds uses around 75,000–100,000 gallons of water per watering. To keep up the green lawns on campus then costs the amount paid to Aqua Systems, plus the cost of the water itself, which is about $47,500/year. All campus water use figures were gathered and analyzed by the author, using Augsburg College utility bills.

38. Peter Gleick, *Bottled and Sold: The Story behind Our Obsession with Bottled Water* (Washington, DC: Island Press, 2011); Beverage Marketing Corporation, *Bottled Water Reports*, accessed May 18, 2022, https://www.beveragemarketing.com/shop /bottled-water.aspx; Peter Gleick and Heather Cooley, "Energy Implications of Bottled Water" *Environmental Research Letters* 4 (2009): 1–6, http://iopscience.iop.org/1748 -9326/4/1/014009/pdf/erl9_1_014009.pdf.

39. My thanks to Matt Rumpza, in Augsburg's Purchasing office, for his help in getting these figures and for his patient engagement with my research.

40. Gleick and Cooley, "Energy Implications of Bottled Water," 6.

41. College Catalog, 1970, 48, 74.

42. College Catalog, 1971, n.p.

43. A partial ban on bottled water sales on campus was approved in 2017. Implementation of this ban, however, has remained difficult, with various forms of bottled water still making their way onto campus via the Pepsi distributor.

12: MONUMENTAL ENCOUNTERS

1. Dewey Albinson, "Communion with Nature," in "The Grand Portage Story and Some Other Tales from the North Country," 1963, 106, Manuscripts Notebooks, Call no. P2386, Minnesota Historical Society, Saint Paul (hereafter Albinson Stories).

2. Albinson, "Foreword," 3, Albinson Stories.

3. In his stories about Grand Portage, Albinson wrote about Ojibwe people at Grand Portage who avoided him, protested his presence, or refused to be painted. See Albinson, "First Trip to Grand Portage," 5; "Old Cramer," 13; "Mrs. Spruce and Mrs. Tamarack," 16, all in Albinson Stories.

4. Mary Towley Swanson, "Dewey Albinson: The Artist as Chronicler," *Minnesota History* 52 (Fall 1991): 269, 276.

5. Swanson, "Dewey Albinson," 265.

6. Ron Cockrell, *Grand Portage National Monument, Minnesota: An Administrative History* (Omaha, NB: National Park Service, Midwest Regional Office, Office of Planning and Resource Preservation, Division of Cultural Resources Management, 1983), 11.

7. For works covering the fur trade history of Grand Portage, see Carolyn Gilman, *The Grand Portage Story* (Saint Paul: Minnesota Historical Society Press, 1992); Grace Lee Nute, *The Voyageur's Highway: Minnesota's Border Lake Land* (Saint Paul: Minnesota Historical Society Press, [1941] 1976); and Grace Lee Nute, *The Voyageur* (Saint Paul: Minnesota Historical Society Press, [1955] 1987).

8. See David Glassberg, *Sense of History: The Place of the Past in American Life* (Amherst: University of Massachusetts Press, 2001). Glassberg argues that the environmental value that we attach to specific places is intertwined with the historical importance that we attribute to them (16).

9. Karl Jacoby, *Crimes against Nature: Squatters, Poachers, Thieves, and the Hidden History of American Conservation* (Berkeley: University of California Press, 2001); Robert H. Keller and Michael F. Turek, *American Indians and National Parks* (Tucson: University of Arizona Press, 1999); James Feldman, *A Storied Wilderness: Rewilding the Apostle Islands* (Seattle: University of Washington Press, 2011).

10. Katrina Phillips, *Staging Indigeneity: Salvage Tourism and the Performance of Native American History* (Chapel Hill, NC: University of North Carolina Press, 2021); Jean M. O'Brien, *Firsting and Lasting: Writing Indians Out of Existence* (Minneapolis: University of Minnesota Press, 2010); Coll Thrush, *Native Seattle: Histories from the Crossing-Over Place* (Seattle: University of Washington Press, 2008); Paige Raibmon, *Authentic Indians: Episodes of Encounter from the Late-Nineteenth-Century Northwest Coast* (Durham, NC: Duke University Press, 2005); Philip Deloria, *Indians in Unexpected Places* (Lawrence: University Press of Kansas, 2006).

11. Donald J. Auger and Paul Driben, eds. *Grand Portage Chippewa: Stories and Experiences of Grand Portage Band Members* (Grand Marais, MN: Grand Portage Tribal

Council and the Sugarloaf Interpretive Center Association, 2000), 11; Gilman, *Grand Portage Story*, 38.

12. Gilman, *Grand Portage Story*, 51.

13. Cockrell, *Grand Portage National Monument*, 6.

14. White Oak Society, White Oak Learning Center, and White Oak Fur Post website, "Grand Portage: An Emporium of Goods and Cultures on the North Shore of Lake Superior," January 11, 2013, http://www.whiteoak.org/historical-library/fur-trade/grand-portage/.

15. Gilman, *Grand Portage Story*, 72.

16. In the 1830s and 1840s the American Fur Company established a short-lived fishing venture at Grand Portage that employed Ojibwe and Métis fishermen. For more information on this venture, see Grace Lee Nute, "The American Fur Company's Fishing Enterprises on Lake Superior," *Mississippi Valley Historical Review* 12, no. 4 (March 1926): 483–503.

17. Article 11, United States, Treaty with the Chippewa, September 30, 1854; Ojibwe leaders designated these rights in the treaties of 1837 and 1842.

18. David Wilkins defines tribal sovereignty as "the spiritual, moral, and dynamic cultural force within a given tribal community empowering the group towards political, economic, and most important, cultural integrity, and toward maturity in the group's relationship with its own members, with other peoples and their governments, and with the environment"; David Wilkins, *American Indian Politics and the American Political System* (Lanham, MD: Rowman and Littlefield, 2002), 339.

19. Gilman, *Grand Portage Story*, 118.

20. Gilman, *Grand Portage Story*, 118.

21. The federal government imposed a broad array of policies on American Indians, which were designed to assimilate them into Euro-American society while attempting to erase tribal cultures and identities. The federal government–initiated allotment policy in the 1880s was their attempt to abolish Indian reservations by the allotment of communally held reservation lands to individual Indians for private ownership. After allotment, surplus plots of land were sold to non-Indians. For more information on how allotment affected Ojibwe communities, see Melissa Meyer, *The White Earth Tragedy: Ethnicity and Dispossession at a Minnesota Anishinaabe Reservation, 1889–1920* (Lincoln: University of Nebraska Press, 1999).

22. Paul Bliss, "Back Two Centuries over Minnesota's Oldest Highway to Oldest Fort," *Minneapolis Journal*, July 16, 1922.

23. Bliss, "Back Two Centuries."

24. Theodore Blegen, "The State Historical Convention at Duluth," *Minnesota History* (August–November 1922): 345.

25. Gilman, *Grand Portage Story*, 128; Cockrell, *Grand Portage National Monument*, 12.

26. Theodore Blegen, "Solon Justus Buck—Scholar-Administrator," *American Archivist* 23, no. 3 (July 1960): 261. Blegen preceded Buck as the superintendent of the Minne-

sota Historical Society in 1931. He also developed a career as a "scholar-administrator" and continued to expand the reach and collections of the MHS. For more information on Blegen's career, see John Flanagan, *Theodore C. Blegen: A Memoir* (Northfield, MN: Norwegian American Historical Association, 1977).

27. Flanagan, *Theodore C. Blegen.*

28. Flanagan, *Theodore C. Blegen.*

29. Jacoby, *Crimes against Nature*, 6.

30. Jacoby, *Crimes against Nature*, 6.

31. See Jacoby, *Crimes against Nature*; and Joseph F. Cullon, "Landscapes of Labor and Leisure: Common Rights, Private Property, and Class Relations along the Bois Brule River, 1870–1940" (PhD diss., University of Wisconsin, Madison, 1995).

32. Paul Bliss, "Back Two Centuries."

33. William Cronon and James Feldman describe how conservation focused on revitalizing a pristine wilderness and disregarded the fact that the America's environment had been altered by humans for thousands of years. See William Cronon, "The Trouble with Wilderness; or Getting Back to the Wrong Nature," in *Uncommon Ground: Toward Reinventing Nature* (New York: W. W. Norton, 1995); and Feldman, *Storied Wilderness.*

34. Bruce White, *Early Game and Fish Regulation and Enforcement in Minnesota, 1858–1920: A Report Prepared for the Mille Lacs Band of Ojibwe*, December 1995, viii, copy given to author by Bruce White.

35. Charles Cleland, "Preliminary Report of the Ethnohistorical Basis of the Hunting, Fishing, and Gathering Rights of Mille Lacs Chippewa," in *Fish in the Lakes, Wild Rice and Game in Abundance: Testimony on Behalf of Mille Lacs Ojibwe Hunting and Fishing Rights*, comp. James M. McClurken (East Lansing: Michigan State University Press, 1991), 104.

36. Cockrell, *Grand Portage National Monument*, 13.

37. The Indian Service was a part of the Bureau of Indian Affairs, the federal agency established in 1824 and moved to the Department of the Interior in 1849. Originally BIA personnel served as diplomatic corps responsible for overseeing trade and other relations with Indian tribes. By the 1860s the BIA had become the leading colonizing agent for the federal government and dominated almost all aspects of tribal life within reservations. Today the BIA is more involved in advocating programs focused on tribal education, social, economic, and cultural self-determination, although it has not entirely separated itself from its paternalistic history.

38. Cockrell, *Grand Portage National Monument*, 14.

39. Ralph D. Brown, "Archeological Investigation of the Northwest Company's Post, Grand Portage Minnesota, 1936," *Indians at Work* 4, no. 18–19 (May 1937): 38–43.

40. Brown, "Archeological Investigation," 43.

41. Ralph D. Brown, "Grand Portage Stockade Excavation, 1937," Sound and Visual Collection, call no. I.245, MHS.

42. Jim Wipson, qtd. in Auger and Driben, *Grand Portage Chippewa*, 23.

43. Jim Wipson, qtd. in Auger and Driben, *Grand Portage Chippewa*, 23.

44. Betty Lou Hoffman, qtd. in Auger and Driben, *Grand Portage Chippewa*, 95.

45. "Grand Portage Stockade Discussion," January 19, 1938, Indian Service and Members of the Historical Society, Minnesota State Historical Office Building, Saint Paul; Bureau of Indian Affairs, *Grand Portage Reconstruction Project, 1935–1942*, National Archives Kansas City Branch Record Group 75; Microfilms Collection 92, MHS, 1.

46. Microfilms Collection 92, 5.

47. Microfilms Collection 92, 5.

48. Microfilms Collection 92, 5.

49. Gilman, *Grand Portage* Story, 129; Cockrell, *Grand Portage National Monument*, 15–16, 28–30, 33.

50. Cockrell, *Grand Portage National Monument*, 17.

51. Elmer Albinson, "Grand Portage Footage," 1940–1949, Sound and Visual Collection call no. #III.7, MHS. Elmer Albinson was Dewey Albinson's brother, and he occasionally visited Dewey while he was living at Grand Portage.

52. Brenda Child provides an excellent explanation of the significance of the jingle dress to Ojibwe people in her book *Holding Our World Together: Ojibwe Women and the Survival of Community* (New York: Viking, 2012), 94–96.

53. Child, *Holding Our World Together*, 94–96.

54. Olga Soderberg, *Grand Portage the Great Carrying Place: A Pageant Script* (Grand Marais, MN: Cook County Historical Society, 1949), call no. F612.C89G7.S6, 1, MHS.

55. Soderberg, *Grand Portage the Great Carrying Place*, 1.

56. Soderberg, *Grand Portage the Great Carrying Place*, 2.

57. Soderberg, *Grand Portage the Great Carrying Place*, 3. The Minnesota Chippewa Tribe (MCT) became the joint governing body of the six bands of Ojibwe in Minnesota and was established in 1934 as part of the Indian Reorganization Act. The MCT has power over matters affecting these bands, though each band still retains autonomy over individual reservation matters.

58. Soderberg, *Grand Portage the Great Carrying Place*, 3.

59. Soderberg, *Grand Portage the Great Carrying Place*, 3.

60. Soderberg, *Grand Portage the Great Carrying Place*, 3.

61. Soderberg, *Grand Portage the Great Carrying Place*, 3.

62. Soderberg, *Grand Portage the Great Carrying Place*, 9.

63. Soderberg, *Grand Portage the Great Carrying Place*, 9.

64. Cockrell, *Grand Portage National Monument*, 28.

65. Cockrell, *Grand Portage National Monument*, 29.

66. Wilderness Society, "Mission and Impact," December 30, 2012, http://wilderness.org/article/mission-and-impact; and "History," December 30, 2012, http://wilderness.org/history.

67. Cockrell, *Grand Portage National Monument*, 31.

68. Cockrell, *Grand Portage National Monument*, 32.

69. I have not found information to explain why the Grand Portage Band did not receive more funding. However, it is likely that, as a result of termination policy under the Truman Administration, funding for tribal projects was limited based on the federal government's efforts to eliminate tribal dependency and tribalism in the interest of assimilating Native Americans into mainstream American society.

70. Cockrell, *Grand Portage National Monument*, 32.

71. Cockrell, *Grand Portage National Monument*, 35.

72. Cockrell, *Grand Portage National Monument*, 35.

73. Cockrell, *Grand Portage National Monument*, 35.

74. Cockrell, *Grand Portage National Monument*, 36. According to Cockrell, the land ceded amounted to 258 acres held by the Grand Portage Band, and 50 acres held by the Minnesota Chippewa Tribe.

75. An Act: To Provide for the Establishment of Grand Portage National Monument in the State of Minnesota, and for Other Purposes, Pub. L. No. 85–910, 72 Stat. 1751 (1958), https://www.nativeamericanembassy.net/TREATIES/digital.library.okstate.edu /kappler/Vol6/html_files/v6po858b.html.

76. Gilman, *Grand Portage Story*, 130; Alan R. and Nancy L. Woolworth, "Grand Portage National Monument: An Historical Overview and Inventory of Its Cultural Resources," vol. 1 (Saint Paul: Minnesota Historical Society, 1982), 230–55.

77. Albinson, "Grand Portage," 38, Albinson Stories.

78. Albinson, "Grand Portage," 39, Albinson Stories.

79. Gilman, *Grand Portage Story*, 133.

80. The Indian Gaming Regulatory Act, Pub L. 100–497, 25 U.S.C. & 2701 et seq. (1988). This federal law lays out the regulatory framework for Indian Gaming.

81. Grand Portage National Monument, "Grand Portage National Monument Heritage Center," National Park Service, January 11, 2013, http://www.nps.gov/grpo /planyourvisit/grand-portage-national-monument-heritage-center.htm.

82. "Grand Portage National Monument Heritage Center."

83. "Grand Portage National Monument Heritage Center."

84. Jacoby, *Crimes against Nature*, 193.

13: PITTSBURGH'S COLONY IN SAINT PAUL'S HINTERLAND

1. Nick Wognum, "Six Charged in BWCAW Shooting Incident," *Ely Echo*, September 21, 2007; Larry Oakes, "Ely Distances Itself from a Rage against Visitors," *Star Tribune*, September 30, 2007; Larry Oakes, "Drunken Spree of BWCA Terror Draws Tears, Fourth Guilty Plea," *Star Tribune*, June 2, 2008.

2. "BWCAW Charges: Community Should Show No Tolerance for Such Behavior," editorial, *Tower Timberjay*, September 22, 2007, www.timberjay.com. On the history of

conflicts surrounding the BWCA, see Mark Harvey, "Sound Politics: Wilderness, Recreation, and Motors in the Boundary Waters, 1945–1964," *Minnesota History* 58, no. 3 (2002): 130–45.

3. D. J. Tice, "The Thing on the Hill," *Corporate Report Minnesota*, October 1982, 63.

4. David A. Lanegran, "Minnesota: Nature's Playground," *Daedelus* 129, no. 3 (Summer 2000): 82.

5. Richard White, "'Are You an Environmentalist or Do You Work for a Living?' Work and Nature," in *Uncommon Ground: Rethinking the Human Place in Nature*, ed. William Cronon (New York: Norton, 1996), 174.

6. David Rosner and Gerald Markowitz, *Deadly Dust: Silicosis and the Ongoing Struggle to Protect Workers' Health*, 2nd ed. (Ann Arbor: University of Michigan Press, 2006); Scott Dewey, "Working for the Environment: Organized Labor and the Origins of Environmentalism in the United States, 1948–1970," *Environmental History* 3, no. 1 (1998): 45, 50.

7. See Brian K. Obach, *Labor and the Environmental Movement: The Quest for Common Ground* (Cambridge, MA: MIT Press, 2004); Dewey, "Working for the Environment"; Robert Gordon, "'Shell No!' OCAW and the Labor-Environmental Alliance," *Environmental History* 3, no. 4 (1998): 460–87; and Chad Montrie, "Expedient Environmentalism: Opposition to Coal Surface Mining and the United Mine Workers of America, 1945–1975," *Environmental History* 5, no. 1 (2000): 75–98.

8. On the history of Minnesota's iron ore ranges in the late nineteenth and early twentieth centuries, see David A. Walker, *Iron Frontier: The Discovery and Early Development of Minnesota's Three Ranges* (Saint Paul: Minnesota Historical Society Press, 1979).

9. Duane A. Smith, *Mining America: The Industry and the Environment, 1800–1980* (Lawrence: University of Kansas Press, 1987), 86, 103.

10. Jeffrey T. Manuel, "Mr. Taconite: Edward W. Davis and the Promotion of Low-Grade Iron Ore, 1913–1955," *Technology and Culture* 54, no. 2 (2013): 317–45; Edward W. Davis, *Pioneering with Taconite* (Saint Paul: Minnesota Historical Society Press, 1964).

11. For examples of hunting and firearms safety clubs on the Iron Range at mid-century, see Calumet Golden Jubilee, 1909–1959 (1959), 54–55, Community Histories: Calumet, Minnesota, Collection, Iron Range Research Center, Chisholm, MN (hereafter IRRC).

12. On the Reserve Mining lawsuit, see Robert V. Bartlett, *The Reserve Mining Controversy: Science, Technology, and Environmental Quality* (Bloomington: Indiana University Press, 1980); Thomas F. Bastow, *"This Vast Pollution . . .": United States of America versus Reserve Mining Company* (Washington, DC: Green Fields, 1986); Thomas R. Huffman, "Exploring the Legacy of Reserve Mining: What Does the Longest Environmental Trial in History Tell Us about the Meaning of American Environmentalism?" *Journal of*

Policy History 12, no. 3 (2000): 339–68; Huffman, "Enemies of the People: Asbestos and the Reserve Mining Trial," *Minnesota History* 59, no. 7 (2005): 292–306; Frank Schaumburg, *Judgment Reserved: A Landmark Environmental Case* (Reston, VA: Reston Publishing, 1976).

13. Huffman, "Exploring the Legacy of Reserve Mining," 340–41.

14. Paul Brodeur, *Expendable Americans* (New York: Viking, 1974).

15. Bastow, "*This Vast Pollution*," 104–6.

16. Bartlett, *Reserve Mining Controversy*, 126–29; Jane E. Brody, "Iron Ore Company vs. the Changing Times: U.S. Court Will Decide on Right to Pollute," *New York Times*, August 8, 1973; Huffman, "Enemies of the People," 299; Wade Green, "Life vs. Livelihood," *New York Times Magazine*, November 24, 1974.

17. Paul H. Landis, *Three Iron Mining Towns: A Study in Cultural Change* (Ann Arbor, MI: Edwards Brothers, 1938), 51.

18. Clay Blair Jr. "Minnesota Grows Older," *Saturday Evening Post*, March 18, 1961, 85. See also Aaron Shapiro, *The Lure of the North Woods: Cultivating Tourism in the Upper Midwest* (Minneapolis: University of Minnesota Press, 2013).

19. Educational Research and Development Council of Northeast Minnesota, *Economy of Northeast Minnesota: Current Economic Activity* (Duluth: Educational Research and Development Council of Northeast Minnesota, 1968), 33. A copy of this report is held at the IRRC.

20. Homer Bigart, "Poverty Blights Iron Ore Region," *New York Times*, January 13, 1964.

21. Dana Miller, "Public Policy and Economic Development: The Iron Range Experience," paper presented to Minnesota Historical Society Annual Meeting, Minneapolis, 1990, 9. A copy of this paper is held at the IRRC.

22. Ray Svatos, "Fishing Minnesota's Abandoned Iron Pits," *Minnesota Volunteer*, July-August 1986, 14–18.

23. State of Minnesota, Iron Range Resources and Rehabilitation Board, *Biennial Report, 1976–1978*, Saint Paul, 1978, 8.

24. D. J. Tice, "The Thing on the Hill (Part II)," *Minnesota Corporate Report*, November 1982, 96.

25. Bill Cook, oral history interview, August 9, 1986, Keewatin, MN, Unemployment on the Iron Range Oral History Collection, A-86-694, IRRC.

26. Midwest Research Institute, *Assessment of the Tourism Development Components of the Iron Range Interpretative Program* (Minnetonka, MN: Center for Economic Studies, 1980), 3–4.

27. Darryl Rice, qtd. in "Winter of Our Discontent," *Star Tribune* (Minneapolis), February 17, 2010.

28. John E. Haynes, "Reformers, Radicals, and Conservatives," in *Minnesota in a Century of Change: The State and Its People since 1900*, ed. Clifford E. Clark (Saint Paul:

Minnesota Historical Society Press, 1989), 383–88; Jennifer A. Delton, *Making Minnesota Liberal: Civil Rights and the Transformation of the Democratic Party* (Minneapolis: University of Minnesota Press, 2002), 1–18.

29. Haynes, "Reformers, Radicals, and Conservatives," 388–89, 392.

30. Izaak Walton League, "Public Information and Conservation Education Honor Roll," June 25, 1960, Izaak Walton League of America Inc. Folder, Box 4; Hubert Humphrey to John Blatnik, July 28, 1964, Personal Correspondence January 24–April 20, 1964, Folder, Box 107, both in Blatnik Papers, MHS.

31. Paul Milazzo, *Unlikely Environmentalists: Congress and Clean Water, 1945–1972* (Lawrence: University Press of Kansas, 2006), 17–37.

32. Huffman, "Exploring the Legacy of Reserve Mining," 354.

33. Megan Boldt, "Chip Cravaack Gets Boost from Key Democrats in Race vs. Rick Nolan," *St. Paul Pioneer Press*, September 21, 2012.

34. Richard Francaviglia, *Hard Places: Reading the Landscape of America's Historic Mining Districts* (Iowa City: University of Iowa Press, 1991), xviii.

35. Smith, *Mining America*, 86.

36. Steven C. High and David W. Lewis, *Corporate Wasteland: The Landscape and Memory of Deindustrialization* (Ithaca, NY: IRL Press of Cornell University Press, 2007), 2.

37. "Minnesota Taconite," *Fortune*, August 1957, 118–27.

38. Thomas J. Baerwald, "Forces at Work on the Landscape," in *Minnesota in a Century of Change: The State and Its People since 1900*, ed. Clifford E. Clarke Jr. (Saint Paul: Minnesota Historical Society Press, 1989), 19.

39. Peter Goin and C. Elizabeth Raymond, *Changing Mines in America* (Santa Fe, NM: The Center for American Places, 2004), 8.

40. Jim Hendrickson, oral history interview by Mike Schomer, July 18, 1990, Aurora, MN, LTV Steel Oral History Project, IRRC.

41. "$300 Million Expansion Planned for Keetac Taconite Operation," *Duluth News Tribune*, February 1, 2008.

42. On the environmental legacies of copper mining in other American regions, see Timothy J. LeCain, *Mass Destruction: The Men and Giant Mines That Wired America and Scarred the Planet* (New Brunswick, NJ: Rutgers University Press, 2009); and Duncan Maysilles, *Ducktown Smoke: The Fight over One of the South's Greatest Environmental Disasters* (Chapel Hill: University of North Carolina Press, 2011).

14: A HOUSE DIVIDED

1. For demographic data, see US Bureau of the Census, "Aitkin County, Minnesota," accessed November 13, 2021, https://data.census.gov/cedsci/profile?g=0500000US27001.

2. Minnesota Session Laws, 1971 Regular Session, Chapter 849, House File No. 189.

3. Minnesota Session Laws, 1971 Regular Session, Chapter 849, House File No. 189, sec. 1.

4. Minnesota Session Laws, 1971 Regular Session, Chapter 849, House File No. 189, sec. 4, subds. 1, 6.

5. Minutes of the Minnesota Experimental City Authority, November 17, 1972, box 1, Minnesota Experimental City Authority Records, 1963–1988, Minnesota State Archives (hereafter MSA), Minnesota Historical Society, Saint Paul (hereafter MHS).

6. Memorandum by township officers of Lake Johanna, Gilchrist, Bangor, and Chippewa Falls, Pope County, State of Minnesota, November 27, 1972, box 2, MHS.

7. Minnesota Pollution Control Agency, "Public Meeting on the Proposed Minnesota Experimental City," transcript of public meeting, February 13, 1973, board room of the Department of Health Building, Minneapolis, MN, included as an enclosure in Charles E. Carson to Homer C. Luick et al., March 6, 1973, folder: "12/72 Experimental City, 1973," Minnesota Pollution Control Agency Records, 1968–1978, MSA, MHS.

8. Richard (Dick) Pemberton's prepared statement, in Minnesota Pollution Control Agency, "Public Meeting on the Proposed Minnesota Experimental City, 18–19.

9. Donald Worster, *A River Running West: The Life of John Wesley Powell* (New York: Oxford University Press, 2001), 467–532.

10. Donald Worster, *Dust Bowl: The Southern Plains in the 1930s* (New York: Oxford University Press, 1979), 182–97.

11. The Tulsa phrase was Daniel Patrick Moynihan's and occasionally appropriated by MXC proponents; see memorandum by Archie Chelseth, March 27, 1969, box 1, Minnesota Experimental City Papers, Northwest Architectural Archives, University of Minnesota Libraries, Minneapolis (hereafter NAA). However, strictly speaking, the phrase was inaccurate for the MXC proponents' purposes: the 1970 population of the city of Tulsa was 366,481 and the 1970 population of the Tulsa metropolitan area was closer to 475,000, whereas the MXC was most consistently pitched as a city of 250,000. See US Bureau of the Census, *1970 Census of Population and Housing, General Demographic Trends for Metropolitan Areas, 1960 to 1970, Oklahoma*, PHC(2)-38, July 2, 1971, 5. Had America grown by an MXC of 250,000 persons per month from 1970 to 2000, the US population in 2000 would have been approximately 295 million; had the nation grown by a 1970 Tulsa (366,481 persons) per month, the US population in 2000 would have been about 337 million. The actual US population in 2000 was approximately 281 million; see US Bureau of the Census, "Intercensal Estimates of the Resident Population by Sex and Age for the United States: April 1, 2000 to July 1, 2010," https://www2.census.gov /programs-surveys/popest/tables/2000-2010/intercensal/national/us-estooint-01.xls.

12. On Ebenezer Howard and the Garden City, see Robert Fishman, *Urban Utopias in the Twentieth Century: Ebenezer Howard, Frank Lloyd Wright, and Le Corbusier* (Cambridge, MA: MIT Press, 1977), 23–90. On Lewis Mumford, the Regional Planning Association of America, and the regional city, see Carl Sussman, ed., *Planning the Fourth Migration: The Neglected Vision of the Regional Planning Association of America* (Cambridge, MA: MIT Press, 1976); Edward K. Spann, *Designing Modern America: The Regional Planning Association of America and Its Members* (Columbus: Ohio State Uni-

versity Press, 1996); John L. Thomas, "Holding the Middle Ground," in *The American Planning Tradition: Culture and Policy*, ed. Robert Fishman (Washington, DC: Woodrow Wilson Center Press, 2000), 33–64.

13. The intellectual debt of the Experimental City idea to either the Garden City idea or the regional city idea was never acknowledged by Spilhaus or Silha. Even when Otto Silha showed Lewis Mumford's film *The City* (1939) to an audience, Silha shared only the first segment of the film that dealt with pollution—stopping short of the later segments dealing with the federal greenbelt towns, the planned community of Radburn, New Jersey, and other examples of decentralized living that Mumford and his colleagues advocated; see memorandum by Walter K. Vivrett, April 16, 1970, box 1, NAA; Otto Silha, "Minnesota's Great Opportunity—The Experimental City," lecture given at the AFL-CIO Conference on Pollution Problems, Duluth, MN, April 25, 1970, box 7, MHS.

14. The new towns in America during the late 1960s and early 1970s took on many shapes, were driven by varied motivations, and often mixed elements of public and private development. Ann Forsyth has identified at least five different types of American new towns during this period: freestanding new towns (such as the Experimental City), satellite new towns, city-center new-town-in-towns, add-on new towns (or "growth centers"), and company new towns; see Ann Forsyth, *Reforming Suburbia: The Planned Communities of Irvine, Columbia, and The Woodlands* (Berkeley: University of California Press, 2005), 27–29.

15. On the push for a national land-use planning act in the 1970s, see Noreen Lyday, *The Law of the Land: Debating National Land Use Legislation 1970–1975* (Washington, DC: Urban Institute, 1976). On the history of national land-use planning in the United States generally, see Todd A. Wildermuth, "National Land Use Planning in America, Briefly," *Journal of Land, Resources, and Environmental Law* 26 (2005): 73–81.

16. *Lake Region (Alexandria, MN) Echo*, December 6, 1972; Otto Silha, lecture given at the Alexandria Rotary Club, Alexandria, MN, November 30, 1972, box 7, MHS.

17. Memorandum by James Alcott, December 7, 1972, box 7, MHS.

18. "New City in Brandon Area Meets Disfavor," *Fergus Falls (MN) Daily Journal*, December 21, 1972; memorandum by Louis Gelfand, "Meeting 2, Minnesota Steering Committee," December 22, 1972, box 3, MHS.

19. Minutes of the Minnesota Experimental City Authority, January 7, 1972, box 1; memorandum by James Alcott, December 7, 1972, box 7; memorandum by Louis Gelfand, "Meeting 2, Minnesota Steering Committee," December 22, 1972, box 3, all in MHS.

20. Memorandum by James Alcott, January 6, 1973, box 1, MHS.

21. Memorandum by James Alcott, January 6, 1973.

22. "Rep. Sherwood Cites Model City Opposition," *Brainerd (MN) Daily Dispatch*, January 31, 1973.

23. John W. Johnson to Otto Silha, January 18, 1973, box 7, MHS.

24. Archie Chelseth to James Alcott, November 3, 1972, box 7, MHS.

25. Exchange between Milton J. Fellowes and James Alcott, in Minnesota Pollution Control Agency, "Public Meeting on the Proposed Minnesota Experimental City," 11–13.

26. For Merritt quotes, see exchange between Grant Merritt and James Alcott, in Minnesota Pollution Control Agency, "Public Meeting on the Proposed Minnesota Experimental City," 14–15; for further insight on the battle between the Reserve Mining Company and the MPCA (among others), see Thomas R. Huffman, "Exploring the Legacy of Reserve Mining? What Does the Longest Environmental Trial in History Tell Us about the Meaning of American Environmentalism?" *Journal of Policy History* 12, no. 3 (2000): 339–68.

27. "Letters Regarding Minnesota Experimental City," Governor: Anderson, Wendell, MSA, MHS.

28. Sound recording of Minnesota Congressional Dinner, April 12, 1967, reel-to-reel tape, box 5, MHS. Silha was citing "A Glance behind the Curtain," by James Russell Lowell (*The Poetical Works of James Russell Lowell*, complete ed. [Boston: Houghton Mifflin, 1882], 44–50):

> The time is ripe, and rotten-ripe, for change;
> Then let it come: I have no dread of what
> Is called for by the instinct of mankind;
> Nor think I that God's world will fall apart,
> Because we tear a parchment more or less.
> Truth is eternal, but her effluence,
> With endless change is fitted to the hour;
> Her mirror is turned forward to reflect
> The promise of the future, not the past.

15: DISSECTING A NATION-LEADING LEGACY

1. I was a research scientist for the MPCA beginning in 1984. I conducted long-range transport modeling studies, analyzed rain chemistry data, wrote portions of the *Statement of Need and Reasonableness*, and testified at the administrative hearings. I have also recently published a modified version of this chapter, supplemented by additional interviews with J. David Thornton, Judge Allan Klein, David Schindler, Ann Cohen, Kevin Proescholdt, and Peter Ciborowski: "The Minnesota Acid Rain Story: A Legacy of Leadership," *Minnesota History* 67, no. 8 (Winter 2021–2022).

2. Toxic Substances Deposition, Minnesota State Statutes, sections 116.42–116.45, 1982, https://www.revisor.mn.gov/statutes/?id=116.42.

3. Fred Hoyle, qtd. in Martin Redfern, "Science: A New View of Home," *Independent*, April 20, 1996.

4. Mark Hertsgaard, "Saving Earth Day," *The Nation*, May 7, 2012. Nixon vetoed the Clean Water Act, but Congress passed it over his veto. He then impounded half of the funds for implementation.

5. James Gustave Speth, *Red Sky at Morning: America and the Crisis of the Global Environment*, 2nd edition (New Haven, CT: Yale University Press, 2004), 329.

6. I use the terms *acid deposition* and *acid rain* interchangeably for the sake of convenience, although the process includes a spectrum of depositions from wet to dry, and deposition is often in the form of acidifying substances rather than strong acids per se.

7. Earl Barrett and Gunnar Brodin, "The Acidity of Scandinavian Precipitation," *Tellus* 7 (1955): 251–57; Eville Gorham, "On the Acidity and Salinity of Rain," *Geochimica et Cosmochimica Acta* 7 (1955): 231–39.

8. Eville Gorham and Alan G. Gordon, "The Influence of Smelter Fumes upon the Chemical Composition of Lakes near Sudbury, Ontario, and upon the Surrounding Vegetation," *Canadian Journal of Botany* (1960): 38, 477–97.

9. For example, Gene E. Likens, F. Herbert Bormann, and Noye M. Johnson, "Acid Rain," *Environment* (1972): 14, 33–40; and James N. Galloway, Gene E. Likens, and Mark E. Hawley, "Acid Precipitation: Natural versus Anthropogenic Components," *Science* (1984): 226, 829–31.

10. Governor Ronald Reagan, speech given opposing expansion of Redwood National Park, September 12, 1965, qtd in Martin Smith, "Brown Cites Successes; Reagan. Cut State Costs, Governor Campaigns in Redwoods," *Sacramento Bee*, March 3, 1966.

11. Ronald Reagan, qtd. in "Killer Trees," in *There He Goes Again: Ronald Reagan's Reign of Error*, ed. Mark Green and Gail MacColl (New York: Pantheon Books, 1983). Reagan later amended this figure to 93 percent. These statements were later shown to be inaccurate.

12. Gorsuch Burford resigned under pressure in 1983 in the face of a scandal involving mismanagement of toxic waste cleanup. James Watt became embroiled in controversy over his banning of a Beach Boys concert on the National Mall and eventually resigned.

13. Clean Air Act Amendments of 1990, 104 Stat. 2468, P.L. 101–549, http://www.epa.gov/air/caa/title4.html. Codified as Subchapter IV-A—Acid Deposition Control in Chapter 85—Air Pollution Prevention and Control, under Title 42 of the Code of Federal Regulations, http://www.gpo.gov/fdsys/pkg/USCODE-2008-title42/pdf/USCODE-2008-title42-chap85.pdf.

14. Since 1913 Minnesota legislators were elected without party designation; however, they ran and caucused as "liberals" or "conservatives," essentially equivalent to Democratic-Farmer-Labor and Republican Parties, respectively. Beginning in 1974 the House, and two years later, the Senate again ran with party designation.

15. Minnesota Legislature Reference Library, "Party Control of the Minnesota House of Representatives, 1951–present," accessed November 13, 2021, http://www.leg.state.mn.us/lrl/histleg/caucus.aspx?body=h.

16. Eville Gorham testified before the House Subcommittee on Oversight and Investigations of the Committee on Interstate and Foreign Commerce. This committee took testimony owing to the interest of representatives serving on the committee and

the potential international implications of acid rain. Gorham was one of the leading international researchers in atmosphere–biosphere interactions, later became a University of Minnesota Regent's Professor, and was elected to the National Academy of Sciences.

17. Minnesota politicians were divided on the BWCA issue. The original 1975 wilderness bill was authored by Congressman James Oberstar and called for dividing the area into a wilderness section and a recreation area open to logging and motorized vehicles. A year later Representative Donald Fraser introduced a competing bill giving wilderness designation to the entire area. Following Congressional hearings in Saint Paul and Ely, Minnesota, Representative Bruce Vento coauthored a full wilderness bill that eventually became law.

18. The state of Minnesota entered the lawsuit opposing the wilderness designation due to concerns about state jurisdiction over the water bodies and that the federal government was overstepping its authority by imposing regulations on motor vehicles in the BWCA; personal communication with Kevin Proescholdt, coauthor with Miron Heinselman and Rip Rapson of *Troubled Waters: The Fight for the Boundary Waters Canoe Area Wilderness* (Saint Cloud, MN: North Star Press, 1995).

19. "Quie Could Make Acid Rain Law Nation's First," *St. Paul Pioneer Press Dispatch*, March 13, 1982.

20. Now Xcel Energy.

21. The executive head of the MPCA was titled "director" until 1986 when the title was changed to "commissioner."

22. Minnesota Statutes, "Administrative Procedure," Chapter 14, accessed November 13, 2021, https://www.revisor.mn.gov/statutes/?id=14.

23. Office of the Legislative Auditor, Program Evaluation Division, "Administrative Rulemaking," 1993, https://www.auditor.leg.state.mn.us/ped/1993/pe9304.htm.

24. All statistics in this paragraph are from Minnesota Pollution Control Agency, *Statement of Need and Reasonableness in the Matter of the Proposed Adoption of Minnesota Rules parts 7005.4010 to 7005.4050, Relating to an Acid Deposition Standard and Control Plan*, 1985, Minnesota Office of Administrative Hearings, accessed June 11, 2022, https://mn.gov/oah/assets/22007921.93_tcm19-159080.pdf.

25. A pH value of 7 is considered neutral. Unpolluted rainfall is slightly acidic (~pH 5.6) due to the presence of carbon dioxide in the atmosphere. pH is measured on a logarithmic scale, so a pH of 4.7 is nearly 10 times more acidic than unpolluted rain.

26. Speth, *Red Sky at Morning*.

27. MPCA offices were located in Roseville, MN, until the fall of 1986, when the agency moved to its current location in Saint Paul.

28. As an MPCA researcher who authored major portions of the SONAR, I presented one half-day of affirmative testimony and was cross-examined by NSP, Minnesota Power, and environmental group attorneys for a full day.

29. State of Minnesota, Office of Administrative Hearings, *Report of the Administrative Law Judge, In the Matter of the Proposed Adoption of Minnesota Rules, Parts*

7005.4010—7005.4050, Relating to an Acid Deposition Standard and Control Plan, PCA-85-002-AK, 6-2200-34-1, 1986, author's possession.

30. Personal interview with J. David Thornton, Minnesota Pollution Control Agency assistant commissioner and director of the MPCA acid rain program in the 1980s.

31. For a fuller description of the history of the environmental movement, see Benjamin Kline, *First along the River: A Brief History of the U.S. Environmental Movement,* 4th ed. (Lanham, MD: Rowman & Littlefield, 2011).

32. For example, see "Koch Brothers behind Environment Killing Measures," Sustainable Business, March 2, 2011, http://www.sustainablebusiness.com/index.cfm/go/news .display/id/21977; and Jonathan Owen and Paul Bignell, "Think-Tanks Take Oil Money and Use It to Fund Climate Deniers," *The Independent,* February 7, 2010, http://www .independent.co.uk/environment/climate-change/thinktanks-take-oil-money-and-use-it -to-fund-climate-deniers-1891747.html.

33. For example, see Bill McKibben, "Cynicism Is for Suckers: How We Can Fight the Greedy Elites That Run Our Lives," AlterNet, January 5, 2012, http://www.alternet.org /environment/153670/cynicism_is_for_suckers%3A_how_we_can_fight_the_greedy_elites _that_run our lives; and David Loy, "The Religion of the Market," 1996, Un Zen Occidental, http://www.zen-occidental.net/articles1/loy8.html.

34. Philip Shabecoff. *A Fierce Green Fire: The American Environmental Movement,* rev. ed. (Washington, DC: Island Press, 2003).

35. The Riverside and High Bridge plants were converted to natural gas as part of the MERP. The Black Dog plant was partially converted before the MERP, and a complete conversion has been authorized. Scrubbers were installed at the Alan King plant.

36. The conservation movement dating to the early twentieth century is sometimes called the "first wave" of environmentalism, which was followed by a second wave that peaked during and after the first Earth Day in 1970. For a description of the evolution of environmentalism from its roots in conservation, see Samuel P. Hays, *Beauty, Health, and Permanence: Environmental Politics in the United States, 1955–1985* (Cambridge: Cambridge University Press, 1987); and Shabecoff, *Fierce Green Fire.*

16: THE URBAN ROOTS OF MILITANT INDIAN PROTEST

1. One study reports over seventy takeovers between 1969 and 1978. See Troy Johnson, Joane Nagel, and Duane Champagne, eds., *American Indian Activism: Alcatraz to the Longest Walk* (Urbana: University of Illinois Press, 1997), 9, 32. On Indian activism in textbooks, see Paul Boyer, *Promises to Keep: The United States since World War II,* 3rd ed. (Boston: Houghton Mifflin, 2005), 329–30; and John Mack Faragher et al., *Out of Many,* 4th ed., vol. 2 (Upper Saddle River, NJ: Pearson Prentice Hall, 2006), 814–15.

2. The history written by Paul Chaat Smith, a Comanche, and Robert Allen Warrior, an Osage, is the best study of these protests. But even this excellent book emphasizes the major occupations, with much less attention paid to community programs. See Paul

Chaat Smith and Robert Allen Warrior, *Like a Hurricane: The Indian Movement from Alcatraz to Wounded Knee* (New York: New Press, 1996).

3. On termination and relocation policies, see Donald L. Fixico, *Termination and Relocation: Federal Indian Policy, 1945–1960* (Albuquerque: University of New Mexico Press, 1986). See also Kenneth R. Philp, *Termination Revisited: American Indians on the Trail to Self-Determination, 1933–1953* (Lincoln: University of Nebraska Press, 2002); and Charles F. Wilkinson, *Blood Struggle: The Rise of Modern Indian Nations* (New York: W. W. Norton, 2006), 57–86.

4. On coal miners' struggles to adapt to cities, see Alessandro Portelli, *They Say in Harlan County: An Oral History* (New York: Oxford University Press, 2011), 259–67.

5. For overviews of AIM's origins, see Peter Iverson, *"We Are Still Here": American Indians in the Twentieth Century* (Wheeling, IL: Harland Davidson, 1998), 148–54; Peter Matthiessen, *In the Spirit of Crazy Horse* (New York: Viking, 1983), 34–41; Smith and Warrior, *Like a Hurricane*, 127–48; and Wilkinson, *Blood Struggle*, 137–49.

6. On the legal claim that Indians were entitled to abandoned federal land, see Vine Deloria Jr., *Behind the Trail of Broken Treaties: An Indian Declaration of Independence* (New York: Delta Books, 1974), 34–41. See also Smith and Warrior, *Like a Hurricane*, 10–12; Johnson, Nagel, and Champagne, *American Indian Activism*, 25.

7. On support for urban Indian centers, see Donald L. Fixico, *The Urban Indian Experience in America* (Albuquerque: University of New Mexico Press, 2000), 130–31.

8. Robert D. Bullard, *Dumping in Dixie: Race, Class, and Environmental Quality* (Boulder, CO: Westview Press, 1994), xiii.

9. On environmental problems facing Indian nations, see Winona LaDuke, *All Our Relations: Native Struggles for Land and Life* (Cambridge, MA: South End Press, 1999). On uranium, see Peter Eichstaedt, *If You Poison Us: Uranium and Native Americans* (Santa Fe, NM: Red Crane Books, 1994). See also Ward Churchill and Winona LaDuke, "Native North America: The Political Economy of Radioactive Colonialism," in *The State of Native America; Genocide, Colonization, and Resistance*, ed. M. Annette Jaimes (Boston: South End Press, 1992).

10. Proclamation qtd. in Matthiessen, *In the Spirit of Crazy Horse*, 37.

11. Bullard's *Dumping in Dixie* was the seminal work, and his other works include *Unequal Protection: Environmental Justice & Communities of Color* (New York: Random House, 1994); and *Confronting Environmental Racism: Voices from the Grassroots* (Cambridge, MA: South End Press, 1999). For an overview, see Luke W. Cole and Sheila R. Foster, *From the Ground Up: Environmental Racism and the Rise of the Environmental Justice Movement* (New York: NYU Press, 2000). Early scholarship includes Bunyan Bryant and Paul Mohai, eds., *Race and the Incidence of Environmental Hazards: A Time for Discourse* (Boulder, CO: Westview Press, 1992); and Richard Hofrichter, ed., *Toxic Struggles: The Theory and Practice of Environmental Justice* (Gabriola Island, BC: New Society, 1993). Early monographs include Andrew Hurley, *Environmental Inequalities: Class,*

Race, and Industrial Pollution in Gary, Indiana, 1945–1980 (Chapel Hill: University of North Carolina Press, 1995); and Laura Pulido, *Environmentalism and Economic Justice: Two Chicano Struggles in the Southwest* (Tucson: University of Arizona Press, 1996).

12. Robert Gottlieb, *Forcing the Spring: The Transformation of the American Environmental Movement* (Washington, DC: Island Press, 1993), 252–53. On uranium, see 250–52, 383–84.

13. On relocation policy, see Fixico, *Termination and Relocation*. On resulting cultural problems, see Jack O. Waddell and O. Michael Watson, eds., *The American Indian in Urban Society* (Boston: Little, Brown, 1971); Fixico, *Urban Indian Experience in America*; Susan Lobo and Kurt Peters, eds., *American Indians and the Urban Experience* (Walnut Creek, CA: Altamira Press, 2001).

14. Testimony of Clyde Warrior before the President's National Advisory Commission on Rural Poverty, February 2, 1967, in Alvin M. Josephy Jr., *Red Power: The American Indians' Fight for Freedom* (New York: McGraw Hill, 1971), 72–73.

15. Clyde Warrior qtd. in Smith and Warrior, *Like a Hurricane*, 37.

16. Dennis Banks with Richard Erdoes, *Ojibwa Warrior: Dennis Banks and the Rise of the American Indian Movement* (Norman: University of Oklahoma Press, 2004), 61, 60.

17. George Mitchell interview with Steven L. Couture, qtd. in Steven L. Couture, "The American Indian Movement: A Historical Perspective" (PhD diss., University of Saint Thomas, 1996), 50, 39.

18. Banks and Erdoes, *Ojibwa Warrior*, 63, 64. See also Fay G. Cohen, "The Indian Patrol in Minneapolis: Social Control and Social Change in an Urban Context" (PhD diss., University of Minnesota, 1973); and James LaGrand, *Indian Metropolis: Native Americans in Chicago, 1945–1975* (Urbana: University of Illinois Press, 2002), 205–7.

19. Vernon Bellecourt interview, in *Native American Testimony: A Chronicle of Indian-White Relations from Prophecy to the Present, 1492–1992*, ed. Peter Nabokov (New York: Viking, 1991), 375. Bellecourt said an Urban League of Minneapolis grant bought the radios and cameras.

20. Clyde Bellecourt testimony in "Public Forum before the Committee on Urban Indians in Minneapolis-St. Paul, Minnesota of the National Council on Indian Opportunity," manuscript, March 18–19, 1969, 171, Illinois State University, Normal.

21. Russell Means with Marvin J. Wolf, *Where White Men Fear to Tread* (New York: St. Martin's Press, 1995), 162–63.

22. Means and Wolf, *Where White Men Fear to Tread*, 19.

23. Means and Wolf, *Where White Men Fear to Tread*, 18–21, 28.

24. Clyde Bellecourt testimony, 171, 170.

25. Means and Wolf, *Where White Men Fear to Tread*, 163. See also Banks and Erdoes, *Ojibwa Warrior*, 64; and "The Angry American Indian: Starting down the Protest Trail," *Time Magazine*, February 9, 1970, 14–15.

26. Vernon Bellecourt interview, in Nabokov, *Native American Testimony*, 375.

27. Leonard Crow Dog and Richard Erdoes, *Crow Dog: Four Generations of Sioux Medicine Men* (New York: HarperPerennial, 1996), 164.

28. Anonymous AIM member interviewed by Couture, qtd. in Couture, "American Indian Movement," 50–51.

29. Banks and Erdoes, *Ojibwa Warrior*, 64.

30. Pat Bellenger interview by Steven Couture, qtd. in Couture, "American Indian Movement," 44–45.

31. On growing social programs and their funding, see Smith and Warrior, *Like a Hurricane*, 132–36; and Matthiessen, *In the Spirit of Crazy Horse*, 35–37. On Means, see Means and Wolf, *Where White Men Fear to Tread*, 136–39.

32. On survival schools, see Fixico, *Urban Indian Experience in America*, 146–49; and LaGrand, *Indian Metropolis*, 207.

33. Gregory Craig, Arthur Harkins, and Richard Woods, *Indian Housing in Minneapolis and St. Paul*, Training Center for Community Programs, University of Minnesota, 1969, 1.

34. Arthur Harkins and Richard Woods, *Attitudes of Minneapolis Agency Personnel toward Urban Indians*, Training Center for Community Programs, University of Minnesota, 1968, 16, see 10–11 on public housing.

35. "Long Range Objectives," "Short Range Objectives," "Indian American Resolution," and "Resolution," *American Indian Movement News*, April 1970, 8, 8, 5, and 6, microfilm 1555, Minnesota Historical Society, Saint Paul (hereafter MHS).

36. Clyde Bellecourt was listed as AIM's executive director at the National Indian Ad-Hoc Board of Inquiry, with John Red Horse and female AIM members Florence Hill and Winnie LaPrairie. Russell Means and Allen Warrior were listed as affiliated with Cleveland American Indian Center. Mary Arpan, "National Indian Ad-Hoc Board of Inquiry," *American Indian Movement News*, April 1970, 9, microfilm 1555, MHS.

37. Mary Arpan, "National Indian Ad-Hoc Board of Inquiry," 10.

38. "This Is Not the Indian Way," *American Indian Movement News*, April 1970, 16, microfilm 1555, MHS.

39. On the Naval Air Station occupation, see "35 Indians Evicted from Naval Station; Sioux Treaty Cited," *New York Times*, May 22, 1971. See also Johnson, Nagel, and Champagne, *American Indian Activism*, 32; and Rex Weyler, *Blood of the Land: The Government and Corporate War against First Nations* (Philadelphia: New Society, 1992), 46.

40. "Indians Seize Air Station; School, Housing Demanded," *Minneapolis Star*, May 17, 1971, 1.

41. "Panthers to Change Tactics, Newton Says," *Minneapolis Star*, May 21, 1971, 3; "Antiwar Rally Draws 27,000," *Minneapolis Star*, May 24, 1971, 1.

42. "Indians Seize Air Station," 3.

43. "Indians Lose Identity in City," *Minneapolis Star*, May 18, 1971, 9.

44. "The Tale of Two Protests," *Minneapolis Tribune*, May 20, 1971, 14.

45. "Indian Leader Seeks Parley," *Minneapolis Star*, May 18, 1971, 11.

46. "Indian Occupation of Naval Station Ends with Arrests," *Minneapolis Tribune*, May 22, 1971, 1, 12; "Lawmen End Indian Air Station Seizure," *Minneapolis Star*, May 22, 1971, 1–2.

47. "Lawmen End Indian Air Station Seizure," *Minneapolis Star*, May 22, 1971, 1.

48. Homer Bigart, "Militancy of Urban Indians Spurs Hope for Change," *New York Times*, February 10, 1972, 1, 26.

49. Chue Kong Thao, "Phillips Neighborhood Lead Collaborative," Neighborhood Planning for Community Revitalization, Center for Urban and Regional Affairs, University of Minnesota, 2005.

50. R. L. Polk, Minneapolis City Directory, 1971, MHS.

51. Sing-Wei Ho, "The History and Health Consequences of the Arsenic Contamination in and around the CMC Heartland Lite Yard Site in South Minneapolis," Neighborhood Planning for Community Revitalization, Center for Urban and Regional Affairs, University of Minnesota, 2005; US Environmental Protection Agency, "EPA to Restart Work on South Minneapolis Superfund Site," Environmental Protection Agency Newsroom, April 14, 2011.

52. Will Delaney, "'More Than Just Bricks and Mortar': A History of Redevelopment Efforts along the East Franklin Avenue Corridor, 1982–2007," Neighborhood Planning for Community Revitalization, Center for Urban and Regional Affairs, University of Minnesota, 2007, 4.

53. Harkins and Woods, *Attitudes of Minneapolis Agency Personnel toward Urban Indians*, 62–63; Craig, Harkins, and Woods, *Indian Housing in Minneapolis and St. Paul*, 3–4.

54. Harkins and Woods, *Attitudes of Minneapolis Agency Personnel toward Urban Indians*, 16.

55. Craig, Harkins, and Woods, *Indian Housing in Minneapolis and St. Paul*, 5.

56. On Means in Cleveland, see Means and Wolf, *Where White Men Fear to Tread*, 155–61. On Powless in Milwaukee, see Banks and Erdoes, *Ojibwa Warrior*, 65; and Weyler, *Blood of the Land*, 46–48. On the school in the Coast Guard station, see Susan Applegate Krouse, "What Came Out of the Takeovers: Women's Activism and the Indian Community School of Milwaukee," in *Keeping the Campfires Going: Native Women's Activism in Urban Communities*, ed. Susan Applegate Krouse and Heather H. Howard (Lincoln: University of Nebraska Press, 2009), 146–59.

57. AIM organized a massive protest in Gordon, Nebraska, just south of Pine Ridge, that forced white officials to seek justice after two white men killed Raymond Yellow Thunder. See "Death of Indian Sparks Protest," *New York Times*, March 8, 1972; "Inquiry Ordered in Indian's Death," *New York Times*, March 9, 1972; "White Man's Town Bows to Angry Indians," *New York Times*, March 20, 1972; "Jury in Nebraska Convicts 2 Brothers in Death of Indian," *New York Times*, May 27, 1972.

58. Mary Crow Dog and Richard Erdoes, *Lakota Woman* (New York: HarperPeren-

nial, 1991), 76. On the partnership between Banks and Crow Dog, see Banks and Erdoes, *Ojibwa Warrior*, 95–104. See also Deloria, *Behind the Trail of Broken Treaties*, 40–41.

59. Banks and Erdoes, *Ojibwa Warrior*, 95.

60. Rick Williams, interview by Charles F. Wilkinson, qtd. in Wilkinson, *Blood Struggle*, 352.

17: RADIOACTIVE WASTE, PUBLIC DEBATE, AND ENVIRONMENTAL JUSTICE AT PRAIRIE ISLAND

1. J. Samuel Walker, *Three Mile Island: A Nuclear Crisis in Historical Perspective* (Berkeley: University of California Press, 2004), 6; Northern States Power Company, "Northern States Power Puts the Accent on Environment," 1967, Box 3, Folder: "Nuclear Radiation, 1958–1968," Minnesota Environmental Citizens' Control Association Files (hereafter MECCA Records), Minnesota Historical Society, Saint Paul (hereafter MHS); Northern States Power Company, news release, February 1, 1969, in "Pamphlets Relating to Nuclear Power in Minnesota," MHS; *Saint Paul Dispatch*, August 21, 1973, in "Newspaper Clippings Regarding Northern States Power Company, 1970–1974," collected by Kenneth Meter, MHS (hereafter NSP Clippings File).

2. Minnesota Committee for Environmental Information, "Scientists Question Reactor Benefits," *Scientist and Citizen* 10 (August 1968): 154–57; "Cooling It in Minnesota," *Environment* 11 (March 1969): 21–25; Arthur Naftalin, "Statement before the Pollution Control Agency," April 8, 1969, Box 3, Folder: "Nuclear Radiation, 1968–1970," MECCA Records; J. Samuel Walker, *Containing the Atom: Nuclear Regulation in a Changing Environment, 1963–1971* (Berkeley: University of California Press, 1992), 310–16.

3. Quotes from Robert H. Engler to members of the Minnesota Pollution Control Agency, August 11 1969, Box 3, folder "Monticello, undated, 1969," MECCA Records; *MECCA Newsletter*, October 1970, MHS; "Minnesota vs. the A.E.C.," *New York Times*, February 16, 1970, 36; Philip M. Boffey, "Radioactive Pollution: Minnesota Finds AEC Standards Too Lax," *Science* 163, no. 3871 (March 7, 1969): 1043–46; Walker, *Containing the Atom*, 310–16.

4. Steve J. Gadler to Grant J. Merritt, July 12, 1971, Box 4, folder 1; and "Statement by Steve J. Gadler before the Congressman Karth and Fraser Congressional Committee," April 20, 1971; and "Closing Argument at Prairie Island Nuclear Power Plant AEC Hearing in St. Paul," November 8, 1973, both in Box 4, folder "Northern States Power Co./ Prairie Island Nuclear Plant, 1968–1979," all in Steve J. Gadler Papers, MHS (hereafter Gadler Papers).

5. MECCA news release, June 14, 1969, Box 3, Folder "Northern States Power Company"; Russell Hatling to Attorney General Douglas Head, August 31, 1970, Box 3, Folder "Monticello," both in MECCA Records; *Saint Paul Dispatch*, May 5, July 22, and November 23, 1971; *MECCA Newsletter*, December 1970, January 1971, and November 1972; *Minneapolis Tribune*, November 20 and 27, 1971, both in NSP Clippings File.

6. Thomas Raymond Wellock, *Critical Masses: Opposition to Nuclear Power in Cali-*

fornia, 1958–1978 (Madison: University of Wisconsin Press, 1998); Brian Balogh, *Chain Reaction: Expert Debate and Public Participation in American Commercial Nuclear Power, 1945–1975* (New York: Cambridge University Press, 1991).

7. See esp. Wellock, *Critical Masses.*

8. J. Samuel Walker, *The Road to Yucca Mountain: The Development of Radioactive Waste Policy in the United States* (Berkeley: University of California Press, 2009), 95–109.

9. This request also led to a contested court case with national implications and eventually to the 1982 Nuclear Waste Policy Act. Walker, *Road to Yucca Mountain,* 169.

10. Jocelyn F. Olson to Howard Kaibel, April 13, 1980, and prefiled testimony of Dale M. Vincent, April 15, 1980, both in Folder 1, Papers of the Minnesota Environmental Quality Board, MHS (hereafter MN EQB).

11. Northern States Power Company, *Certificate of Need Application: To Increase Storage Capacity of the Spent Fuel Pool at the Prairie Island Generating Plant* (1979), MHS, quotes on 9, 88, 22.

12. The NRC had conducted a generic EIS on all spent fuel pool expansions in general and deemed this sufficient. *MECCA Newsletter,* November/December 1979; Mike Sweeney, "Prairie Island, Love Canal Ties Told at Hearing," *St. Paul Pioneer Press,* June 18, 1980; Olson to Kaibel, April 13, 1980.

13. The EQB consisted of citizens and state officials, and reviewed projects that influenced the state's environment; although NSP applied to the Minnesota Energy Agency for its certificate of need, the EQB determined whether or not the project required an EIS.

14. State of Minnesota, Environmental Quality Board, transcript of EQB hearing, Red Wing, May 1, 4, and 5 (quotes, 23, 7), 1980, Folders 2 and 4, MN EQB.

15. "Neighbors Not Told of Nuclear Plant Accident," *Minneapolis Star,* October 3, 1979; "Plant Cool: But Debate Heating Up," and "Official Response? Confusion," *Minneapolis Star,* October 4, 1979; and "Prairie Island Plant Ready," *Minneapolis Star,* October 18, 1979.

16. Mark Mason, "Findings of Fact, Conclusions, and Decision," 1981, 44; and "Memorandum," in Minnesota Attorney General, Environmental Protection Division, Nuclear Waste Disposal and Regulation Files, MHS.

17. See Walker, *The Road to Yucca Mountain*; James Flynn et al., *One Hundred Centuries of Solitude: Redirecting America's High-Level Nuclear Waste Policy* (Boulder, CO: Westview Press, 1995); James W. Hulse, *Nevada's Environmental Legacy: Progress or Plunder* (Reno: University of Nevada Press, 1999).

18. "NSP Agrees to Study on Effects of Its Plan for Spent Nuclear Fuel," *Star Tribune* (Minneapolis), December 22, 1989; quote from "Showdown on NSP Nuclear Waste Starts," *Star Tribune,* November 11, 1991; Northern States Power Company, *Application for Certificate-of-Need for Prairie Island Spent Fuel Storage,* Docket NO. E002/

CN-91–19, Minneapolis, 1991, https://www.lrl.mn.gov/webcontent/lrl/guides/nuclear%20waste/NSP_1.pdf.

19. Minnesota Environmental Quality Board, *Final Environmental Impact Statement: Prairie Island Independent Spent Fuel Storage Installation* (Saint Paul: Minnesota Environmental Quality Board, 1991).

20. Northern States Power Company, *Application for Certificate-of-Need for Prairie Island Spent Fuel Storage*, 8, 14–15.

21. Northern States Power Company, *Application for Certificate-of-Need for Prairie Island Spent Fuel Storage*, 17–18.

22. Allan W. Klein, "Findings of Fact, Conclusions and Recommendation in the Matter of the Application of Northern States Power Company for a Certificate of Need for the Construction of an Independent Spent Fuel Storage Facility," April 10, 1992, 42–43, https://www.lrl.mn.gov/webcontent/lrl/guides/nuclear%20waste/oahnsp41092.pdf.

23. "State Supreme Court Hands Legislature a Hot Issue," *Star Tribune*, July 20, 1993.

24. "Full Senate, House Approve NSP Plan for Nuclear Waste," *Star Tribune*, May 7, 1994; "Raitt and Pirner Sing Different Tunes, but Antinuclear Message Is the Same," *Star Tribune*, February 24, 1994; "Committee Backs NSP Waste Storage," *Star Tribune*, March 4, 1994; "23 Arrested in Protest of NSP Nuclear Storage," *Star Tribune*, October 13, 1994; "National Nuclear Waste Day at Prairie Island Nuclear Plant," Prairie Island Coalition newsletter, September 1994, Wisconsin Historical Society, Madison, Wisconsin; "Senate OKs NSP Plan," *Star Tribune*, March 31, 1994.

25. Dane Smith, "Prairie Island: A Nuclear Fight Full of Fears," *Star Tribune*, April 3, 1994.

26. See Joni Adamson, Mei Mei Evans, and Rachel Stein, *The Environmental Justice Reader: Politics, Poetics, & Pedagogy* (Tucson: University of Arizona Press, 2002); Eileen Maura McGurty, "From NIMBY to Civil Rights: The Origins of the Environmental Justice Movement," *Environmental History* 2, no. 2 (1997): 301–23; Eileen McGurty, *Transforming Environmentalism: Warren County, PCBs, and the Origins of Environmental Justice* (New Brunswick, NJ: Rutgers University Press, 2007); Gerald R. Visgilio and Diana M. Whitelaw, *Our Backyard: A Quest for Environmental Justice* (Lanham, MD: Rowman & Littlefield, 2003).

27. "Showdown on NSP Nuclear Waste Starts," *Star Tribune*, November 18, 1991; "Minnesota Runs Out of Space for Its Nuclear Waste," *Morning Edition*, National Public Radio, March 10, 1994; "Leaders See Racism in NSP's Choices," *Star Tribune*, April 6, 1994.

28. Minnesota Environmental Quality Board, *Final Environmental Impact Statement*, Comment Letter 14.

29. Nancy B. Collins and Andrea Hall, "Nuclear Waste in Indian Country: A Paradoxical Trade," *Law & Inequality* 12 (June 1994): 267–350; Conrad L. Huygen, "Clouded Vision: The Mescalero Apache and the Nuclear Legacy," *Environmental Law and Policy*

Journal 15 (December 1994): 1–8; Tom Meersman, "NSP, Mescaleros OK Nuclear Storage Talks," *Star Tribune*, February 4, 1994.

30. Joint Religious Legislative Coalition, "Joint Religious Legislative Coalition Position on Prairie Island Spent Fuel Storage," 1993, MHS.

31. "Full Senate, House Approve NSP Plan for Nuclear Waste," *Star Tribune*, May 7, 1994; "Senate Panel OKs Amended Nuclear Bill," *Star Tribune*, March 25, 1994; *Laws of Minnesota, 1994*, Chapter 641-S.F. No. 1706.

32. Prairie Island Coalition, *Confronting Nuclear Racism: A Prairie Island Coalition Report* (Lake Elmo, MN: Prairie Island Coalition, 1996), 4–5.

33. An excellent summary of the ongoing controversy over nuclear energy at Prairie Island and Monticello can be found at https://www.lrl.mn.gov/guides/guides ?issue=nuclearwaste.

34. Minnesota Department of Health, "Prairie Island Nuclear Generating Plant April 2022," https://www.health.state.mn.us/communities/environment/radiation/docs /monitor/pi/m0422.pdf.

AFTERWORD: MINNESOTA'S MANY INTERSECTING CROSSROADS

1. James E. Sherow, "Introduction: An Evening on Konza Prairie," 5; and Dan Flores, "Spirit of Place and the Value of Nature in the American West," 31, both in *A Sense of the American West*, ed. Sherow (Albuquerque: University of New Mexico Press, 1998).

2. Thomas Andrews and Flannery Burke, "What Does It Mean to Think Historically?" *Perspectives on History*, January 1, 2007, at https://www.historians.org/pub lications-and-directories/perspectives-on-history/january-2007/what-does-it-mean-to -think-historically.

3. The concept of the *longue durée* emerges from the French Annales School that emphasizes long-term historical structures and recurring cycles, especially those grounded in the environment. See, for example, Fernand Braudel, *The Mediterranean and the Mediterranean World in the Age of Philip II*, vol. 1 (New York: Harper and Row, [1966] 1972), 20.

4. Anne Kelly Knowles, "GIS and History," in *Placing History: How Maps, Spatial Data, and GIS Are Changing Historical Scholarship*, ed. Knowles (Redlands, CA: ESRI Press, 2008), 3.

5. Richard White, "The Nationalization of Nature," *Journal of American History* 86, no. 3 (December 1999): 977. There are exceptions, such as Richard White, *Railroaded: The Transcontinentals and the Making of America* (New York: W. W. Norton, 2012); and William Cronon, *Nature's Metropolis: Chicago and the Great West* (New York: W. W. Norton, 1991). Other scholars implicitly address intersecting scales, such as Kathleen A. Brosnan, *Uniting Mountain and Plain: Cites, Law and Environmental Change along the Front Range* (Albuquerque: University of New Mexico Press, 2002); and Kathryn Morse,

The Nature of Gold: An Environmental History of the Klondike (Seattle: University of Washington Press, 2003).

6. Richard White, discussing Henri Lefebvre's *Production of Space*, trans. Donald Nicholson-Smith (Cambridge, MA: Harvard University Press, 1991), 28–31, in White, "Nationalization of Nature," 977–79.

7. White, "Nationalization of Nature," 978.

8. See, for example, Martin V. Melosi, *The Sanitary City: Environmental Services in Urban America*, abridged ed. (Pittsburgh: University of Pittsburgh Press, 2008), chap. 8; and Sarah S. Elkind, *Bay Cities and Water Politics: The Battle for Resources in Boston and Oakland* (Lawrence: University Press of Kansas, 1998), chap. 2.

9. Lefebvre, *Production of Space*, 33–39.

10. See, for example, Roy Rosenzweig and Elizabeth Blackmar, *The Park and the People: A History of Central Park*, rev. ed. (Ithaca, NY: Cornell University Press, 1998), 284–306.

11. See, for example, Kent Blansett, "San Francisco, Red Power, and the Emergence of an 'Indian City,'" in *City Dreams, Country Schemes: Community and Identity in the American West*, ed. Kathleen A. Brosnan and Amy L. Scott (Reno: University of Nevada Press, 2013), 261–83.

12. Lorenzo Veracini, *Settler Colonialism: A Theoretical Overview* (New York: Palgrave Macmillan, 2010), 8. Also see Lynette Russell, ed., *Colonial Frontiers: Indigenous-European Encounters in Settler Societies* (Manchester: Manchester University Press, 2001).

13. See David P. Billington, Donald C. Jackson, and Martin V. Melosi, *The History of Large Federal Dams: Planning, Design, and Construction in the Era of Big Dams* (Denver: Department of the Interior, Bureau of Reclamation, 2005), 353–82, 399–405.

14. Jaskiran Dhillon, "What Standing Rock Teaches Us about Environmental Justice," Items: Insights from the Social Sciences, December 5, 2017 (emphasis in original), https://items.ssrc.org/just-environments/what-standing-rock-teaches-us-about-environmental-justice/

15. For example, see Traci Brynne Voyles, *Wastelanding: Legacies of Uranium Mining in Navajo Country* (Minneapolis: University of Minnesota Press, 2015).

16. Lorenzo Veracini, *The Settler Colonial Present* (New York: Palgrave Macmillan, 2015), 68–87.

17. Lefebvre, *Production of Space*, 33–39.

18. Lefebvre, *Production of Space*, 35–39.

CONTRIBUTORS

John O. Anfinson is the retired Mississippi National River and Recreation Area superintendent, National Park Service, and a longtime Mississippi River historian.

William C. Barnett is professor of history at North Central College.

Kathleen A. Brosnan is the Paul H. and Doris Eaton Travis Chair in Modern American History and director of graduate studies at the University of Oklahoma.

Kevin C. Brown is a research associate in the environmental studies program at the University of California-Santa Barbara.

James W. Feldman is professor of environmental studies and history at the University of Wisconsin-Oshkosh and the director of the environmental studies program.

Thomas Finger is associate professor of history at Northern Arizona University.

David A. Lanegran is emeritus John S. Holl Professor of Geography at Macalester College.

James Longhurst is professor of history at the University of Wisconsin-La Crosse.

Jeffrey T. Manuel is professor of history at Southern Illinois University-Edwardsville.

Michael D. McNally is the John M. and Elizabeth W. Musser Professor of Religious Studies at Carleton College.

Chantal Norrgard is assistant professor of First Nations and Indigenous studies at the University of Wisconsin-Superior.

Gregory C. Pratt is a retired research scientist for the Minnesota Pollution Control Agency and adjunct assistant professor in environmental health sciences at University of Minnesota.

Aaron Shapiro is executive director of Patapsco Heritage Greenway.

Robert S. Thompson is a historian with the Defense POW/MIA Accounting Agency.

Joseph Underhill is associate professor of political science and director of environmental studies at Augsburg University.

George Vrtis is professor of history and environmental studies at Carleton College.

Karen Wellner teaches in the Biological Sciences Department at Chandler-Gilbert Community College.

Christopher W. Wells is professor of environmental studies at Macalester College.

Todd A. Wildermuth is associate teaching professor at the University of Washington School of Law, where he is director of the environmental law program and policy director of the Regulatory Environmental Law and Policy Clinic.

INDEX

Note: References in *italics* refer to figures and tables.